Analysis with
LOCAL CENSUS DATA
Portraits of Change

Analysis with
LOCAL CENSUS DATA

Portraits of Change

Dowell Myers

School of Urban and Regional Planning
University of Southern California
Los Angeles, California

ACADEMIC PRESS, INC.
Harcourt Brace Jovanovich, Publishers
Boston   San Diego   New York
London   Sydney   Tokyo   Toronto

ACADEMIC PRESS, INC.
1250 Sixth Avenue, San Diego, CA 92101

United Kingdom Edition published by
ACADEMIC PRESS LIMITED
24–28 Oval Road, London NW1 7DX

Library of Congress Cataloging-in-Publication Data

Myers, Dowell.
    Analysis with local census data : portraits of change / Dowell
Myers.
        p.   cm.
    Includes bibliographical references (p.   ) and index.
    ISBN 0-12-512308-6 (acid-free paper)
    1. Social sciences — Statistical methods.   2. United States —
Census.   I. Title.
HA29.M94   1992
304.6′0723 — dc20   91-29642
                                                                CIP

Printed in the United States of America
92 93 94 95   9 8 7 6 5 4 3 2 1

For Sue

Contents □□

□ **Part 3**
Extracting Information from Local Microdata

Preface □□□

The modern census of population and housing was established in 1940, with the incorporation of the housing component and the introduction of sampling techniques for use with a supplementary, long-form questionnaire. For 50 years, that basic format has persisted, but today there is growing discussion of alternative formats that are better attuned to the realities of a changing society. Some of these alternative procedures may be adopted for use in the year 2000 census, but, given the long lead times required for planning and testing, wholesale change is not likely until 2010. Thus, the current era continues.

My first experience with the census grew out of a project using the 1960 data in conjunction with the newly released 1970 data. Under the leadership of Paul Davidoff at the Suburban Action Institute in White Plains, New York, we were trying to measure increasing income segregation in suburban communities outside New York City. Like most census users, I learned on the job, receiving off-the-cuff training and various pointers from co-workers, especially Virginia Gordon. Davidoff's "quintile method" was good enough to stand the test of time and is demonstrated here in Chapter 11.

That applied experience gave me a taste of what could be accomplished with local census data. When I returned to school at the University of California, Berkeley, I proceeded to acquire theory to bolster my newfound interest, reading the works of greats in the fields of demography, urban geography, and urban planning. Meanwhile, I was carrying out small research projects on population and housing changes in the suburbs of San Francisco and Oakland. My professor, Donald L. Foley, was helpful in guiding me, but the published literature revealed a glaring gap between the grand theories of the national thinkers and the practical realities of local researchers. In subsequent training at MIT and Harvard University, David Birch and George Masnick were instrumental in showing me how deeper insights could be extracted from census data.

It is often said that authors only write books they wish they could have read. This book fits that truism exactly. How much more efficient and accurate my work would have been if I had found a guide to census data and its analysis. I was never quite sure where my work fit it, what the limitations were, or how many useful opportunities I was missing. Doubtless, a great many other local census data users have felt the same way. The small minority who have mastered the use of local census data can rightfully take pride in their exclusive accomplishment.

In the past decade, census use has been democratized. Once sophisticated research with census data was restricted to programmers using large mainframe computers. The proliferation of microcomputers has changed all that. Now it is possible for anyone to make use of large numbers of figures from the census, by entering them into spreadsheets or other programs.

The microcomputer has also made this book possible, for reasons of the increased speed it lends to the research and writing process. Widespread interest in the census is short-lived, lasting three or four years, and a book must be timed to coincide with that interest if it is to be of use. How to research and produce a book in a timely fashion is a major problem.

If an author waited for the new census data to arrive, analyzed it, and then proceeded to write a book, that work would not appear until mid-decade or later. (Witness the 1987 and 1988 publication dates of the 1980 Census Monograph series.) By then, the data from the latest census would already be cold, and many researchers would have already completed their work.

The alternative strategy is to begin a book long in advance of the release of new census data. The difficulty in this strategy is that many details concerning the format of the data are not finalized until very late in the process, even after the first results of the census are starting to be released. Thus, the advance strategy is effective only within a fairly brief window of opportunity. Consider that the decision whether or not to adjust the census for undercount was not announced until July 15, 1991, creating an especially hazardous environment for the advance planner.

Given the necessarily constricted time frame, this book could not have been researched, written, and published without a microcomputer. The 1990 census was the first for which this was possible. In previous censuses, even in 1980, the whole process would have been slowed down by the turnaround time consumed in submitting, and then correcting, a series of tasks for each chapter: computer programs, manuscript drafts, and exhibits. (This Preface was produced in final form for the publisher, for example, in one-third or one-half of the elapsed time formerly required.) Under the pre-microcomputer system, these functions also would require a staff and substantial resources to pay them. For all these reasons, a census book simply may not have been feasible prior to 1990, even if it was sorely needed.

Despite the author's relative freedom of production, a host of individuals provided support for the work in various capacities. Critical review of drafts to various chapters was provided by colleagues at the University of Southern California: Dana Cuff, Pini Herman, Curt Roseman, and Christopher Williamson. In particular, William Baer reviewed the entire early draft, provided encouragement, and helped direct my course. I also benefited from reviews by other Los Angeles area colleagues, notably Leo Estrada and Ira S. Lowry, both of whom shared the wisdom of their unique expertise. Harvey Choldin, Andrew Isserman, and David Sawicki also provided valuable written comments on key portions of the text. Also, Greg Lipton deserves credit for volunteering one of the computer programs listed in Appendix B. Finally, my mother, Ruth Dowell Myers, applied her skills to editing some of the work, urging me to an even simpler and clearer standard of writing.

It bears emphasis that the book would be of much lower quality without the support of the U.S. Bureau of the Census in providing access to preliminary information and in reviewing the data chapters of the book. I appreciate the support of key officials William Butz, Paula Schneider, and Daniel Weinberg, and I thank Michael Batutis, Carmen Campbell, William Downs, Richard Griffin, Don Hirschfeld, Howard Hogan, Jane Ingold, J. Gregory Robinson, and Andrea Rowland for their advice and review comments.

For assistance with the data, I thank Richard Lovelady, of the California State Data Center, and Muriel Kotin, a DBase programmer, both of whom generously shared their time. I also benefited from use of the library holdings and support services of the Population Research Laboratory at USC, headed by Don van Arsdol and David Heer. A number of student assistants also helped with various data chores, including Deuk Ho Cho, Seung Youn Choi, Yeol Choi, and Chuck Hong. Scott Allen provided quality control as my designated student critic of the book's presentation. I wish to thank the School of Urban and Regional Planning, and Dean Alan Kreditor and Associate Dean Peter Gordon, for making those assistants available and for providing a mainframe computer account when needed.

Direct research funding was received from two sources. The USC Faculty Research and Innovation Fund supported a related project, parts of which transformed themselves into sections of the book. The Education Foundation of the International Council of Shopping Centers, which has long had an interest in small-area data, provided a timely grant for which I am very grateful.

My editors at Academic Press, Charles Glaser and Elizabeth Tustian, also deserve credit for recognizing the importance of this work and moving it along on an accelerated production schedule. The reader may join me in thanking them for the timeliness of their efforts.

These are the individuals and organizations who have directly contributed to this book. However, as other authors have attested, the largest contributors may be the family members who are forced to give up valuable family time to the cause. My two young sons, Ben and Jesse, showed remarkable tolerance for "Daddy's deadline," wondering if it would ever be done, and when could we go outside and play ball. (It was nearly dinner time on the weekend, after all.) I've got some catching to do now.

Most of all, I wish to acknowledge my wife, Susan Marie Tuemmler. She showed remarkable strength and resilience in holding it all together for the duration. I needed her support more than anyone's, and she is my partner in this work, as in everything. I only hope that sometime soon we can trade roles in support. This book is dedicated to her.

D. M.
Altadena, California

1 □□□

Introduction

Everyday life takes us down streets full of cars and pedestrians going to work, to school, or to the shopping center. We may sometimes wonder where these people come from, where they live, or what they are like. These questions arise most often when we sense a change. Any one local area is different from another, and different from itself 10 years ago. The population may have grown and houses may have been built, but incomes have also changed and people have grown older. More minority groups exist than before and also a wider variety of family types. All these changes have occurred in just a decade.

Census data provide the most reliable and detailed information for describing local areas—our neighborhoods, cities, and counties. Businesses, government agencies, universities and nonprofit organizations all feast on these data, not only when they are newly released, but for years afterward. However, knowledge about appropriate analysis is not as widespread as the data themselves. For lack of guidance, most census users only skim the surface of the data, missing the rich and exciting discoveries that can be made about each local area.

Local area analysis with census data has not received the scholarly and professional attention it deserves. The vast majority of census users seek information about local areas, yet these local users lack examples or methodology texts prepared by experts in the field. Experts on census data who have explained basic methods all tend to focus at a much higher level of geography—at the level of the nation and its regions.[1] However, local level analysis requires somewhat different theory and dif-

---

[1]The classic book by Pittenger (1976) may be the only exception to this rule; yet even that work addressed only state and county populations, not subcounty areas such as cities and neighborhoods.

ferent methods from those applicable to larger areas. Local analysis is bottom-up in orientation (not the top-down view of national researchers), and it follows a different line of causality, often the reverse of that typical at the national level.

To unlock the power of local census analysis, users require guidance about the content of the data and strategies of analysis. The aim of this book is twofold: to promote greater understanding of the data, and to provide methods to help users more fully exploit the rich potential of local census data.

☐ 1. Use of This Book

This book provides a comprehensive introduction to the scope of local census data and carefully explains the different variables and sources of data. The major effort is to show how analysis can be conducted to extract maximum use from local data. The initial skill requirements are relatively minor—multiplication and division, and perhaps a familiarity with microcomputer spreadsheets. With that background, the guidance offered by this book will unlock considerable analytic power.

This book was written to be available in time for use with 1990 census data. It explains those data in detail and contrasts them with the 1980 census. However, because the actual 1990 census results were not released until 1991 and 1992, the analytical examples make use of 1980 and 1970 data. The analyses shown with those data are designed to be repeated with 1990 and 1980 data in the reader's own community.

Many types of users will benefit from this book. Anyone wishing a basic orientation to the census, or to local data availability, will profit from this work. In addition, with the release of the 1990 data, most local users will share a common question: How has this area changed since the last census? Academics, government analysts, market researchers and many others will share this same quesion. Specific users may need to focus on specific topics, and they can find concentrated guidance by locating their topic in the table of contents or the index.

Most readers are unlikely to read the book from beginning to end. Many readers may simply desire to undertake a particular analysis and seek guidance for a limited task. Other readers may have extensive experience with census data and wish to fill some minor gaps. Still others may seek only a basic introduction to census data and may not wish to conduct much analysis.

For each of these reader groups we offer the following advice.

1.1. Whole Book Readers

Those reading the whole book through can usefully read the chapters in sequence, paying particular attention to the introductory discussions and skimming over the more detailed method descriptions. To truly understand the latter, readers will need to get out of their armchairs and "get dirty with the numbers." Trying the methods is the only way to learn them.

1.2. Single Task Readers

Readers seeking guidance to complete a single task, such as an age or income analysis, should turn to those chapters and read them in full. Each gives enough background about the data to get started. Readers can choose the method or methods that seem most useful and then go to work.

1.3. Experienced Analysts

Those who already have extensive experience with census data analysis should certainly read this introduction and then browse through the book. Experienced analysts will know how to pick out subjects of interest. Although the microdata section is tailored more specifically for experts, each chapter contains points of potential interest to expert users, especially the latter parts of each chapter.

1.4. Readers Desiring a Basic Census Introduction

Those who are beginners may wish to follow the same strategy as experts, simply browsing through the book to see what is of interest. Chapters 2, 3, and 4 are most relevant to them. In addition, beginners should read the early portions of subsequent chapters but avoid the microdata section entirely.

☐ 2. Major Emphasis of the Book

This book is designed for the needs of local census data users, providing a comprehensive introduction to the data and teaching useful methods of analysis. More than a reference work or a how-to manual, the emphasis throughout is on developing a sophisticated, critical understanding of the census, thus turning sheer numbers into exciting insights. Beginners will acquire both immediate skills and a firm foundation for future development. More experienced analysts will learn how their expertise fits into a broader picture and can round out their skills.

Four major themes are the focus of the chapters that follow:

- mastering access to the full range of census data;
- measuring changes over time;
- comparing places; and
- demonstrating how different aspects of change fit together.

So vast is the output of census data produced each decade that users often drift in a sea of reports and computer tapes. This book condenses the expertise of many into a brief guide that quickly brings the reader up to speed. Even experts can benefit from such considerations as the *hierarchy principle* governing data access in larger and smaller areas. Users may wonder also just how accurate *are* local census data and how accuracy of use can be maximized. Which reports come from just a sample and not from a 100-percent count? Which reports offer the most detailed data and for what kind of places? The book provides an overview of all the data offerings from the census and then gives detailed guidance about how to use the different data for specific analyses.

The particular analytical focus with these data is on *changes over time* in local areas. The major question many local analysts will want to address with the 1990 census data is "How has this local area changed since the 1980 census?" Answers to the question include not only the total growth of the area, but more importantly its changing characteristics.

A second common question asked by local users is "How does this local area differ from other areas or from the surrounding region?" The simple answer to this question compares places at one point in time using the most recent census data. The stronger answer *compares patterns* of change in different places. For example, two areas may have the same income level, but what is more important is that one area's income has been moving up and the other down.

Methods to be presented lead toward *integrated portraits* of change that describe and compare local areas in depth. Changes in age structure, housing type, income and labor force, among other factors, are part of a single combined reality. Much like the proverbial blind men feeling separate parts of the elephant, we must learn to envision different parts of the census data as fitting together. Combined, they tell a unified story that is much bigger than suggested by a single variable alone. For this reason, the book places major emphasis on the linkages between different types of variables in the census. Such analysis, of course, requires the broad understanding of census data sought at the outset. It also requires some theoretical orientation, as discussed in the next section. To date, many of the linkages within census data have not been well researched in local areas, and there is a rich body of questions here that deserves greater analysis.

Overall, the methods to be described support *value-added analysis* of census data, extracting far greater meaning from census figures than is revealed by the raw data alone. Most census users can profit from access to such simply crafted but powerful tools.

2.1. Some Limitations

The comprehensive scope of this book is not without limits. We do not cover several important topics. First, although stressing trends over time, we do not delve into forecasting methods. Our emphasis on understanding change between censuses is essentially historical, laying important groundwork for future projections but stopping short of that endeavor.

A second restriction in scope is that we offer a limited degree of spatial area analysis, only briefly introducing the TIGER files and the developing technology of geographic information systems (GIS). Those topics are so vast that they deserve comprehensive treatment in a separate work. The emphasis of the present book is on concentrated analysis of individual places. GIS users who learn to construct in-depth local portraits may be able to turn those lessons into more informed spatial analysis cutting across many communities.

A third limitation of this book stems from our sole focus on census data. A full understanding of local changes requires attention to major forces not recorded in the census, such as the local economy, politics, or roadways, shopping malls, and other major developments. Our emphasis on census-measured changes should not be seen as failure to recognize the importance of these additional factors. Rather, our goal is to extract maximum analytical value from the census data available in localities.

☐ 3. Theoretical Basis

Local area analysis has special qualities that demand specialized theory. Three differences stand out:

- bottom-up orientation;
- reversed causal order, from housing to population to labor force; and
- opportunistic and fuller use of local census data.

3.1. Bottom-Up Perspective

One of the barriers keeping national researchers from contributing to local area analysis is that their methods and theories are not suitable for local application. Their approaches impose a top-down causal thinking that assumes local areas largely mirror changes in the nation or region as a whole. Although this is true to some extent, a bottom-up approach toward local analysis stresses the unique history of present population and housing characteristics in each locality. With a stronger grasp of local dynamics, analysts can more realistically link the effects of growth drawn from outside the locality.

In general, the interaction between local changes and regional changes is of substantial interest. No local area exists in isolation. At the same time, every local area has both unique characteristics and a history that create a momentum carrying from one period to the next. The approach taken in this book is to stress the bottom-up perspective while keeping the regional context closely in mind.

3.2. Reversed Causal Order

In a bottom-up perspective, the dynamic linkage of variables in local areas is the opposite of that found in the nation or region. The direction of linkage is raised by the question "What comes first, population or housing?" At the national level, it is clear that population growth creates housing needs that lead to housing construction. However, at the local level, houses must be built first before they can be filled with population. Once an area is fully developed, the number of persons that can live in that locality is largely constrained. Only through expansion of household size can the area's population continue to grow, and it may even shrink.

For reasons of housing's precedence in local areas, practitioners of small-area demography have learned to rely on the *housing unit method* as one of their primary tools for estimating population. Indeed, the housing stock forms the primary basis for collecting census data because this collection is rooted in local areas. Whereas housing may be relatively unimportant for explaining population in the nation or regions, housing forms the essential substructure for small-area demography.

Until recently, information on the housing stock was not well integrated with local population dynamics. Only in local area analysis does housing rise to prominence, and so there was little guidance from national-level researchers on appropriate theory. The emergence of a new interdisciplinary theory, termed *housing demography*, has begun to remedy that deficiency (Myers 1990a). This theory stresses the detailed interconnections between populations and their housing stocks.

The present work extends this concept for linking population and housing to include linkage of labor force and employment. Housing stocks may be filled with population, but these persons have labor force behavior necessary for paying the housing bills. Labor force participation is well known to depend on the age and sex of residents, and we will show how it is also related to their type of housing.

3.3. Exploiting Local Census Data

Local census data may be underused because scientific theories have not been designed to exploit the data opportunities available. The sophisticated models developed for analyzing change at the national level are not usable at the local level because the requisite data are not available locally. At the same time there is a wealth of local census data that lies relatively untapped.

Theoretical barriers posed by a top-down orientation and a nationally oriented causal order may have limited more creative and effective use of the rich local data. The goal of this book is not to teach a single, theoretically correct mode of analysis. Rather, the goal is to describe a wide variety of possible analyses with varying degrees of complexity or technical sophistication.

The key to fuller use of locally available census data is to recognize the possibilities. The aim of the book is to promote a flexible set of skills and research approaches that more fully exploits the rich potential of local census data.

☐ 4. Plan of the Book

This book strives for comprehensive coverage of census data and their analysis. The scope entails all forms of census data available locally, as well as skills useful for both beginners and experts. This ambitious scope has led to a particular structure for the book, which we describe in terms of skill levels and topic content.

4.1. Background Required

On the whole, the book's aim is to be readable by any person with some college education and a familiarity with numbers. The methods employed here require no more math than addition, subtraction, multiplication, and division. Some of the methods require only a calculator, most a microcomputer spreadsheet, and others a mainframe computer. (See the discussion on computers below.)

4.1.1. Layered Sequencing of Two Skill Levels

To accommodate the needs of both beginners and experts, the book has adopted a *two-skill-level strategy*. The presentation is layered in sequence from easier to more difficult. In general, the earlier sections of each chapter are devoted to simpler concepts and procedures, while later sections grow more difficult. The later chapters of the book are also more attuned to the interests of experts.

A range of variations are often described for particular methods, some of which are more difficult or time consuming than others. Most of the explanation is devoted to the simpler methods, with paragraphs or footnotes provided to guide analysts toward more expert variations. With this format, census users can choose the method that either matches their skill level or their data and time available.

The one major exception to this beginner-expert sequencing of topics occurs in the first part of the book. The review of fundamental concepts and definitions presented in Chapters 3 through 5 is designed to hold equal interest for the beginner and those already experienced in the field. Those chapters, although simply stated, reach for a deeper understanding than found in most other sources.

4.1.2. Computer Requirements

Many of the methods that follow can be executed with a calculator. However, it is advantageous for the analyst to have access to a microcomputer and to be familiar with spreadsheet software such as Lotus 1-2-3 or Microsoft Excel, although any spreadsheet will suffice. Microcomputers speed up the calculations and make possible a number of reruns where new assumptions or corrections are introduced. Modern spreadsheets are also linked to graphing programs, permitting detailed illustrations of the data in addition to producing tables.

Rapid changes in technology since 1980 have created a revolution in census analysis. Analysis that once required days of FORTRAN programming and a mainframe computer now can be carried out quickly on the desktop of the average person. At the time of the 1980 census, very few persons had skills with microcomputers. However, 10 years later, these skills have become almost as common among younger persons as typewriting skills. This book capitalizes on the new technology and its accompanying new talent.

Spreadsheets are sufficient for almost all of the calculations in the book because their row-and-column format is well suited to census data in tables. For the analysis of only a few places, it is often easiest to key in the data from published sources. Users with access to CD-ROM drives or mainframe computers have the advantage of being able to process the data for many places at the same time, permitting comparison of the results from hundreds of census tracts. However, the in-depth analysis of a single place with a spreadsheet may often be preferred to a more superficial analysis of 100 places using a mainframe computer.

4.2. Sequence of Topics

This book consists of three sections: Definitions and Strategies, Analysis of Local Changes, and Extracting Information from Local Microdata. The first section covers foundation concepts and introduces census data. The second shows how census data in summary tables can be analyzed. The third addresses the more complex, but infinitely more flexible, analysis of microdata records.

4.2.1. Definitions and Strategies

In the first section we present basic concepts and definitions, surveying ideas about census data and exploring the structure of the data.

As a leadoff to the first section, Chapter 2 introduces case example neighborhoods in Los Angeles County that will be analyzed in later chapters. Key characteristics of each neighborhood are identified and questions for subsequent research are raised. These examples include the richest and poorest parts of Beverly Hills, singles neighborhoods in Hollywood, suburban family neighborhoods in the San Fernando Valley—home to the teenage "Valley girls" and boys—and minority neighborhoods near the Pasadena Rose Bowl, one changing from white to black, the other from black to Hispanic. This chapter also serves as a practical introduction for beginners to some basic census data and concepts.

Even experienced census users are often unaware of the full variety of available data and the alternative uses. Both beginners and experts can benefit from the concise overviews of fundamental concepts and data structures offered in Chapters 3 and 4. Readers might decide initially to skip all of the explanations about the data and move right to the analysis. But sooner or later, most readers will very likely come back to those chapters in search of clarification. This book does not replace the need for the definitive documentation provided by the Bureau of the Census. Rather, the book provides a concise, overall explanation that many readers may find more accessible.

Attention is also given in the introductory section to strategies for presenting results in tables and graphs. Such visual aids are very useful for conveying local portraits of change to an interested audience. Chapter 5 summarizes the recommendations of leading thinkers on tabular and graphic presentation. With these techniques, the results of census analysis can be made to look more professional, and hard-earned analytical results can have their communication needs served more effectively.

4.2.2. Analyzing Local Changes

In the second section of the book we turn to examples of local area analysis and focus on separate topics in each of six chapters. These topics include the housing stock, household relationships and marital status, age structure, migration and residential mobility, racial change, and income. Each of these topics is examined with data in summary tabulations—most of it published in census tract books and accessible to the widest range of researchers.

Specific methods of analysis are introduced in each chapter and illustrated with examples drawn from the case example neighborhoods. Most of the methods presented in these chapters should be accessible to beginning analysts with spreadsheet familiarity (or a fast calculator).

4.2.3. Extracting Information from Local Microdata

The third section of the book turns to the analysis of microdata records on computer tape or CD-ROM. Chapters in this final section cover more advanced topics and require greater computer access than earlier sections. Custom tabulation with microdata allows for substantial creativity but requires a sound, disciplined grasp of the underlying logic.

Chapter 12 begins with coverage of the basic skills required for structuring microdata in tables. Choosing alternative denominators, disaggregation, and manipulation of marginals are all essential skills. This chapter also introduces how the microdata files can be extracted into different databases for different purposes. With customized programming, major sets of census variables can be linked to one another: employment to population, housing to employment, or population to housing. Each is illustrated in subsequent chapters.

In Chapter 13 we analyze detailed data on labor force participation by age, sex, and race, illustrating both a detailed, population-based analysis, and also showing how employment is linked to population. A further purpose of the chapter is to demonstrate how the detailed picture drawn from a custom microdata tabulation, selected from larger area data, can be used to augment the more limited data available for small areas. Widely used techniques of iterative proportional scaling by marginals are introduced.

Chapter 14 then demonstrates alternative tabulation bases. In addition to population-based analysis, we can carry out housing-based or employment-based analysis, and we can also link these three together. In one application, the local age structure is related to characteristics of the housing stock the population occupies. In a second application, employment-based analysis is used to analyze characteristics of workers in relation to their type of employment.

☐ 5. Maximizing Use of Local Census Data

The book concludes with a summary of major lessons about census data use. The orientation given here provides a platform for maximum use of local census data. Readers will find many ways to extend the basic concepts and procedures outlined. We do not exhaust all the possibilities, but merely seek to open the door into some of the richest areas. The book's goal is to create the opportunity for more exciting and rewarding local analyses. In every community, a wealth of data beckons.

Part 1

Definitions and Strategies

2 □□□

Using Census Data for Local Portraits

Census data can tell us a great deal about a local area. We can learn how large a place is, whether in terms of population, number of housing units, or land area in square miles. By comparison with previous censuses we also can learn how rapidly the area has grown (or declined) in population or housing.

The richness of census data is revealed when we break down the total into different categories. Some of the most popular breakdowns are population by age group, housing by size of unit, and households by income level. Growth can be calculated for many separate categories, not just the total size of the place.

In crafting a portrait of a local area, it makes a great difference what proportion of the total falls into different brackets. One place may be an elderly retirement community, another a young singles neighborhood. With census data we can identify key characteristics of each local area and see how they are changing.

This chapter introduces some case example neighborhoods that will be analyzed in later chapters of the book. Through short, overall descriptions of these places, we also aim to give the flavor of what census data can tell us. Only summary profiles will be addressed in this chapter; more detailed dynamics will be delved into later. To better orient beginning census users, the chapter also offers a brief introduction to some basic census concepts. The following chapters develop these issues in greater depth.

☐ 1. Strategy of Analysis

Most analyses are motivated by the desire to learn simple answers to basic questions. As we learn the answers to a series of separate questions, we begin to build a portrait of the community.

One set of questions concerns *overall growth*. How much did the locality grow over the decade? Did it grow faster or slower than other places? Different measures of growth may yield different results, depending on whether the focus is on housing, households, population, or economic factors.

The growth question quickly leads to a more complicated analysis: if the locality grew this much over the past decade, how much will it grow over the coming decade? A desire to answer this forecasting question requires us to examine closely the dynamics of growth over the past decade. Without a detailed understanding of past growth—obtainable from census data—any venture into forecasting remains superficial and very uncertain.

A second general set of questions concerns the *composition* or makeup of the local population. How rich is this neighborhood? How many minorities are there? Is this a place for young singles? To really answer these questions, we must compare the local area to others. Otherwise we will not know if $30,000 income or 30 percent singles is a lot or a little. The key is to find out whether this local area is above or below average for tne region.

Questions about composition become even more intriguing when we ask how the composition has changed over time. Essentially all answers must be relative to some standard. "How rich" depends on the rest of the city, county, or region. And change over time may involve a relative standard that is also shifting. If the average income in the county rises from $20,000 to $30,000 between censuses, and our local income has held constant at $20,000, what does this say? In this example, local income has not kept pace with the average change in the county, and now it is below average.

Reasons for local changes can often be found by comparing different data for the same area. For example, it is possible that income was stagnant for an area because it holds a very large share of elderly population, most of whom live on fixed incomes. Or income may have been held down because a great many apartments were built in the area and renters tend to have lower incomes than owners.

1.1. Integrating Pieces of Census Information

The strategy recommended here is to build local portraits of change that combine and compare different aspects of census data. Local areas are measured relative to the county average, and different census variables are linked to one another. In this manner we build an integrated portrait of change in the community.

The census contains such a mountainous wealth of information that no one ever mines it completely. At best we can only tunnel in here and there, hoping to pull out some of the riches. This book offers a guide to some of the most important census information, suggesting how different parts of the data can be fit together into a total portrait that is greater than the sum of the parts.

In our analysis of growth and changing composition over time, we investigate four general areas with local census data:

- housing—the physical dwellings people live in;
- households—the person or persons who occupy a housing unit (this includes families as well as persons who are unrelated to one another);
- population—the number, age, sex, or race of persons living in the area; and
- economic characteristics—income levels and employment behavior.

These four topic areas are distinct but also related to one another. Housing provides the place where people live, and the population living in housing units forms the areas households. At the same time, different population groups have different economic characteristics. In turn, the amount of money people have (and their lifecycle status) determines what kind of housing people occupy.

A host of more specific topics might be investigated by different census users. For example, transportation analysts would probably focus on the number of vehicles at the residence and commuting patterns. School authorities might focus on the number of persons enrolled at different grade levels.

For a generalized usefulness, however, we must address broader topics that underlie many different specific concerns. Few census users can avoid addressing the four general topics identified above. Methods demonstrated in the context of these broad topics can be applied to more specialized questions if desired.

1.2. Measuring and Describing Change Over Time

Many census users have a hard time describing the amount of change in a local area. Even with the data in hand for two different census years, it is not obvious which calculations are most relevant. A number of different calculations can be made, each of which tells a slightly different story. The analyst must choose which calculation to report and then select the appropriate words to describe the story. With a little guidance, users can avoid the common traps that catch so many census users.

After measuring we must explain in words what meaning we find in the numbers. Adherence to some basic terminology can help to clarify distinctly different concepts. "Population change" is not a specific enough term to represent a calculation. Does that mean the numerical change in population or does it mean the percentage change in population? A local area may have grown by a total of 2,000 people, representing a 10 percent increase over its former size of 20,000 persons.

The use of percentages to describe change seems so simple, but this often leads to a quagmire of confusion. The need for clear terminology becomes especially important when different areas are being compared or when describing how the parts of a total have changed. Consider the example of Table I for an area made up of only two population subgroups—blacks and Hispanics. Given these data, what measurements—and what words—should we use to describe the changes in the number of blacks versus the Hispanics?

The black population increased from 1980 to 1990 by 1,000 persons, or by 5.9 percent, and the Hispanics by 33.3 percent. Alternatively, we might state that blacks

Table I.

	1980	1990	Numerical Change	Percentage Change
Blacks	17,000	18,000	1,000	5.9
Hispanics	3,000	4,000	1,000	33.3
Total	20,000	22,000	2,000	10.0

and Hispanics shared equally in the locality's growth because each accounted for a 50 percent share of the total growth. In still another view, blacks can be said to have declined as a percentage share of the area's total population, falling from 85 percent of the total in 1980 to 81.8 percent in 1990. Conversely, the Hispanic share of the total increased between 1980 and 1990 because they grew more rapidly in number, by 33.3 percent, even though they experienced the same amount of growth as blacks.

The key to making clear these different interpretations is to spell out the meaning with complete sentences. Shorthand statements about "percent of change" are not sufficient and often misleading. Does that imply the percentage change,[1] the percentage share of the total change,[2] or the change in the percentage share of the total (i.e., the difference in percentage points)?[3] The conventional rule is that any statement using percentages must specify *percent of what*. The alternative measurements of change each have a different conceptual meaning, as demonstrated in various applications throughout the book.

☐ 2. Identifying Places to Analyze

Probably the most outstanding feature of census data is that the same types of data are provided in the same format for so many different places. We can choose to analyze any state in the nation, or any of its 3,141 counties, or any city or neighborhood. Data are available in a standardized format for every place, so the same analysis can be repeated easily anywhere.

For most of this book we emphasize the most commonly available data for small areas—at the level of the municipality or the neighborhood, termed a *census tract* in urban areas and a *block numbering area* in rural areas. A census tract is a large neighborhood, generally with a population between 1,500 and 6,000. Every metropolitan area has been mapped with census tracts, and in rural areas data are provided for the equivalent spatial unit, the block numbering area. Because these small areas are so numerous, they are identified by numbers such as Tract 1373. When development

[1]Percentage change (also known as the growth rate) is the numerical change in each subgroup divided by the original number in that subgroup, times 100; e.g., 1,000 / 17,000 × 100 = 5.9 percent.

[2]Percentage share of growth is the numerical change in each subgroup divided by the total growth for all subgroups combined, times 100; e.g., 1,000 / 2,000 × 100 = 50 percent.

[3]Percentage point change (also referred to as the change in percent or the change in share) is a subgroup's percentage share of the total at one point in time minus its share of the total at an earlier point in time; e.g., 81.8 percent − 85.0 percent = −3.2 percentage points.

causes population to grow, the tracts are subdivided and the numbers grow decimals (Tract 1373.02). These are real places bounded by streets and identified on census tract maps by their number IDs.

2.1. Example of Census Geography

A later section describes census geography in more detail, but it is helpful to illustrate the basic concepts here. Figure 1 shows an example of the geographic subdivisions in a metropolitan county. The example, prepared by the Census Bureau, makes clear the hierarchical structure of the census geography. The geography builds from the block level, at the bottom of the diagram, on up to block groups and census tracts (or block numbering areas). In turn, the census tracts and block numbering areas are nested within counties. Counties then combine to the level of states and the states make up the nation.

Cities and small municipalities can overlap the census tracts and even spill across county boundaries. At a larger scale, metropolitan statistical areas (MSAs) are formed out of groups of whole counties, but the notion of an urbanized area is defined more specifically to exclude rural portions of MSAs (as shown in Fig. 1).

The importance of Fig. 1 is that it shows how a specific block, e.g., Block 914 that lies between 2nd and 3rd avenues, makes up a piece of Census Tract 1014 and a piece of the City of Altoona, as well as a piece of Blair County. Maps published by the Census Bureau allow users to find what block, census tract, or other statistical geography pertains to each specific place in the nation. With the geography identified, users can then look up the relevant data in books or in computerized files.

For most of this book we focus at the subcounty level, mainly using census tract data for illustration, although the same data may be obtained for block numbering areas, municipalities, or counties. The Census Bureau makes these small-area data available in published books or in computerized form. Somewhat more detailed data are found in the computerized versions, but the published books are quite adequate for most purposes. All of these data resources are described at length in Chapter 4.

The census tract (and block numbering area) books are especially useful because they contain data about larger-sized areas as well: cities, counties, and the metropolitan area. In one source we gain access to comprehensive data in a standardized format for all geographies—except individual blocks and block groups—in the local area.

□ 3. Case Example Neighborhoods

The standardization of data format permits us to choose any place as an example for analysis. Methods explained in this book are demonstrated with data from the Los Angeles area, principally 10 neighborhoods located in the city of Los Angeles and surrounding area. These neighborhoods were selected because they are well known—some of international fame—and because they illustrate dynamics of change in a variety of ways.

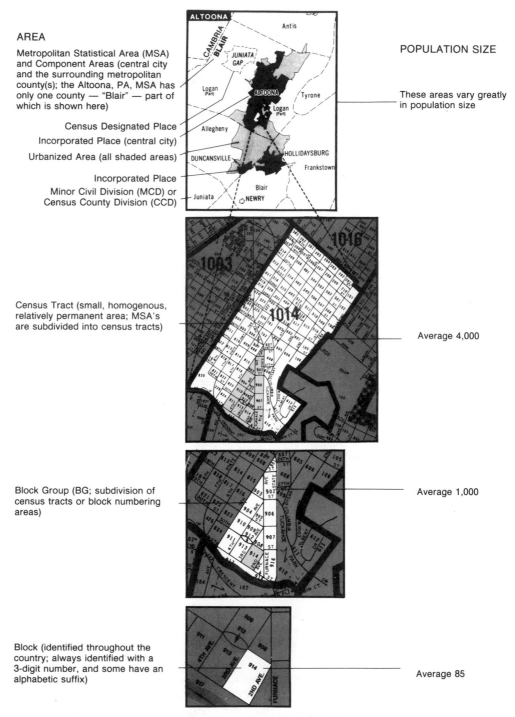

AREA

Metropolitan Statistical Area (MSA) and Component Areas (central city and the surrounding metropolitan county(s); the Altoona, PA, MSA has only one county — "Blair" — part of which is shown here)

Census Designated Place
Incorporated Place (central city)
Urbanized Area (all shaded areas)

Incorporated Place
Minor Civil Division (MCD) or Census County Division (CCD)

Census Tract (small, homogenous, relatively permanent area; MSA's are subdivided into census tracts)

Block Group (BG; subdivision of census tracts or block numbering areas)

Block (identified throughout the country; always identified with a 3-digit number, and some have an alphabetic suffix)

POPULATION SIZE

These areas vary greatly in population size

Average 4,000

Average 1,000

Average 85

Figure 1. Example of geographic subdivisions in a metropolitan county. (Taken from the U.S. Bureau of the Census.)

Why not choose examples about which there is widespread curiosity and that are fun to explore? The same analyses can be carried out for any locality or neighborhood in the nation. The author's preference is for the Pasadena area where he lives, but the reader's own home town may hold even greater interest than the famous places identified here.

The case example neighborhoods are drawn from four general parts of the Los Angeles region. Many of the following places are well known and are located on the accompanying map (Fig. 2):

- San Fernando Valley;
- Hollywood;
- Beverly Hills; and
- Pasadena.

First, as a representation of suburban development, two census tracts are identified in the San Fernando Valley, the home of middle-aged families and teenage "Valley girls" (and boys). One of these examples illustrates a case in which household size drops markedly as teenagers grow up and leave home. The other illustrates a case in which apartment construction transforms a neighborhood once wholly composed of detached single-family houses.

Examples of inner city, singles neighborhoods are two census tracts in Hollywood and one tract that is a gay neighborhood in West Hollywood. These neighborhoods illustrate a variety of different situations, including a mixture of old and new housing and the mixture of other population types among the young singles.

Two contrasting neighborhoods in Beverly Hills are identified for showing methods of measuring income. One is the wealthy enclave most expect to see in Beverly Hills, but the other is a much more modest neighborhood. How poor can you be and still live in Beverly Hills?

Finally, to explore the complex subject of racial and ethnic change, we have selected three census tracts extending northeast from the Pasadena Rose Bowl. These include a core black area, one mixed black and Hispanic, and one mixed black and white (non-Hispanic), all of which are changing rapidly.

The San Fernando Valley and Hollywood tracts all fall within the city limits of Los Angeles. The Beverly Hills tracts are contained within Beverly Hills City. The Pasadena area tracts all fall within that city, except for Tract 4602, which lies in an unincorporated area, a census-designated place called Altadena that borders Pasadena along the northern mountains.

All of these tracts will be compared to the whole of Los Angeles County so that we can gain a better perspective on their similarities and differences. The county is synonomous with the Los Angeles–Long Beach MSA and provides a good regional basis for comparison.

Because of the long lead time needed for publication, please bear in mind that our analysis of these places was conducted before the release of the 1990 census data. However, we have advance information about the forthcoming data products and their formats, enabling us to guide the reader from 1980 to 1990. With this book in hand, readers will be able to repeat our 1970 to 1980 analysis with 1990 data when released.

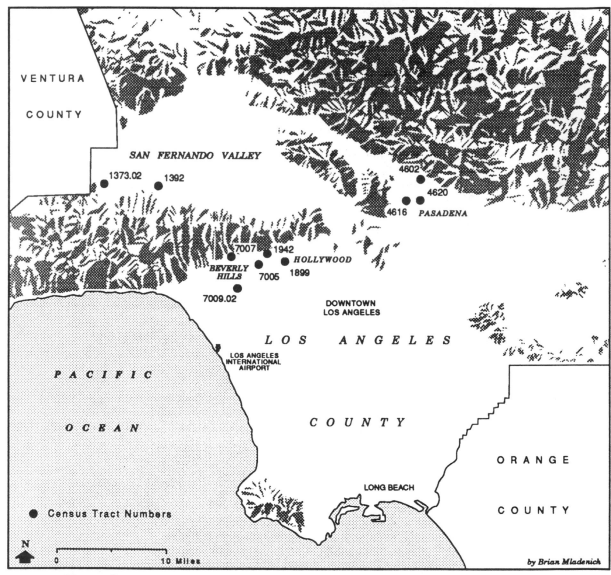

Figure 2. Locations of example census tracts and selected cities in Los Angeles County.

3.1. Comparison of Population Growth

As an initial examination of the data for these tracts, let us compare their rate of population growth. For this first analysis, it may prove helpful to show each single step in the calculation because there is more than one way to measure growth. Amplifying the distinctions made previously, consider the basic options for measuring growth from an earlier period to a later period (e.g., from 1980 to 1990):

- absolute growth or numerical change: (the number later minus the number before);
- percentage growth: (absolute change / number before × 100); and
- change in percentage of the total or percentage point change: (a subgroup's later percentage of the total minus the subgroup's percentage of the total before).

3.1.1. Interpreting the Different Measures

Each of the methods has different advantages and is used at different points in the chapters to follow. The absolute growth measures the actual numeric increase—how many people are added. Often we need to know the actual number change because that is what counts for purposes such as political redistricting or for planning the size of local government services.

The advantage of percentage change is that this measures the rate of growth, such as 20 percent growth over the decade. Conversion of absolute growth to percentage growth is essential for comparing places of different sizes. For example, absolute growth of 2,000 persons is 20 percent of an initial population of 10,000 residents but only 5 percent of an initial population of 40,000 residents. Thus the same absolute growth can be part of either a faster or slowing growing area, depending on the initial size of the place.

The third method—change in percentage of the total—measures a completely different concept. This measure is more useful for describing change in composition, i.e., how a subgroup's percentage of the total has changed over time. Nevertheless, we can illustrate this method in the context of growth analysis.

These three methods are illustrated with our tract data in Table II. Column A reports the 1970 population size and column B the 1980 size. The first method of measuring change—numerical change—is shown in column C. The second method—percentage change—is shown in column D. Results of the third method are then presented in the last column, following some intermediate calculations in columns E and F.

3.1.2. Overall Numerical Growth and Decline

The amount of numerical change ranges widely, from growth of 2,344 in West Hollywood Tract 7005 to a loss of 1,199 in San Fernando Valley Tract 1373.02. In fact, six of the tracts lost population over the decade. Examination of additional data for each tract may reveal whether that population loss was due to loss of housing units, fewer occupied units, or a shrinking number of persons per household. Similarly, additional data may reveal the causes of the substantial population growth in other areas.

Table II. Measurement of Population Growth in Example Neighborhoods

	Total Population 1970 A	Total Population 1980 B	Numerical Change (B−A) C	Percentage Change (B−A)/A × 100 D	Percentage of Total Sample 1970 (A/total.A) × 100 E	Percentage of Total Sample 1980 (B/total.B) × 100 F	Change (F−E) G
San Fernando Valley							
1373.02	5,909	4,710	−1,199	−20.3	11.4	8.5	−2.9
1392	3,315	4,341	1,026	31.0	6.4	7.8	1.5
Hollywood Area							
1899	8,003	7,814	−189	−2.4	15.4	14.1	−1.3
1942	5,236	5,092	−144	−2.8	10.1	9.2	−0.9
7005	5,125	7,469	2,344	45.7	9.9	13.5	3.6
Beverly Hills							
7007	3,846	3,555	−291	−7.6	7.4	6.4	−1.0
7009.02	6,371	6,101	−270	−4.2	12.3	11.0	−1.3
Pasadena Area							
4616	3,487	4,477	990	28.4	6.7	8.1	1.4
4620	4,559	5,983	1,424	31.2	8.8	10.8	2.0
4602	6,141	5,925	−216	−3.5	11.8	10.7	−1.1
Total of Sample Tracts	51,992[a]	55,467[b]	3,475	6.7	100.0	100.0	0.0
Los Angeles County	7,032,075	7,477,503	445,428	6.3			

Notes: [a]Total.A (i.e., the sum of the case example neighborhoods in column a, or 1970).
[b]Total.B (i.e., the sum of the case example neighborhoods in column b, or 1980).

3.1.3. Comparing Growth Rates

The percentage population growth also varies considerably, from a positive 45.7 percent to a negative 20.3 percent. The advantage of this method is seen when we compare these population growth rates to that for the entire sample or the whole of Los Angeles County (bottom of Table II). Those larger areas have much greater numerical change, but relative to their larger size this works out to around six percent growth over the decade.

From this slow rate of growth, the county accelerated to an 18.5 percent increase between the 1980 and 1990 censuses, according to early returns from the 1990 census. That is a numerical increase in the county that is nearly one million greater than experienced in the 1970–80 period. Although much of this growth was accommodated in newly developed parts of the county, accelerated growth was experienced throughout the county's neighborhoods.[4]

3.1.4. A Changing Share of the Total

Finally, change can be measured by the third method, showing the change in each subarea's share of the total population in our sample. This illustrates the type of analysis census users can conduct with the subareas that make up their own commu-

[4]County of Los Angeles Department of Regional Planning (1990).

nity. For example, if our 10-sample census tracts were all in one community, we would find that some areas were accounting for a larger share of the total over time. West Hollywood Tract 7005 has increased its share by 3.6 percentage points, whereas San Fernando Valley Tract 1373.02 decreased its share by 2.9 percentage points. In Table II note how these percentage *point* changes (column G) are different from the percentage change measured by the second method (column D).

All of the above methods address changes in the total population. This focuses only on the very tip of the iceberg of local change. Many more dimensions can be analyzed, including not only the size of each place but also changes in its characteristics.

3.2. Profiles of Change for Each Place

A large number of indicators have been collected to profile each of these neighborhoods. Data are presented for both 1970 and 1980, and these can be extended to 1990 once those census data are released. These indicators address a large number of characteristics in summary fashion. More detailed analysis of changes will be carried out in subsequent chapters. Our purpose here is merely to provide a summary profile for each of the sample places.

The indicators are given in four groupings that correspond to our four basic sets of variables. Three of the indicators measure percentage change in total population, households (occupied units), and housing. Others measure change in the tract's composition, as shown by change in the tract's percentage that falls in a specified category (e.g. change in percent owners). The "difference" recorded in the summary profile is simply the change in percentages in a certain category (or change in medians). All of these measures are explained in some detail in later chapters in which that topic is elaborated.

The list of profile measures is repeated for each tract in a series of tables. The first, Table III, presents the summary profiles for Los Angeles County and the two San Fernando Valley tracts. Subsequent tables record the same data for the Hollywood tracts, the Beverly Hills tracts, and the Pasadena tracts. The following sections use these data to describe briefly each of the case example neighborhoods.

3.2.1. San Fernando Valley Tracts

The two Valley census tracts shown in Table III are drawn from different ends of the economic spectrum. Tract 1373.02 is an upper middle-class area with a median family income in the 1980 census equal to 191 percent of the county median. (This method of standardizing income as a ratio to the county median is elaborated in the chapter on income, Chapter 11.) This neighborhood is almost exclusively single-family homeowners in large houses averaging seven or more rooms. Most of the homes were new in 1970, but since then little construction has occurred.

Valley Tract 1373.02 lost a surprising amount of population (a decline of 20.3 percent) between 1970 and 1980. Because occupied housing decreased by only 2.2 percent, the population loss apparently is related to falling household size (from 4.20 to 3.44). In 1980 note that 42.9 percent of homes were occupied by longtime residents (10 or more years). In that time span children may have grown up and left home,

Table III. Summary Profile of Los Angeles County and San Fernando Valley Tracts

	Los Angeles County			San Fernando Valley Tract 1373.02			San Fernando Valley Tract 1392		
	1970	1980	Difference	1970	1980	Difference	1970	1980	Difference
Housing									
% Change in stock	—	12.5	—	—	-0.5	—	—	48.8	—
% Owners	48.5	48.5	0.0	97.5	96.6	-0.9	46.7	43.1	-3.7
% Single-family units	62.0	56.2	-5.8	100.0	98.2	-1.8	64.5	52.6	-11.9
Median rooms	4.5	4.5	0.0	7.0	7.2	0.2	4.0	4.3	0.3
% New units	24.1	15.0	-9.1	95.4	6.4	-89.0	33.4	35.4	2.0
% pre-1940	25.3	17.7	-7.6	0.7	0.0	-0.7	0.7	0.3	-0.3
Median value	$24,300	$87,400	—	$46,000	$166,400	—	$19,500	$83,300	—
ratio to county	100	100	0.0	189	190	1.1	80	95	15.1
Households									
% Change in occupied units	—	12.3	—	—	-2.2	—	—	44.1	—
Median HH size	2.40	2.26	-0.1	4.20	3.44	-0.8	2.20	2.02	-0.2
% One-person HHs	24.3	27.4	3.1	1.4	3.2	1.9	26.6	31.2	4.6
% Occupants < 5 yrs.	60.0	57.4	-2.6	59.7	38.8	-20.9	64.1	71.3	7.3
% Longtime occupants	22.9	26.7	3.8	1.4	42.9	41.5	20.2	17.9	-2.3
% Not in HHs	2.1	1.9	-0.2	0.1	0.0	-0.1	0.0	1.2	1.2
Population									
% Change in population	—	6.3	—	—	-20.3	—	—	31.0	—
% Single	25.1	30.2	5.2	26.5	31.5	5.1	19.1	27.7	8.5
% Children < 15	26.8	22.0	-4.8	39.1	18.9	-20.3	23.6	16.8	-6.7
% Elderly 65+	9.3	9.9	0.7	1.2	2.9	1.6	7.4	10.4	3.0
% Black	10.8	12.6	1.8	0.2	0.8	0.6	0.1	1.2	1.1
% Hispanic	18.3	27.6	9.3	7.6	2.5	-5.1	7.0	11.1	4.1
Economic									
Median family income	$10,972	$21,125	—	$20,515	$40,447	—	$10,678	$21,497	—
ratio to county	100	100	0.0	187	191	4.5	97	102	4.4
% of families in poverty	8.2	10.5	2.3	1.8	2.3	0.5	3.3	4.8	1.5
% Women in labor force	44.6	53.5	8.9	36.6	52.8	16.2	49.5	62.4	12.9

thereby shrinking household size. Note also that the percentage of population under age 15 has fallen from 39.1 to 18.9 percent of the total. The number of Valley girls living in this area is in sharp decline.

In contrast, Tract 1392 is middle to lower middle class in economic status. Fewer than half of its households are owners and just over half of the units are single family. This has fallen since 1970 because of construction additions at the rate of 48.8 percent to the stock in one decade, much of which is apparently apartments. Note that only 35.4 percent of units are described as new (built in the past 10 years) in 1980. A larger percentage for change in stock is calculated because the denominator in 1970 (total housing units at that time) is smaller than the denominator in 1980 (total current housing units).

Population in this area grew by a smaller amount—31.0 percent—because household size declined somewhat (fewer people per house means more houses are needed). The number of singles increased as a share of all persons and children declined. The elderly are a little more common than in the other Valley tract, as are minorities.

3.2.2. Hollywood Tracts

The opposite to the suburban San Fernando Valley may be the bright lights and dense apartment housing of Hollywood. The first tract shown in Table IV, Tract 1899, has only 8.6 percent of its housing in single-family units, and only 7.0 percent of its households are homeowners. The units are very small, averaging only 3.1 rooms in size. The median family income is only 72 percent of the county average, but the percentage of families in poverty is actually lower than the county.

Hollywood Tract 1899 has a high percentage who are single (37.8 percent), and this has grown sharply since 1970. The tract also has a very high percentage who are elderly (23.5 percent). Is the percent single (i.e. never-married) greater than expected with the number of young people in the area? A method for testing this is presented in Chapter 7. At any rate, the prevalence of small housing units and a young singles population mixed with elderly leads to a very small household size, only 1.42 persons per unit. This is less than half that found in Valley tract 1373.02 and well below the county average.

The next tract, Hollywood Tract 1942, lies in the Hollywood Hills. About 60 percent of its housing is single-family and just over half (51.8 percent) of the households are owners. Ownership has actually increased since the last census. The house values are very high, more than twice the county average (with the majority falling into the highest category of $200,000 or more). Prices appear to have risen substantially in relative terms since 1970. There has been a 10.0 percent increase in housing since 1970.

Hollywood Tract 1942 has a median family income that is 135 percent of the county average. The percent single in this area is even higher than in Tract 1899 (39.0 percent), and the percent elderly is lower. Household size averages a little larger than in Tract 1899, but then the housing is larger, also.

The final Hollywood area tract is 7005 in West Hollywood, a recently incorporated city that is the center of the Los Angeles gay community. This area borders on Beverly Hills and is a highly desirable area. Prior to its incorporation, the city expe-

Table IV. Summary Profile of Hollywood Area Tracts

	Hollywood Tract 1899			Hollywood Tract 1942			West Hollywood Tract 7005		
	1970	1980	Difference	1970	1980	Difference	1970	1980	Difference
Housing									
% Change in stock	—	5.4	—	—	10.0	—	—	72.7	—
% Owners	7.6	7.0	-0.7	45.1	51.8	6.7	10.0	20.1	10.1
% Single-family units	10.2	8.6	-1.5	60.0	60.2	0.2	16.8	12.4	-4.4
Median rooms	3.2	3.1	-0.1	4.4	4.4	0.0	3.2	3.2	0.0
% New units	40.5	19.4	-21.1	19.1	8.6	-10.5	43.8	32.2	-11.6
% pre-1940	24.6	15.6	-9.0	29.3	26.2	-3.1	19.5	13.0	-6.5
Median value	$32,400	$117,900	—	$44,100	$200,000+	—	$32,800	$147,600	—
ratio to county	133	135	1.6	181	over 228	—	135	169	33.9
Households									
% Change in occupied units	—	3.6	—	—	9.4	—	—	55.3	—
Median HH size	1.60	1.42	-0.2	1.80	1.54	-0.3	1.30	1.26	0.0
% One-person HHs	47.0	54.6	7.6	39.4	48.3	9.0	59.2	65.5	6.3
% Occupants < 5 yrs.	76.5	69.1	-7.4	62.7	54.8	-7.9	79.9	78.3	-1.5
% Longtime occupants	12.0	14.9	2.9	22.1	28.3	6.3	9.1	10.4	1.3
% Not in HHs	0.8	0.4	-0.3	0.1	0.0	-0.1	0.3	0.0	-0.3
Population									
% Change in population	—	-2.4	—	—	-2.8	—	—	45.7	—
% Single	25.8	37.8	12.0	31.3	39.0	7.7	38.9	50.7	11.7
% Children < 15	6.4	4.9	-1.5	9.1	6.3	-2.8	4.6	2.3	-2.3
% Elderly 65+	22.6	23.5	0.9	15.1	17.3	2.2	11.0	13.0	2.0
% Black	0.9	3.8	2.9	2.1	2.4	0.3	2.2	5.5	3.3
% Hispanic	6.6	6.1	-0.5	5.8	3.1	-2.7	7.1	6.0	-1.1
Economic									
Median family income	$9,976	$15,113	—	$15,690	$28,581	—	$13,869	$22,272	—
ratio to county	91	72	-19.4	143	135	-7.7	126	105	-21.0
% of families in poverty	7.6	9.4	1.8	3.6	5.7	2.1	3.8	7.1	3.3
% Women in labor force	49.3	55.0	5.7	49.1	58.6	9.5	59.6	69.7	10.1

rienced a 72.7 percent increase in housing stock over the decade of the 1970s. That is a tremendously high rate of development for an already built-up community and it has brought a population explosion to a formerly quiet community. Blocks of stylish, small two-bedroom houses were torn down and replaced with condos and apartments. The percentage of housing in single-family units fell to 12.4 percent by 1980, but the percent homeowners doubled to 20.1 percent, reflecting a high degree of condominium construction.

The size of the units tends to be very small—3.2 rooms. Not surprisingly, the median household size is also extremely low—1.26 persons. The smallest allowed household size is 1.00. Without one person you don't have a household. In this area, 65.5 percent of the households have only one person, more than double the average in the county. The bulk of the area's household size is contributed by the third of households with two or more persons. In keeping with the alternative lifestyle of this area, the percent single is also extremely high—50.7 percent—and the percent married is very low—23.1 percent. Children are almost nonexistent.

3.2.3. Beverly Hills Tracts

Beverly Hills is incorporated as a small city and is known as one of the wealthiest in the nation. But even the richest of areas has highs and lows, as reflected in Table V. Beverly Hills Tract 7007 is the wealthiest part of the city. The median family income is "off the charts," meaning it falls somewhere in the top, open-ended category of the census. In 1980 this median is at least 355 percent of the county average. Yet some families—4.8 percent—still live in poverty!

Beverly Hills Tract 7007 is composed almost entirely of single-family homeowners, but there are 7.2 percent who are renters. In fact, about 10 percent of the residents are also minority. The tract has actually lost population over the decade, and over half of the residents have been in their homes for a long duration (10 or more years).

In contrast, Beverly Hills Tract 7009.02 is one of two neighborhoods from the flat lands that looks almost middle income. How poor can you be and still live in Beverly Hills? The median family income in this area is only 116 percent of the county average, and has declined since the previous census. Surprisingly, only 19.6 percent of households are homeowners, but the value of these homes is more than twice the county average. Overall, the units are fairly small—4.0 rooms—less than half the median size of units in Tract 7007.

Household sizes are also small in Tract 7009.02, only 1.56 persons, and the percent elderly is high—24.5 percent. Singles are also high (28.0 percent), but children are almost as common here as in the wealthier Beverly Hills tract. The percent minority is actually lower in this tract of more average income—less than 5.0 percent. These differences within the Beverly Hills neighborhoods require some sorting out and are addressed in Chapter 11.

3.2.4. Pasadena Tracts

Pasadena is one of the historic centers of black residence in the Los Angeles region, as well as being famous for its historic architecture, the Rose Bowl, Cal Tech, and the

Table V. Summary Profile of Beverly Hills Tracts

	Beverly Hills Tract 7007			Beverly Hills Tract 7009.02		
	1970	1980	Difference	1970	1980	Difference
Housing						
% Change in stock	—	4.7	—	—	4.4	—
% Owners	92.7	92.8	0.2	17.6	19.6	1.9
% Single-family units	99.2	92.8	−6.4	15.0	15.5	0.5
Median rooms	8.5	8.5+	—	4.0	4.0	0.0
% New units	11.1	8.3	−2.8	23.0	6.2	−16.8
% pre-1940	59.8	47.9	−11.8	43.1	36.2	−6.9
Median value	$50,000+	$200,000+	—	$50,000+	$200,000+	—
ratio to county	over 205	over 228	—	over 205	over 228	—
Households						
% Change in occupied units	—	1.4	—	—	3.2	—
Median HH size	3.10	2.79	−0.3	1.80	1.56	−0.2
% One-person HHs	8.0	12.9	4.9	40.4	48.1	7.6
% Occupants < 5 yrs.	40.0	32.6	−7.4	62.8	61.2	−1.6
% Longtime occupants	36.5	50.6	14.1	21.5	23.0	1.5
% Not in HHs	0.2	0.0	−0.2	0.1	0.0	−0.1
Population						
% Change in population	—	−7.6	—	—	−4.2	—
% Single	27.6	26.0	−1.6	21.8	28.0	6.2
% Children < 15	19.3	15.4	−3.9	12.7	10.9	−1.8
% Elderly 65+	14.6	17.3	2.7	24.0	24.5	0.5
% Black	2.9	2.4	−0.5	0.3	1.6	1.3
% Hispanic	9.9	7.4	−2.5	1.7	3.3	1.6
Economic						
Median family income	$50,000+	$75,000+	—	$14,411	$24,555	—
ratio to county	over 455	over 355	—	131	116	−15.1
% of families in poverty	0.5	4.8	4.3	5.9	8.3	2.4
% Women in labor force	34.7	35.6	0.9	42.3	54.1	11.8

Jet Propulsion Laboratory, a center for space research. Unlike the massive concentration of minorities in south central Los Angeles, including the Watts area, the Pasadena minority population is much smaller scale and somewhat better integrated. The Pasadena area is a good laboratory because diverse racial and economic groups live within short distances of one another. This area holds special interest for the author because this book was written in a residence bordering one of the example neighborhoods. We will examine three neighborhoods extending northeast from the Rose Bowl toward the mountains.

Tract 4616 lies in the heart of Pasadena's black community. In 1980, this neighborhood was 75.8 percent black and 13.1 percent Hispanic (see Table VI). The percentage black has declined seven percentage points from the previous census. Fully 20.7 percent of families have incomes under the poverty level, and the median family income is only 64 percent of the county-wide average (a slight increase from the previous census). Right around 43 percent of households are homeowners and

Table VI. Summary Profile of Pasadena Area Tracts

	Pasadena Area Tract 4616			Pasadena Area Tract 4620			Pasadena Area Tract 4602		
	1970	1980	Difference	1970	1980	Difference	1970	1980	Difference
Housing									
% Change in stock	—	16.8	—	—	6.9	—	—	1.4	—
% Owners	43.8	42.9	-0.9	23.5	20.8	-2.7	77.6	80.1	2.5
% Single-family units	76.7	73.5	-3.2	49.8	38.2	-11.7	93.1	89.3	-3.7
Median rooms	4.7	4.8	0.1	3.9	3.8	-0.1	5.5	5.6	0.1
% New units	9.2	20.7	11.5	26.6	12.7	-13.9	16.0	1.5	-14.5
% pre-1940	63.0	32.9	-30.1	35.0	34.8	-0.2	21.6	29.3	7.7
Median value	$17,400	$64,600	—	$16,700	$57,100	—	$23,300	$77,100	—
ratio to county	72	74	2.3	69	65	-3.4	96	88	-7.7
Households									
% Change in occupied units	—	16.3	—	—	11.6	—	—	1.6	—
Median HH size	2.20	2.65	0.5	2.10	2.64	0.5	2.80	2.60	-0.2
% One-person HHs	31.7	23.2	-8.5	32.2	27.3	-4.9	14.4	18.8	4.4
% Occupants < 5 yrs.	47.8	50.1	2.3	69.8	66.1	-3.7	47.9	45.9	-2.0
% Longtime occupants	37.5	31.4	-6.0	17.4	18.4	0.9	25.8	30.7	4.9
% Not in HHs	0.3	0.3	0.1	1.3	1.5	0.2	1.1	0.9	-0.2
Population									
% Change in population	—	28.4	—	—	31.2	—	—	-3.5	—
% Single	27.7	36.1	8.4	27.1	39.0	11.9	22.8	29.6	6.7
% Children < 15	28.2	29.2	1.1	28.3	32.8	4.5	27.7	23.4	-4.3
% Elderly 65+	14.1	9.7	-4.4	10.0	7.3	-2.8	8.8	7.6	-1.1
% Black	83.0	75.8	-7.2	74.2	63.2	-11.0	29.0	57.4	28.4
% Hispanic	8.7	13.1	4.4	10.9	29.4	18.5	7.6	6.6	-1.0
Economic									
Median family income	$6,367	$13,622	—	$6,451	$9,836	—	$12,446	$24,346	—
ratio to county	58	64	6.5	59	47	-12.2	113	115	1.8
% of families in poverty	21	20.7	-0.3	19.5	30.5	11.0	4.2	6.2	2.0
% Women in labor force	47.6	53.1	5.5	51.8	53.4	1.6	48.7	63.6	14.9

nearly three-quarters of the housing is single-family structures. In 1980, the median value of the owned homes was 74 percent of the county median.

In contrast to the other areas we have described, population has grown substantially faster (28.4 percent) than the housing stock (16.8 percent). The proportion living in one-person households has declined and the median household size has increased (from 2.20 to 2.65 persons). At the same time, the proportion married or elderly has declined, but the proportion children and the proportion single has risen.

The second Pasadena tract is 4620, bordering Tract 4616 on the east. This area is especially interesting because of its greater mix of Hispanic and black population. Blacks accounted for 63.2 percent of the population in 1980, an 11 percentage point decline from the previous census. Instead, Hispanics have increased their share of the population from 10.9 to 29.4 percent.

This area is even poorer than Tract 4616. Just over 30 percent of families were in poverty in 1980, up from 19.5 percent in 1970, and the median family income has fallen from 59 percent to 47 percent of the county median. This area has many more apartments (only 38.2 percent of the stock is single family) and many fewer homeowners—only 20.8 percent.

Like Tract 4616, population in this area also grew faster from 1970 to 1980 than occupied housing units—31.2 percent versus 11.6 percent. Similarly, median household size in this tract grew from 2.10 to 2.64 persons. Children and singles increased substantially, while marrieds and elderly declined. Overall, it is unclear how much of the change in this tract is due to changes in the black population, changes in the Hispanic population, or changes in the balance between the two. We will give this tract close examination in the chapter on race, Chapter 10.

The final neighborhood, Tract 4602, lies two miles to the north, up against the flank of the San Gabriel mountains. This is in the Altadena area north of the Pasadena city limits, although it is closely tied to Pasadena by roadway access and by membership in the Pasadena school district.

Tract 4602 has a substantially lower proportion black in its population, but this has grown from 29.0 percent in 1970 to a majority, 57.4 percent, in 1980. Unlike the other Pasadena cases, the area is composed of 80.1 percent homeowners and 89.3 percent single-family housing units. Median house values fell substantially from 96 percent in 1970 to 88 percent of the county level in 1980. However, median family income held steady at 115 percent of the county level. With a low percentage in poverty, the area is clearly middle class in economic status. Also unlike the other Pasadena cases, the area has experienced slower growth in population than in housing. Median household size fell from 2.80 to 2.60 persons.

The magnitude of racial change in this middle-class area also beckons closer examination. What kinds of houses are the black families moving into? Which of the white families are leaving? The results are fairly surprising and form an interesting contrast to the pattern of black to Hispanic racial change in Tract 4620.

☐ 4. Conclusion

This overview of neighborhoods has highlighted some interesting cases. We can easily compare the different places in a broad-brush analysis, yet this inquiry raises more interesting questions than it answers. We have only scratched the surface of data available for each location. More detailed analysis clearly would be useful for spelling out the dynamics of change in each place.

Subsequent chapters delve into different facets of a detailed analysis. Methods of analysis will be illustrated with the cases that are most interesting for that subject. Before launching into those specifics, we turn now to some crucial background information about the census data. What is the difference between such things as household and family income? And how can we better link the data together to gain more satisfactory local portraits? When to worry about the accuracy of the data is another concern we must also address.

3 □□□

Concepts, Definitions, and Linkages

Effective use of census data requires a grasp of key concepts and an understanding of basic definitions. These are not separate ideas, but are related to one another in a coherent system. With this coherent system, census users can tie different bits of data into integrated, local-level portraits of change.

We begin with a review of the purpose of the census and its evolving governmental uses. The enormous political importance of the census has exposed it to increasing criticism in recent years. A little background on that controversy will help users evaluate the accuracy of the data more realistically.

The second section explains the foundations on which census data are collected. Basic concepts used in the census are defined. We distinguish the long- and short-form questionnaires and summarize the content of the four major groupings of census questions: population, household, housing, and economic.

Next we explore different concepts and measures that are used to link different variables within the census data base. The focus is on grasping the overall structure of census variables and concepts. Particular attention is given to understanding household data because these link the population and housing portions of the census data. More specific details on individual variables are addressed in later chapters where data are analyzed.

The last section of this chapter addresses special qualities of analysis required for understanding local areas. How do we think about the causal structure that ties together different bits of data for local areas? In important respects, local analysis may be seen as the reverse of what we expect for national-level analysis. A specialized mode of analysis is required for researching local community change.

1. Purpose and Evolving Uses of the Census

If the census weren't important, people wouldn't argue over it. The growing criticism about the census in recent years has paralleled the growing weight that is placed on it. This may be a healthy sign of democracy at work in a modern era increasingly dominated by statistics. The Census Bureau has been long recognized as the most professional statistical agency in the world, and, over the decades, the Census Bureau has continued to refine its techniques and become ever more proficient. Yet, at the same time, changing conditions in the U.S. population have made taking a census ever more difficult.

1.1. Political Representation

Article 1, Section 2 of the U.S. Constitution mandates the taking of a census every 10 years. This was required in order to apportion political power in the new Congress and for periodic reapportionment. Every decade, census counts are used to assign seats in the House of Representatives to the states. Each state is guaranteed one seat. After that, seats are distributed in proportion to population size, with a total of 435 seats to be allocated.[1] With some states growing more rapidly, and others stagnating or declining, the decennial census defines political winners and losers.

At the state and local levels, similar procedures are carried out. Ultimately, every new legislative seat, whether in the U.S. House of Representatives, the state legislature, or a city council district,[2] must be related to a specific piece of geography and its census count. Census data, detailed to the block level, are used for drawing boundaries for those local districts. By law, the necessary data must be reported to the president and Congress by the January after the census.[3]

A number of events since 1960 have focused attention more acutely on the census's apportionment impacts (Anderson 1988). Prior to 1960, reapportionment at the substate level was carried out more casually, leaving districts of highly unequal size, most often with small rural districts and large urban districts. This practice violated the principle of equal representation and was overturned in the courts. In addition, the civil rights struggle of the 1960s, particularly the Voting Rights Act of 1965 and a series of judicial decisions, brought added awareness of minority representation (Thernstrom 1987). More than just an adequate overall headcount was required. Minority advocates demanded a full minority count in order to gain equal representation in new districts being formed.

---

[1] A *method of equal proportions* is used for reapportionment. Legislative seats are assigned by a priority value in a series of allocation rounds. A state's priority value for each round is determined from division of the state's population by the geometric mean of the number of seats assigned through previous rounds and the next seat that would be assigned. First, every state receives one seat. Then priority values are calculated and the state with the highest priority value gets the next seat.

[2] Some council districts are *at large*, meaning that the council members represent the whole city and not a specific subarea.

[3] The specific data for this redistricting are the Public Law 94-171 tables. These were the first results to be released from the 1990 census and included a minimum of detail (total population and housing, voting age population, and crosstabulation by race and Hispanic origin).

1.2. Revenue Implications

A second use of census data has contributed economic urgency in addition to the political impacts. Since World War II, federal aid programs and revenue sharing have increased steadily. These funds have been allocated by formulas that rely on census data, including not only the total population size, but also characteristics such as per capita income, unemployment rates, and age of housing. A larger population entitles a local area to more aid, adding fuel to local arguments over the adequacy of census counts. In particular, mayors of cities with many minorities argued that the whole city was cheated of needed funds if their minorities were not fully counted.

In 1989, a total of $58.7 billion, or an average of $236 per capita, was transferred from federal to state and local governments through such formula grants. (In addition to these federal transfers, many states distribute additional funds to their cities by using census data in formulas.) Adjustment for census undercount would not increase local revenue shares by $236 per capita, because the total is fixed and would be merely redistributed. However, among those places that were undercounted by more than the average, it is estimated that federal funds would be increased by $56 per capita (Murray 1991). Those funds would be taken away from other places that were undercounted to a lesser degree. In general, the fixed size of the federal (or state) pie creates controversy over the alternative distribution formulas and their underlying census data, often leading to battles over computer printouts (Nathan 1987).

1.3. Difficulties in Counting the Population

Despite the growing importance placed on the census, counting the population has become increasingly difficult. The population has been growing more heterogeneous and contains more persons who lack a command of the English language. Also, many cities have poor districts that are home to difficult-to-count persons who may be disinclined to have any contact with government representatives. Furthermore, even among the middle class, cooperation has been eroded by the large increases experienced in "junk mail." The result is that public cooperation in filling out the census questionnaire has been lagging. The best indicator of the worsening environment for census taking is the response rate to the census mail questionnaire that is sent out in March. By the time of scheduled followup for nonresponse, only 63 percent of questionnaires had been returned in 1990. That was down from 75 percent in 1980 at the same point in the process (Jones 1991). Some states had higher cooperation than others, but all states fell by about the same amount from their 1980 levels. The public is becoming harder to reach through the mail questionnaire, and the Census Bureau must visit more addresses in person, requiring more staff time and much greater expense.

1.4. Protecting Against Undercount

Concerns about census undercount, particularly of minorities, have been mounting since the 1970 census. Evidence about the degree of undercount and methods of adjustment are discussed in the next chapter as part of a review of the data's accuracy.

The Census Bureau has taken the lead role in uncovering and correcting the undercount.

After each decade's census, the Census Bureau conducts a detailed "post mortem," a series of studies of what worked well and what could be improved (Butz 1991; Jones 1991). An estimate of the undercount is part of that process, as discussed in Chapter 4. Planning efforts for recent censuses have made reduction in undercount a primary goal. As part of that improvement, local governments were invited to review housing unit counts for each block and to identify blocks where there were discrepancies. Bureau staff then rechecked cases of significant discrepancy. The ironic outcome was that, although the local review process was intended to enhance the quality of the count, local political leaders' comments in the news media about these discrepancies may have generated greater negative perceptions about the accuracy of the census.

1.5. Political Economy of the Census

The issues cited above illustrate a general tension inherent in the census. The Census Bureau is a statistical agency that is immersed in a political context. Although the census is conducted in accordance with exacting scientific procedures, political influences exert themselves in a myriad of ways. A broad definition of political influence is used here, ranging from suggestions and criticism to persuasion, all of which are backed by a threatened loss to the Census Bureau of public credibility, popular support and participation in the census, or loss of financial support within the federal government. This influence may be wielded by formal advisory committees or outside advocates that represent census users or interest groups, as well as by congressional leaders and key decision makers in the executive branch, principally the Department of Commerce and the Office of Management and Budget. These influences do not necessarily *taint*, or detract from, the quality or accuracy of the census; rather, political guidance is essential for making sure that the census provides data that are more *relevant* to public purposes.

With the growing debates surrounding the 1980 census, it became clear that census numbers could not be separated from politics (Mitroff, *et al.*, 1983). This led the National Committee for Research on the 1980 Census to sponsor a monograph on *The Politics of Numbers*, in which the editors strive to "open up a field that scarcely exists—the political economy and sociology of statistics" (Alonso and Starr 1987: 4).

Here we call attention to four main opportunities for political influence on the census: 1) determining questionnaire content; 2) determining how data will be reported; 3) deciding the adequacy of census coverage (how much tolerance for undercount); and 4) allocating the budgetary resources required to finance different programs and levels of effort by the Census Bureau.

1.5.1. Determining Questionnaire Content

The questionnaire is the source of all census data products, and therefore *content determination* about what questions are to be included in the questionnaire may be the single most important issue about the census. This matter is vital because it is irreversable. Information not collected by the questionnaire in one census cannot be discovered, especially for smaller areas, until the next census 10 years later.

Given this importance, the Census Bureau has organized an explicit decision making process for receiving suggestions, the details of which are documented in a special series of reports (U.S. Bureau of the Census 1990c). Overall, there is a conservative bias toward retaining the same questionnaire. This is due to restrictions on expanding the length of the questionnaire, as well as the desire to maintain comparability over time for local trend analysis (a major emphasis of this book). In addition, existing questionnaire items tend to develop user constituencies who lobby for their retention.[4]

Suggestions on content for the 1990 census were received in a detailed procedure involving essentially three stages. First, the needs of users outside the federal government were solicited through a series of local public meetings, special conferences, and written comments. In a second stage, the Census Bureau formed 10 Interagency Working Groups, chaired by Census Bureau staff members and organized along questionnaire content lines, to learn directly what the needs of federal agencies were. In a final stage, the Office of Management and Budget (OMB) organized a Federal Agency Council (FAC) to advise OMB on census data needs. The FAC took the results of prior stages as input in evaluating Census Bureau plans for developing the 1990 questionnaire. The OMB held a strong negotitating position with the Census Bureau over the final content of the questionnaire.

1.5.2. Reporting the Data

The decision process on how the data are to be reported is much less documented, but, as mentioned above, this is also less critical than the content of the questionnaire. User input seems to be the primary influence, as received in 10 regional census product meetings and 65 local public meetings. Census Bureau staff reviewed current and emerging federal legislation that could impact data needs, and they also evaluated the new potential of CD-ROM disks for use with microcomputers in analysis of census data (U.S. Bureau of the Census 1989b:1–2). The usefulness of products from the 1980 census was also evaluated, and certain report series were

---

[4]The Census Bureau describes five selection criteria for questionnaire items: "First, only essential data were considered—those with a broad demonstrated societal need and those needed to meet Federal, State and local statutory data requirements. These data would have to be needed for relatively small areas (local governments and small statistical areas) or numerically small population groups. If data were required only at the national or regional level, sample surveys were the more appropriate vehicles.

Second, many of the questions asked in 1980 were repeated in 1990 because they provided a continuum of vital socioeconomic and housing trend data.

Third, there would be no significant increase in the number of questions the Census Bureau would ask in 1990, relative to the 1980 census. Public cooperation—essential for a successful census—could be undermined by a questionnaire that respondents found too burdensome.

Fourth, questions would not be used that were intrusive, offensive, or widely controversial. Controversial subjects could influence or reduce response to the census and were to be avoided.

Fifth, the Census Bureau had to be able to formulate a clear, concise question on each subject that would yield accurate data. Wording and format were especially important because the census is conducted primarily by mail, using a self-administered questionnaire." (U.S. Bureau of the Census 1990c: 4)

discontinued in 1990, but generally there was an effort toward maintaining relative comparability of products over time.

It is less clear how specific tables were designed for published reports or summary tape files, but these formats are intended to respond to users' needs, particularly those related to federal agency programs. Those needs tend to emphasize disadvantaged classifications that are the subject of federal programs. More space is devoted in the data products to poor persons than middle-income or higher, more to elderly than younger-aged persons, and more to race and Hispanic origin than other demographic factors. As noted in Chapter 4, the increase in the number of racial categories and the expansion of tables crosstabulated by race in the 1990 census products, often in preference to age or other descriptors, may also reflect the strong lobbying of congressional leaders and minority advisory committees.

1.5.3. Adequacy of Coverage

Debates over the census undercount have been a major political issue. The principal advocates for undercount adjustment are big city mayors with larger-than-average minority populations. Their political influence is wielded through congressional leaders and through threats of lawsuits, some of which were filed even before the results of the 1990 census were known. However, in this case, the major political influence is wielded after the questionnaire content and reporting plan already have been determined. The dispute over adequacy of coverage, unlike debates over content or reporting, which are about relevance and usefulness, is a political dispute over the *accuracy* of the data. This debate bears more acutely on the fundamental purpose of the census than the other debates.

1.5.4. Budget

A final opportunity for political shaping of the census is especially important—the budget of the Census Bureau. The budget varies over the decennial cycle, with special needs for census planning, field operations, and data coding and reporting in selected years. The entire cost of the 1990 decennial census amounted to approximately $2.58 billion. The necessary funds for this massive operation, as well as the Census Bureau's other ongoing programs, are allocated through congressional appropriations.

With greater funds the Bureau can report data in more detail and with greater accuracy. Yet it is also true that funding beyond a certain level may have diminishing additional benefits for specific activities. Because different political constituencies may value the additional benefits differently, the Bureau's budget allocation can be politically contentious. Overall, it is important to emphasize that budget decisions underlie all other political decisions about the census, limiting the content, reporting, and coverage of the census.

For example, if it were sufficiently valued, coverage could be increased—at a price—by greater staffing for pre-census planning and follow-up for cases of nonresponse. With greater funding, the potential accuracy of post-census adjustments could also be enhanced. Whereas the Census Bureau proposed in fiscal year

1986/87 to field a 1990 post-enumeration survey (PES) of 300,000 housing units to estimate the necessary adjustment factors for population subgroups in local areas, funds were made available for a survey of only 170,000 units. The entire cost of the PES fieldwork and analysis ran approximately $60 million, and the larger survey could have been fielded for another $30 to $40 million. With those additional funds, and the larger sample size, the sampling error would have been reduced by approximately 40 percent, greatly enhancing the accuracy of the adjustment process. Given the terms of the decision by the Secretary of Commerce (Mosbacher 1991), the larger survey might have contributed sufficient accuracy to warrant adjusting the census (see Chapter 4). However, the Secretary of Commerce had indicated early on that statistical adjustment for undercount was not a political or budgetary priority of the Administration.

1.5.5. Politics and Statistics

In summary, the collection and reporting of census statistics is subject to numerous political influences. It is naive to assume that *truth* exists in a vacuum waiting to be discovered. There are different truths to be found, not all of which can be found at the same time or for the same price. The Census Bureau is the major statistical agency by which our democracy discovers and describes itself. Following the pioneering concept of Innes (1990), we must view statistics as being *negotiated interactively* among users and providers, and among competing viewpoints and purposes. Census Bureau statisticians, census users, and political leaders all have a hand in shaping the nature of the census. As this book is oriented toward the needs of census users, it bears emphasis that the more informed users become, the greater will be their influence in planning the next census, a process already commenced in 1991.

☐ 2. Organized Content of the Data

The census collects a large amount of information about demographic characteristics, economic behavior, and persons' living arrangements. This information has been organized in approximately the same manner for the past several censuses, permitting useful analysis of changes over time. To make the most of the data, it is necessary to understand some key organizing concepts and to view the content of the census questionnaire in a structured context.

2.1. Four Major Universes

The census collects information from four broad universes: population, workers, housing, and households. These four bases have a high degree of overlap, as many persons are employed in the labor force, almost all people live in households, and those households occupy housing units. But not all persons are in the labor force, in households, or in housing. And not all housing units are occupied.

In a general view, a universe consists of any identifiable set, such as school children or mothers of children.[5] However, the census data base consists of some major sets that are not merely subsets of one another. Persons, households, housing units, and workers have special characteristics defining them as distinct universes. A synonym often used for universe in this book is *base*, as when a universe serves as the base for a tabulation or base for a ratio. Each of the major universes is discussed below.

2.1.1. Basic Definitional Units

Each of the four universes has a fundamentally different unit of definition (U.S. Bureau of the Census 1982). For housing it is the *housing unit*, defined as any room or group of rooms intended to be occupied as separate living quarters. Separate living quarters are those in which the occupants live and eat separately from other persons in the building and which have direct access from the outside of the building or through a common hall.

The definition of a *household* is the *occupied* housing unit, and so the number of households equals the number of occupied housing units. The household is the central concept linking population and housing. Households are made up of one or more persons sharing a housing unit, defining the household's size. The distinction between households and families is important. *Families* are a subset of households. Families are defined as persons living together and related by blood, marriage, or adoption. A married couple is a family; two unrelated lovers are not. A woman and a baby living in an apartment are a family household, as are a brother and sister living together. If their mother lives by herself elsewhere, she is not a family household.

Individual *persons* are the fundamental unit of population. The key point to observe is that persons are counted at their *place of usual residence*. Most often this is the location of the housing unit where their household resides. However, a small percentage of persons (less than 3 percent) live in institutions or group quarters or are homeless on the street. These persons would be counted at the point where they usually sleep. For persons with more than one residence, the place-most-often-slept rule also applies.

Economic status or behavior is more diffuse and has two broad categories: one is *income*, the other *labor force participation*. Individuals' incomes may be summed for all family or household members, yielding *family* or *household income*, and the total of individuals' income in an area can be divided by total population, yielding *income per capita*. In contrast, labor force behavior is defined solely in terms of individual persons and cannot be aggregated like income. Substantial data on the type of job, and commuting behavior, is collected for *persons at work* the week prior to the census.

By way of summary, it is helpful to conceptualize the four definitional units in a hierarchy. Households are a subset of housing units. Persons are subunits within households, but a very small, extra proportion of persons reside outside households.

---

[5]The Census Bureau defines a *universe* as "the set of entities of which the characteristics are studied, or about which an inference is made" (U.S. Bureau of the Census 1982: 51).

Finally, economic behavior is practiced by persons, largely restricted to a subset of those over age 15.

2.1.2. The Housing Sample Universe

Census data are collected primarily through a *housing unit sample frame*. Census questionnaires are mailed to a list of addresses that includes every housing unit known to exist. The Bureau strives for a 100 percent sample of housing units. People who live in the housing unit are expected to fill out the questionnaire, answering questions about themselves and about the housing unit.

If a questionnaire is not returned from the housing unit after several weeks, census takers pay a personal visit to collect the information. Three efforts are made to speak directly with some adult individual who lives in the housing unit. If unsuccessful, the census taker will try to collect some information about the household from neighbors as a last resort. Similar efforts will be made if the housing unit is not occupied. Missing details about persons or housing units are estimated later through computerized procedures of allocation and substitution (see Chapter 4).

The housing unit is the primary means of locating population in space, and different types of people and households tend to live in different types of housing. In small areas, housing patterns vary markedly from neighborhood to neighborhood, and those patterns tend to persist because housing units are both immobile and long-lived. Accordingly, housing provides a crucial foundation for understanding local patterns and changes.

2.1.3. The Household Universe

Occupied housing units (excluding those that are vacant) define the household universe. This universe actually consists of three intersecting universes: (1) the occupied housing units; (2) the household groups living in those units; and (3) the persons in those households. A separate set of characteristics can be defined for each of these universes, e.g., occupied housing units have number of bedrooms, households can have number of persons, and persons in households can have different ages.

The household is a complex concept that forms the bridge between population and housing and interacts with both of these other sample bases. It also forms the crux of the census data structure and therefore deserves greater attention.

The characteristics of households are often used to classify both housing units and population. For example, we may identify black housing units or renter-occupied housing units. Alternatively, population members can be classified by their household status: for example, children living in female-headed households or the renter population.

In turn, households are often classified based on the characteristics of their population members. Examples include black households, elderly households, or less educated households. Less often, households are also classified by the characteristics of their housing units, e.g., apartment households or public housing households.

So important is the household concept that we will return to it for special attention in a later section.

2.1.4. The Population Universe

The ultimate goal of the census is to gain a complete count of the population. The household population can be reached through their housing units, but persons not living in housing units much be reached in other ways. Census takers also visit institutions and group quarters such as nursing homes, prisons, and college dormitories. In 1990, they also made a special effort on *S* (for shelter) night to locate homeless persons living on the street or in improvised shelters.

Despite their best efforts, a small percentage of persons remains uncounted—around two percent in 1990. The evidence on census undercount, and a brief explanation of how it is estimated, is presented in Chapter 4.

The population universe consists of individuals abstracted from their household and housing contexts. Most often these data are tabulated by age, race, and sex. Two key population variables encode information on individual's relationships to others. One is marital status and the other is a person's relationship in a household. The latter will be described in detail when we return to the household concept.

2.1.5. Labor Force and the Employment Universe

As defined earlier, labor force behavior is a key aspect of economic behavior. This universe is defined on the basis of the population universe: It is persons who seek and hold jobs. However, labor force data in the census are more than just a subset of population data. The key rationale for defining them as a separate universe is that jobs are usually held at a *different location* from the place of residence.

The labor force includes both employed persons and unemployed persons who are available for work. As such it is helpful to think of the labor force as a population-based concept that generates workers from a population in a residence place. These workers then turn into employees at a work place. Thus the labor force concept forms a bridge between the population universe and the employment universe.

The separation of employment from the housing unit, and place of residence, poses difficulty for local analysis of this universe. Because of commuting, workers in local firms may not be local residents, who are also in the labor force but work elsewhere. A further complication is that there is no definite number of workers per household. (There is always at least one *person* per household.) Finally, job definitions are much less clear-cut than variables in the other universes and require many more categories. For all these reasons, the employment universe is the most difficult to analyze with census data.

2.1.6. Overlap of the Different Universes

By way of summary, the overlap of three of the universes is depicted in Fig. 1. The shaded rectangles at the center of the diagram represent the universe of population living in different situations. The top of the diagram depicts housing units that are vacant and without households; the bottom depicts persons without housing units. Labels on the right-hand side describe the household population and the total population. Labor force is enclosed in the total population universe and is not otherwise shown.

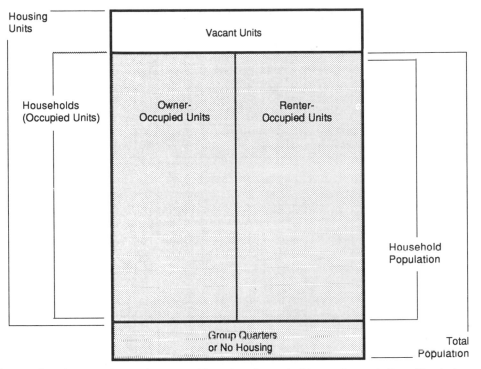

Figure 1. Overlapping universes of housing, households, and population. Shaded area denotes the population universe.

2.2. Content of the Census Questionnaire

The census questionnaire asks questions pertaining to all four universes. These are explained below. First, it is useful to overview the set of questions and explain the critical difference between the sample and 100-percent count data.

2.2.1. Sample and 100-Percent Questions

Census data are collected by means of two questionnaires. A *short-form* questionnaire asks a short list of questions and is distributed to the majority of the population. Supplementing these data, a *long-form* questionnaire includes the questions on the short form (the 100-percent component) and adds a larger number of questions that are answered by only a sample of the population.[6] The sample component of the data has a certain amount of sampling error, making information for those variables less accurate than the 100-percent data. It is also important to know which publica-

---

[6]This strategy was first developed with the 1940 census, when a supplementary schedule of questions was distributed to 5 percent of the population. Introduction of a sampling strategy required a substantial increase in statistical methodology and marks the 1940 census as the first of the modern era (Jenkins 1987; Anderson 1988). The tactic proved to be a cost-effective means of expanding the information collected by the census. In 1950, the sampling rate was expanded to 20 percent, and it has stayed around that level since.

tions and computer files contain one set of data or the other. Those issues are addressed in the next chapter. Here, we inquire into the content of the questions.

Table I lists the two components of the census data and also identifies the separate population and housing variables (because these appear in separate reports). Observe that all of the social and economic characteristics pertaining to persons are collected on a sample basis, as well as a majority of the housing data. This somewhat weakens their usefulness for analysis in small areas.

Table I.   1990 Census Content from the 100% and Sample Questionnaires

100% COMPONENT	
Population	**Housing**
Household relationship	Number of units in structure[a]
Sex	Number of rooms in unit
Race	Tenure (owned or rented)
Age	Value of home or monthly rent paid
Marital status	Congregate housing (meals included in rent)[b]
Hispanic origin	Vacancy characteristics

SAMPLE COMPONENT	
Population	**Housing**
Social Characteristics:	Condominium status[c]
Education (enrollment and attainment)	Number of bedrooms
Place of birth, citizenship and year of	Year structure built
immigration	Year moved into residence
Ancestry	Farm residence
Language spoken at home	Shelter costs, including utilities[e]
Migration (residence in 1985)	Plumbing and kitchen facilities
Disability[d]	Telephone in unit
Fertility	House heating fuel
Veteran status	Source/method of water and sewer
	Number of vehicles available
Economic Characteristics:	
Labor force status	
Occupation and industry of worker	
Place of work and journey to work	
Work experience in 1989	
Year last worked	
Income in 1989	

Changes from the 1980 content:
[a]Moved from sample to complete count
[b]Entirely new question on complete count
[c]Moved from complete count to sample count
[d]Entirely new question on sample count
[e]New, detailed questions added in this category

Deleted questions:
Number of bathrooms
Air conditioning
Stories in building
Marital history

The 100-percent variables listed at the top of Table I have been given priority by the Census Bureau because of their importance to the most basic uses of the census: allocating representation in the electoral process, measuring growth, and analysis of basic population or housing dynamics.

Table I also summarizes the basic changes made in the questionnaire content between the 1980 and 1990 censuses. Relatively few changes were made, thereby preserving the usefulness of the census for measuring change over time. Questions for 1990 in each of the major topic areas are now reviewed. Readers may wish to examine a copy of the 1990 long-form questionnaire that is provided in Appendix A.

2.2.2. Housing Variables

The most important variable purely about housing is structure type, derived in 1990 from question H2, "Which best describes this building?" Among the 10 response options are such housing structure types as mobile home, one-family detached house, or building with 10 to 19 apartments. In 1990, this essential question was moved from the sample to the 100-percent count, making it more useful for small-area analysis.

Tenure (owner or renter status) is a household characteristic that serves also as an essential housing variable. Question H4 asks: "Is this house or apartment—Owned by you or someone in this household. . . ?" In addition, the census takers also record the vacancy status of housing units not currently occupied.

Other important housing variables include the number of rooms in the unit (from question H3), year the structure was built (H17), and price—house value or monthly rent (H6 and H7). More detailed questions related to housing expenses are asked in questions H20 to H26. The expanded detail in the 1990 questionnaire reflects the prominence of concerns about housing affordability problems.

Additional housing variables include number of bedrooms (H9), whether or not the housing unit is a condominium (H18), and the presence or absence of complete plumbing, constructed from questions H10 and H11. Less important housing variables include telephone (H12), water (H15), and sewer (H16) utilities, as well as the number of vehicles kept at the home—up to seven or more (H13). The latter variable may support transportation analyses, along with commuting variables included in the economic grouping.

A number of other housing variables were entirely eliminated from the census in 1990. In the past, number of bathrooms and air conditioning were especially important as indicators of dwelling amenities and quality. Apparently these variables did not draw enough use to warrant retention.

2.2.3. Household Variables

Almost no direct questions about households are asked in the census questionnaire. Instead, most household information is derived from information about occupied housing or from persons' household relationships. Two key questions are asked in the housing section—the tenure question cited above and a question asking when the householder moved into this unit (H8), a measure of the household's occupancy duration. The latter question measures mobility in an alternative fashion to the migration question (residence five years ago) asked of all persons age 5 and older. (In addition, a question about household relationships is asked of all *persons* and is discussed under population.)

Most household variables must be inferred from a combination of housing variables and population characteristics. The Census Bureau processes the raw data to construct a number of important derived variables. These include the number of persons sharing the same housing unit (household size) and the family status of the household. Presence of a family is inferred if persons related to the householder are living in the same housing unit. Presence of children under age 18 in the family is detected if one or more persons in the same housing unit are under age 18 and have the appropriate relationship to the householder.

2.2.4. Population Variables

The greatest number of questions is asked about characteristics of population. Major variables of importance include age (P5), sex (P3), race (P4), Hispanic origin (P7), and marital status (P6). A question about each person's relationship within his or her household (P1) is the lead question in the entire questionnaire. This matter of household relationship has such great importance for linking population to households, and for structuring the census data base, that we examine it more closely in the next section.

All of the preceding population questions are in the 100-percent component. Additional population questions are listed in the sample component. One set asks about migration, including place of residence five years before the census (P14), place of birth (P8), and year of immigration (P10). Further questions ascertain ancestry or ethnic origin (P13) and language in the home (P15). Still other variables include educational enrollment (P11) and attainment (P12), as well as health conditions (P18 and 19) and children ever born (P20).

2.2.5. Economic Variables

A large group of questions in the population section of the questionnaire addresses economic issues. These include a whole column of multipart questions about the income of persons (P32 and P33). The Census Bureau processes these answers to yield measures of poverty status, as well as measures of total household and family income.

A second group of economic questions concerns the labor force behavior of individuals. The Census Bureau determines labor force status from a number of questions about behavior *last week* (P21, P25, and P26). The type of job is explored through a person's occupation (P29) and industry or employer (P28 and P30). Finally, a set of transportation questions is related to the journey to work, including place of work (P22), mode of transportation (P23), and exact time of day and length of commute (P24).

☐ 3. Relationships and Linkages in the Data

The questionnaire is designed to collect data useful for analyzing relationships, as well as for counting totals. Several important linkages are built into the database. Those connections are highlighted in this section.

3.1. Spatial and Temporal Relationships Identified in the Data

Census data are collected at a single point in time and at a single point in space (the usual place of residence). Nevertheless, to be most valuable, it is useful to gather information reflecting the dynamics of change over time and movement over space. Several variables achieve that effect and we call attention to them here.

3.1.1. Spatial Relationships

The precise location or address of the place of residence is coded by the Census Bureau but not released for reasons of confidentiality. Instead, location is summarized at different levels of specificity depending on the report or data series. (See Chapter 4 for details of census geography and data series.)

Additional locations are coded as well. The place of residence five years before the census provides essential information for studies of migration and residential mobility (see Chapter 9). Similarly, data on place of birth provide insight into patterns of lifetime migration. Finally, for persons who are currently employed, the place of work is coded in a detailed fashion that assists transportation planners (COMSIS Corporation 1983) or others interested in journey-to-work movements.

3.1.2. Temporal Relationships

An even greater number of variables in the census embody time. These time variables range in scale from the daily (length of commute and time of day) to the lifetime (persons' age). Two migration questions center on time: place of residence five years before and year of immigration. Two housing questions also address time: year householder moved into the unit and year the unit was built.

Observe that some of the time variables measure duration, others a date in history, and still others age. With knowledge of the current census date, all of these variables can be converted to a common scale that measures either duration or history. The different temporal variables can then be interrelated in dynamics of change. One example has been to show how the age of the housing stock restricts how long people have lived there, and then investigate how that affects the age composition of the local population (Myers 1978).

The aforementioned temporal relationships are all discoverable in census data from a single year. When data are collected from two census years, the prospects for temporal analysis expand even more.

3.2. Key Linkages Among Major Groupings

Relationships among census variables are overviewed in Table II. The major groupings are related to one another in a systematic fashion, as shown by the linkages via key variables.

Housing units are vacant or occupied. The occupied portion of the housing stock defines the presence of households. This linkage may be measured in the aggregate as the percentage of housing units that are occupied (or the complement to those that are vacant).

Households are linked to housing, not only through their definition as occupied units, but also through their tenure (owner- or renter-occupied). In principle, any number of housing characteristics can be used to classify households, but the owner/renter distinction is the fundamental classification used throughout the published census data.

Population is linked to households most simply through the percent of population living in households and through the average household size, or persons per household. More fundamentally, population is linked to households through individuals' household relationship status. Of particular importance is the concept of the *householder*, the reference person used to describe all other household members' relationships. Every household has only one householder, such that the number of householders in the population equals the number of households. The aggregate relationship between population and households is expressed as the percentage of the population that are householders (the headship rate). The householder concept is

Table II. Major Census Data Groupings and Linkages

	Housing	Household	Population	Economic
Definitional unit	Housing unit	Occupied housing unit	Persons at usual place of residence	Economic status or behavior of persons
Major classifications linking to other bases	Occupied or vacant	Owner or renter occupied	Relationship in household, especially the householder	Total household income and poverty status Person-specific rates of behavior
Variables describing aggregate linkage	% occupied (or vacant)	% owner-occupants (ownership rate)	% in households % householders (headship rate) Persons per household (average HH size)	Median HH income Per capita income % families or persons below poverty level % in labor force (LF particip. rate)
Major classifications internal to each category	Structure type Tenure Size of unit (rooms) Year built Price	Family or nonfamily —marital status —presence of children Size (number of persons) Duration of occupancy	Age Sex Race and Hispanic origin Marital status Educational attainment	Labor force status Occupation and industry Individual income Commuting behavior

sufficiently complex—and so important—that it is addressed in greater detail in a following section.

Economic status is linked both to population and households. Individual income is aggregated within households to derive household (or family) income and to determine poverty status. Income is summarized alternately for persons, households, or families. Labor force behavior focuses on a population base, measuring the percent of persons who are in the labor force (employed or unemployed).

3.3. The Central Household Linkage

The heart of the census database is the linkage among persons who share housing units. This is most evident in the hierarchical file structure of the microdata files that sequences person records for each occupant behind a common housing record (see discussion in the next chapter). Most census data publications flatten out this household structure, reporting either population data alone or housing data absent persons. The total omission of household data would divorce the housing characteristics from the resident population, and few would welcome that separation.

The Census Bureau emphasizes a few key household variables, some of which have proven troublesome. The difficulties are partly technical, as household variables reside at the junction between population and housing, and partly political, because the household variables formally express relationships between men and women.

3.3.1. Average Household Size

The simplest household variable is household size—the number of persons living in the housing unit. The most widely used descriptor related to households may be *average household size*, the mean number of people living in each housing unit in a locality. This is calculated simply as the ratio of total persons living in households to the total number of occupied housing units. The denominator of the ratio leaves out the small percentage of housing units that are vacant (roughly 3 to 6 percent), and the numerator leaves out the small percentage of persons who are not part of the household population (roughly 1 to 2 percent).

Average household size is widely used in practice, for it has two reciprocal uses, alternatively determining housing or population. Excepting the small percentage of units that are vacant, the number of housing units times average household size equals population (excluding those living in group quarters):

$$\text{Population} = \text{Occupied Housing Units} \times \text{Persons per Unit}$$

This relationship is the heart of the housing unit method so frequently used by small-area demographers (Smith and Lewis 1983).

Conversely, excepting the small number of institutional residents, population divided by household size equals the number of occupied units:

$$\text{Occupied Housing Units} = \text{Noninstitutional Population} / \text{Persons per Unit}$$

This relationship is widely used by real estate analysts to predict the demand for housing in local areas (Carn *et al.* 1988).

Here we have a tautology. The real estate analysts base their housing calculations on local area population forecasts produced by demographers. The demographers, in turn, base their population data on the assumed future number of housing units in the area. This tautology is recognized by many, but not all, local analysts. The reciprocal uses of average household size create a two-edged sword that can be dangerous if left unguided by a causal model indicating whether population or housing comes first. Such a model is introduced shortly.

A major additional problem is that household size embodies the combined effects of so many different forces. Declining mortality has led to more elderly households, which are typically very small in size, while declining fertility has led to fewer large families with children. At the same time, shifting age structure and marriage patterns have produced many more young adults who are unmarried and disposed to living by themselves or with a roommate. In addition to these compositional changes, the propensity of all age groups and marital statuses to form separate households has increased markedly since 1940.

In local areas, additional effects on household size are created by variations in the size of housing units being built. An area of small apartments will normally attract smaller households than one of larger dwellings. Even in an area with a constant housing stock, household size can plummet because the families who originally moved into the subdivision have grown old enough that their children have left home.

Average household size is affected by all these factors. To make sense of its changes in local areas requires an analysis that, at a minimum, integrates age structure, marital status, household relationships, and housing type. The crucial household relationship is the householder, the person who forms the nucleus of the household around which other members cluster.

3.3.2. Householders and Household Headship

The householder is a potentially confusing concept, yet it is crucial for linking the population and household universes. Therefore, it bears careful consideration. The *householder* is the person listed on the first line of the census questionnaire and, according to instructions printed there, is supposed to be "the household member (or one of the members) in whose name the home is owned, being bought, or rented." If more than one person qualifies as householder according to this criterion, the household members may nominate one of the persons to be the statistical *head* or reference person in the household.

The householder concept plays a central role linking population and housing. Since every household has one householder, the number of householders equals the number of households. And if we know the percent of population that are householders, a measure known as the *headship rate*, we can determine the number of households that will be yielded from the population. For example, a 40 percent headship rate applied to population growth of 10,000 persons yields an expected increase of 4,000 households. Headship rates can be calculated for specific subgroups of the population defined by age, sex, marital status, race, and other attributes, and sharp differences in headship are usually found among the subgroups. Analysis of trends in the subgroup headship rates provides insights into future headship rates.

These subgroup-specific rates can then be applied to projected population in each subgroup, yielding forecasts of households headed by members of the different subgroups.

This procedure hinges on clear identification of the member of the household who is responsible for forming the household. Designation of this key individual, the householder, is clear-cut in many households: persons living alone; single adults living with children; or elderly homeowners with grown children and grandchildren living in the home. Uncertainty arises most often in two particular cases.

One common case of uncertainty is the group of two or more roommates sharing an apartment. Who is the householder or *head* of the household? In practice, it is likely to be whoever starts filling out the census questionnaire or whoever answers the door when the census taker calls. Statistically, in this case it probably doesn't much matter who is designated householder. The roommates are likely of similar age, sex, and educational attainment. Any one of them represents the characteristics of the others. One out of the three roommates will be designated householder, giving their common age-sex group a headship rate of one-third.

3.3.2.1. Designating a Householder in Married Couples. The case of married couples has proven very troublesome. Is the husband or the wife the householder? In reality they are *co-heads* of household, each should receive half credit, but only one spouse can be designated *head* for statistical purposes. Otherwise the principle is violated that the number of householders equals the number of households. This issue is mired so deeply in gender politics that the Census Bureau was forced to revise its procedures for the 1980 census. Prior to that time the terminology *head of household* was used in place of *householder*.

In 1970, the head was to be the person "who is regarded as the head by the members of the household. However, if a married women living with her husband was reported as the head, her husband was considered the head for the purpose of simplifying the tabulations."[7] In other words, the Census Bureau's practice was to disavow the wife's status as head, overruling the couple's choice by substituting the traditional designation of the man as head of house. By the end of the 1970s, this did not sit well.

Prior to the 1980 census, the Census Bureau made two changes. First, the term household *head* was dropped in favor of the more neutral sounding term *householder*. Second, the designation of wives as head, or householder, would no longer be blocked. Either husband or wife could now be termed householder. This is a welcome recognition of modern marriage and gender equality, but it has some statistical consequences.

The characteristics of the householder are assigned to the household. That is how we determine the race of households, among other characteristics. Although wives' marital statuses are the same as their husbands', on average they are a couple of years younger, have a little less education, and earn less income. Under the new rules, if large numbers of married couples suddenly designated the wife as householder, we might see the age of married households fall and a diminishment in their

---

[7]Documentation to published census tract books, 1970 census [PHC(1), pages App-4 and App-5].

education level. This could translate into a surprising rise in homeownership among younger households and among those with less education and lower income. Comparisons between the 1980 and 1990 censuses, which both used the new practice, would be unaffected *as long as* the proportion of wives termed householder remained relatively constant over time.

In 1980, 3 to 4 percent of married couples termed the wife as householder. However, this practice has changed since that time. Figure 2 shows data from four different survey years, depicting the percentage of wives named the householder. Substantial changes have been recorded among younger women in particular. Over 10 percent of the youngest wives were named householder in 1990, and this percentage appears to be rising steadily. Although these data are derived from the national Current Population Survey, comparison of 1980 and 1990 patterns will be possible for specific areas in the nation using public use microdata on computer tape.

One implication of this trend during the 1980s is social—what it may say about changing gender roles or marital relationships. The pattern of change certainly invites further analysis. Yet another implication concerns the growing inconsistency in the historical data series on household headship. Analysts should be forewarned that headship or householder status among men is falling while that of women is rising. However, that says nothing about actual propensities to form households. *Credit* for the household is simply being redistributed more fairly. For most purposes, the simplest and safest solution now may be to combine the two sexes before calculating headship rates.

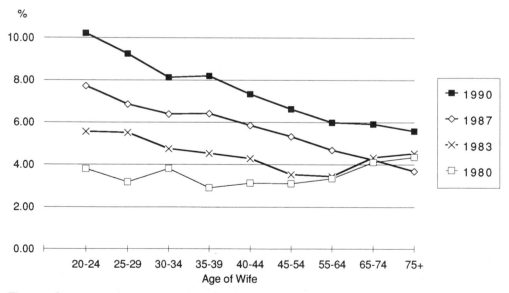

Figure 2. Percent of married couple households where the wife is designated householder, by age of woman and date of survey. (*Source:* Current Population Survey; reported in Table 2 of P-20 Report Series, Nos. 365, 389, 423, and 450.)

3.3.3. Disaggregated Linkage of Population and Housing

A practical illustration will help make clear the relationships between the different concepts we have presented. An illustration of the basic data structure used to link population and housing is depicted in Table III. This tabulation covers a comprehensive five-percent sample of the six-county Los Angeles region, as reported by the 1980 census. Population data are arrayed along the row dimension, here represented by age of persons. Household and housing status is arrayed across the top in columns. Although further variables could be added on each dimension, this table is sufficient to describe the basic structure.

Column H reports the total population in each age group, while column G shows the number who are in group quarters or institutions and not part of the household population. Among the remaining population, household status is defined in three categories (householder, spouse of householder, or other household occupant), each of which is further classified by the type of housing unit in which they live (owned or rented).

The totals of the columns can be used to compute three familiar measures of average household size, total headship rate, and total homeownership rate. Note that three different bases are used in the calculations. The headship rate is derived from a *full population base*, dividing the number of householders (rented + owned) by the total population. The average household size is derived from a *household population base*, dividing the number of persons in households (all columns save G) by the number of householders (representing the number of households). The homeownership rate is derived from a *household base*, dividing the number of householders in owned units by all householders (owned + rented).

Table III. Living Arrangements of Persons: Household Status and Tenure Status
Six-County Los Angeles Region in 1980; 5 Percent Sample

	A	B	C	D	E	F	G	H
	Householder		Spouse of Householder		Other House-hold Occupant		Living Outside Households	Total Persons
	Owned	Rented	Owned	Rented	Owned	Rented		
Age of Person								
under 15	0	0	0	0	75,680	56,565	245	132,490
15 to 19	106	1,511	130	1,203	32,374	15,059	1,696	52,079
20 to 24	1,928	13,710	2,817	6,878	16,535	12,219	2,227	56,314
25 to 34	19,803	31,571	18,553	11,978	9,288	9,744	1,359	102,296
35 to 44	24,487	15,218	18,758	5,116	2,819	2,669	630	69,697
45 to 54	23,273	10,481	17,066	3,291	2,029	1,599	518	58,257
55 to 64	22,134	9,372	14,608	3,010	2,140	1,340	558	53,162
65 to 74	14,085	7,648	6,871	1,943	2,253	994	786	34,580
75 or more	7,632	5,827	2,016	753	2,516	870	2,616	22,230
Total	113,448	95,338	80,819	34,172	145,634	101,059	10,635	581,105

The results of calculating these measures for the Los Angeles sample are reported in Table IV. Of all persons in the region, 98.2 percent are occupants in households and 35.9 percent are householders. Among the household population, there are 2.732 persons per household, and among all households the homeownership rate is 54.3 percent.

Calculations of average household size and headship rates are approximately the inverse of one another. The former is the number of persons per household, while the latter is the number of households per person. The one discrepancy in this symmetry is that the household headship calculation takes in persons from the nonhousehold population as well. Note that if we omitted the nonhousehold population from the denominator of the headship rate this would become 36.6 percent instead of 35.9 percent, the exact inverse of average household size—2.732. Because of this slight difference in sample bases, the headship rate only approximates the inverse of average household size.

The true power of these measures linking population and housing is unlocked when we disaggregate the analysis by subgroups of the population. As shown in Table IV, the household occupancy rate reaches a low of 88.2 percent among elderly persons aged 75 or older, largely due to nursing home residency, but the residency rate also dips between ages 15 and 24 due to military service and college dormitory living. At other ages household residency is very high.

The headship rate changes more dramatically with age, rising to 50.2 percent by age 25 to 34, and then growing slowly to a peak of 62.8 percent among persons aged

Table IV. Descriptive Measures Linking Population and Housing

Six-County Los Angeles Region in 1980; 5 Percent Sample

Calculation Base	Household Occupancy Rate[a]	Headship (Householder) Rate[b]	Average Household Size (Persons per HH)[c]	Homeownership Rate[d]
	All Persons	All Persons	Persons in Households	All Householders
Total	0.982	0.359	2.732	0.543
Age-specific				
under 15	0.998	0.000	0.633	0.000
15 to 19	0.967	0.031	0.241	0.066
20 to 24	0.960	0.278	0.259	0.123
25 to 34	0.987	0.502	0.483	0.385
35 to 44	0.991	0.570	0.331	0.617
45 to 54	0.991	0.579	0.277	0.689
55 to 64	0.990	0.593	0.252	0.703
65 to 74	0.977	0.628	0.162	0.648
75 or more	0.882	0.605	0.094	0.567

Notes: [a]Total or age-specific persons living in households divided by total or age-specific population.
[b]Total or age-specific householders divided by total or age-specific population.
[c]Total or age-specific household residents divided by total households.
[d]Total or age-specific homeowners divided by total or age-specific householders.

65 to 74. The increase at later ages is attributable to the cumulative depletion of married couples, as both divorce and mortality release formerly married persons from sharing household headship with their partners.

Household size also can be broken into age groups. Age-specific household size is not a rate but simply a decomposition of total household size into its age components. Table IV shows that the largest group (.483 persons per household) is age 25 to 34, due to the bulge of the baby boom generation in this age bracket. The 1990 census should show that the largest category is now aged 35 to 44 but continued high migration to the Los Angeles region will keep the numbers high in the younger age bracket.

Finally, the homeownership rate rises most dramatically of all, reaching 61.7 percent by age 35 to 44, before peaking at age 55 to 64 with 70.3 percent homeowners. The upward curve across age groups reflects the impact of marriage, income, and wealth accumulation on homeownership attainment. The decline in the oldest age groups does not reflect exit to nursing homes—those persons have left the household sample and are not part of this calculation. Rather, the decline indicates that more elderly are opting for rented quarters, possibly in retirement complexes. An alternative explanation is that the oldest persons never had very high homeownership, maturing as they did in the Great Depression and war years. Instead, it is the pre-retirement generation in 1980 who has the highest ownership because of their access to VA and FHA loans that stimulated homeownership.

Together, these relationships define a richly structured interface between population and housing. Although some of this information is directly available from the census only at levels of geography including 100,000 or more population, we can make use of this knowledge in building detailed portraits of changes at the neighborhood level. That task requires a causal scenario describing how the different bits of data are tied together.

☐ 4. Causal Dynamics in Local Area Analysis

Which comes first, population or housing? This simple question probes to the heart of a methodological crisis that has been simmering with little debate for years. One group of researchers assumes, naturally, that housing is built to meet the needs of population. Another operates with the opposite outlook: that housing brings people. Forecasting models are more accurate if they drive off the causally prior variable, but we will argue that the causal order differs between local and regional scales. Here we define the regional scale as the area comprising an economic region, often equivalent to a metropolitan area but sometimes larger. Persons both live and work within this relatively self-contained region. In contrast, the local level is a relatively small subarea within that larger region.

A second problem concerns the availability of data to carry out an analytical procedure. Because census data are much more richly detailed at the regional level than for local area analysis, particularly at the subcounty level, the question of data availability interacts with the development of methodologies appropriate at different spatial scales.

While it is readily apparent that differences in outlook are related to geographic scale, the question of suitable methodology is more confused. Local area analysis has special problems and qualities not recognized by researchers trained in methods more appropriate for national- or state-level analysis.

4.1. Two Alternative Scenarios

When the linkage of population and housing is utilized in models of urban growth and change, the key question is which factor comes first, causing the other? The base that is prior can be used to predict the other. If population drives housing demand or needs, then we would prefer population-based measures. Conversely, if housing drives population growth then we would prefer housing-based measures.

Researchers of state and national populations are quite certain that population growth precedes housing. Their typical scenario is that growth in employment induces migration to states or substate economic regions. (Alternatively, retirement migration may precede any employment growth.) Migration combines with fertility and mortality to produce population growth, such as described by the familiar cohort-component projection model. Thus, migration constitutes the link between economic factors and population. Next, population growth leads to household formation, and all these added households require housing. Finally, housing units are built to accommodate needs that began with employment growth. In sum, we may describe this causal sequence as:

$$E \rightarrow P \rightarrow HH \rightarrow HU$$

The above scenario describes well the dominant pattern of causation found at geographic scales at the county or higher levels. But a different causal scenario is played out in local areas consisting of neighborhoods or municipalities. In fact, many local observers subscribe to just the opposite causal direction:

$$E \leftarrow P \leftarrow HH \leftarrow HU$$

In the local view, houses are built and then occupied by households. Next, the occupied units, households, have people living in them of different ages and sexes. Finally, those people engage in employment, as expressed by their labor force participation rates. (The occupants also engage in many other locally based behaviors, ranging from commuting to shopping and school enrollment.) To the national researcher, the reversed causality of this scenario seems absurd, but it makes perfect sense to the local school official forecasting next year's enrollment, or the local transportation planner, or the developer planning a shopping center.

Houses come first, followed by people. This is the model subscribed to by the census taker who begins with a locally based housing unit sample frame and works toward population. This relationship is also the basis for the housing unit method, the most commonly employed method of small-area demographers for estimating population growth (Rives and Serow 1984).

The two opposing views are summarized in Figure 3. One is appropriate at the national or metropolitan level; the other at the subcounty or neighborhood level. The higher geography view is favored by national and regional researchers, those who

Geographic Scale

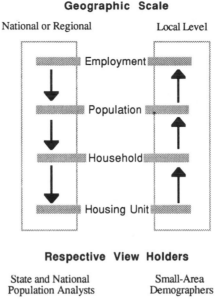

Respective View Holders

State and National Population Analysts	Small-Area Demographers
Regional Economic Modelers	School Officials
	Urban Planners
State and National Housing Policy Analysts	Transportation Analysts
	Census Takers

Figure 3. Implicit order of causation between major sets of census variables at two different geographic scales.

write the most books and set the standards of methodology for others to follow. The local geography view is favored by local analysts and practitioners on the street. The local view has been under-recognized in a published literature dominated by national researchers.

4.2. Top-Down Versus Bottom-Up Structure

How do we reconcile these two very different perspectives? In the terminology of urban and regional modelers, we face a tension between a top-down and bottom-up causal structure. The experience of urban modelers is worth considering because it may shed light on how we should think about local census data.

The *top-down* perspective determines changes at a higher level of geography, then allocates a share of the changes down to each local area. In its pure form, this procedure does not take into account unique characteristics already existing in each local area. This is a simple accounting procedure for dividing the pie at higher geographies and assigning pieces to subareas. In this *step-down* calculation, shares are often allocated based on local areas' past shares of growth, less often by their future capacities for growth (infrastructure, vacant land, etc.).

The alternative is a *bottom-up* perspective in which each local area is analyzed in isolation, then summed to the total of a higher geography. In its pure form, this procedure does not take into account causal forces shaping growth for the whole of the higher geography. Past trends in the locality are simply extrapolated to the future. Great emphasis is placed on the momentum of current characteristics, highlighting the unique circumstances of each area.

An awkward accounting problem occurs whenever a bottom-up analysis is carried out for all local areas and summed to the total of the higher geography. The total rarely matches the results of an independent analysis carried out for the higher geography. The most common solution is to treat the results of the higher level geography as a control total. A scalar is formed by the ratio of the control total to the sum of the bottom-up analysis. When this is multiplied by each of the local areas' bottom-up results, each is scaled up or down proportionately so that their sum now equals the control total. This method patches over the error created by the mismatch between top-down and bottom-up methods. It is a mechanical solution that forces the correct total.

More sophisticated solutions to linking subareas and regions fall into three categories, often involving elaborate computer programming. Higher level growth is integrated into the local analysis with very different priorities and emphases in each of the three methods.

4.2.1. Employment-Based Models

First, employment-based models are favored by regional agencies focused on transportation planning. These are usually carried out by urban economists, regional scientists, or transportation engineers. Ideally, new employment can be forecast for specific locations, with population and housing effects rippling from there via a gravity model. The early prototype of this approach was the Lowry (1964) model. More elaborate versions incorporate the detailed effects of transportation accessibility (Putnam 1983).

The employment-transportation-based models have been the most highly developed means of higher and lower level integration. However, their mathematical complexity is so great that relatively few variables can be incorporated. These models largely surrender any attention to local population and housing characteristics. There is no bottom-up perspective. Local areas have no histories and few current characteristics, save current jobs, total housing, and a current transportation infrastructure. Instead, integration of subareas and regions is achieved by imposing the top-down approach on local areas by stressing the projected location of employment and roadway capacity.

4.2.2. Population-Based Models

A second means of integration stresses detailed analysis of local population dynamics and characteristics. This approach is favored by local health departments and service agencies for whom population characteristics are all important. Carried out by demographers, this approach simulates and projects internal changes in local area populations via births, deaths, and migration.

When multiple subareas are analyzed, results are summed across all areas and compared with higher level control totals for each separate age-sex-race group. Separate adjustments are then made to all subareas for each demographic subgroup. Nevertheless, this is emphatically a bottom-up approach, stressing the momentum of current population structure. This method gives much more attention to who lives in each subarea than the employment-transportation models. However, the demographic method usually pays no attention to the employment, transportation, housing, land, or infrastructure attractions of local areas.

4.2.3. Housing-Based Models

A final method for integrating local area and higher level analysis is less commonly employed, perhaps because it falls in the intellectual domain of neither demographer or economist. Yet this may be among the easier methods of integration. Essentially, this method is a housing-based allocation model that follows a two-tiered linkage of population and housing. This method directly confronts the problem that causation between the two variables reverses between geographic levels.

The two-tiered model has four steps, traveling from the regional scale down to the local scale and switching from one predictive base to another, as outlined in Figure 4. First, a projection is made of a future regional population based on standard demographic methods, incorporating an economic analysis of likely migration rates in response to predicted employment change. Labor force participation rates are applied to each age-sex group to predict the future number of workers. If predicted employment exceeds predicted workers, migration is induced and added to the population.

Second, the future regional population is converted into future housing needs by applying household headship rates to each age-sex group. The resultant number of households headed by each subgroup is then multiplied by the homeownership rate for the group to yield the number of owners and renters. Adding consideration for normal vacancies indicates the total number of rental and owner units required. The existing housing is then subtracted from these totals to yield required or expected additions to the housing stock.

The next step bridges to the local level by allocating shares of the expected housing construction to subareas. These shares are based on the subareas' past attractiveness (measured by past shares of housing permits and/or by proximity to employment concentrations) and their resources to support development, e.g., vacant land or roadway access. These factors hold the key to allocating shares of regional housing construction to local areas. They also hold the key to future local population growth.

The final step is to convert the expected local housing into expected future population in each subarea. At the neighborhood level, population growth and characteristics are based largely on processes of residential mobility. The departure of mover households and aging of the stayer population, supplemented by occupants of new housing construction, combine to create local demographic change. An extended version of the housing unit method, such as that demonstrated in Chapter 14 and Myers and Doyle (1990), can be used to translate housing into local population with detailed characteristics.

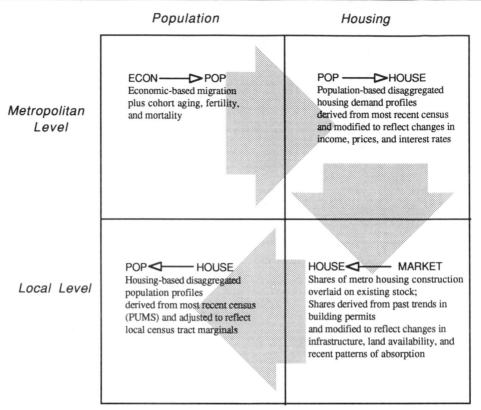

Figure 4. Methods appropriate to each stage of the two-tiered linkage between population and housing.

Overall, the advantage of the proposed four-step model is that it makes explicit in steps three and four the key assumptions implicit in most small-area population forecasts. Made explicit, local analysts can research the assumptions to provide more accurate parameter values. Each step in the model uses different sample bases and different measures for linking population and housing.

4.3. Practical Methods for Local Analysis

Methods in this book are attuned to local analysis and, therefore, stress a bottom-up view. Ideally, we would set each local area in its regional context using a complex computer modeling scheme. However, this degree of integration requires many thousands of hours of data management and computer programming. In practice, only the largest regional planning agencies are able to undertake such extensive research enterprises.

The alternative is to explore a local area in depth, affording it the detailed treatment that is not possible in large scale regional models. Sophisticated analysis can be carried out by using local-area census data to construct integrated local portraits and by comparing these results to county-wide or regional benchmarks. Bearing the regional context in mind will greatly improve the portrait of change constructed for an individual locality. A grasp of alternative causal patterns under different scenarios will help analysts choose variables wisely, and think about their implications more coherently.

☐ 5. Conclusion

The census provides a uniquely rich data resource. No other data source offers coverage as comprehensive, both in the extent of geographic identification and in the topics investigated. Yet to make sense out of the data requires a keen sense of the linkages among the data bits. This chapter has attempted to highlight the connections between different variables and concepts. And we have stressed means of understanding the unique position of local areas within broader regional contexts. Our bottom-up perspective must be checked against broader realities. We next address the types of data available at different levels of geography.

4 □□□

Data Available for Local Area Analysis

The wealth of data produced from the decennial census is enormous. Data are reported for 39,000 jurisdictions and 7,000,000 street blocks nationwide. A wide range of data items are reported for each of these places, often with different versions and formats of the data appearing in different report series.[1] Although only a handful of basic data items are reported for the smallest areas (individual blocks), in somewhat larger geographic areas much richer data are made available—reaching into the thousands of numbers describing each area. In fact, in large counties and cities, even more detailed data series become available, some of which permit unlimited custom tabulations by the user.

The present chapter has three broad goals. The first objective is to present the basic structure of census geography. The spatial units for which data are reported are defined politically, as in municipalities, counties, or states, and statistically, when arbitrary areas—block groups, census tracts, etc.—are defined for data reporting purposes. Census geography is important not only for locating data from the most relevant local area, but also because of the *hierarchy principle* that more detailed data are reported for areas at higher levels in the geographic hierarchy (e.g., the county instead of its many neighborhoods).

A second goal of the chapter is to summarize the different data series that are released from the census for each level of geography. Some data are published in books

---

[1] A comprehensive listing of all current offerings from the Census Bureau is published annually in an inch-thick *Census Catalog and Guide* (U.S. Bureau of the Census, 1991c).

and others are issued in computerized form. A second important distinction is between summary tabulations, whether printed tables or counts recorded in computerized form, and microdata records, individual person and household records with all names and addresses removed. The latter are issued for larger places only and can be used to carry out custom tabulations or conduct analyses of statistical associations between variables.

The third goal is to explain how accurately the census represents the population. Whereas some questions are asked of all persons (the 100-percent component), many of the questions are asked only of a sample of local residents (the sample component). Data from the latter questions are *weighted up* to represent the entire local population, but these estimates have associated sampling error and we must address their statistical significance. A final topic of consideration concerns net undercount of the population and methods of adjusting the census.

With understanding of these three broad issues—census geography, the different forms of data, and accuracy of representation—local data users will be better prepared to take advantage of the rich opportunities available. Although the Census Bureau does an excellent job describing the contents of different data series, a certain amount of background knowledge is required to make intelligent choices from a vast menu.

☐ 1. Data Access by Geographic Level

All census data are reported for some geographic area. These areas are arranged in a pyramid fashion with one area, the nation, at the top and a multitude of increasingly smaller geographic areas at the bottom. In between we have nine census regional divisions, 50 states plus the District of Columbia, and then the counties and municipalities within each state. Metropolitan areas (MAs), consisting of groups of economically related counties, often span the border between two states. At a still lower geographic level reside census tracts (with an average population of around 4,000) and their subunits—block groups and blocks.

Different amounts of data, in greater or less detail, are reported for higher and lower levels of geography. In many cases users want to access the most detailed data available for a particular area. But how does one know what level of data can be accessed or whether the most detailed data have been found? Close attention to census geography is necessary for identifying which data series afford the most detailed data for a particular geographic level.

1.1. Basic Hierarchy of Census Geography

The Census Bureau's geography is structured generally in one basic hierarchy from larger to smaller entities. This basic hierarchy is supplemented by specialized geographic constructs that overlap one or more levels of the basic hierarchy. Table I outlines these elements of census geography.

The basic hierarchy has 10 levels, 11 if we include the smallest unit of geography, the housing unit or institutional living quarters. The upper portion (top four levels)

Table I. Census Geography for 1990 Data Files and Publications

A. Basic Hierarchy (with summary level codes)

010	United States
020	Region
030	Division
040	State (including D.C.)
050	County (or equivalent entities)
060	County subdivisions [minor civil division (MCD), census county division (CCD), and unorganized territory]
070	Place (both incorporated and census designated)
080	Census tract and block numbering area (BNA)
090	Block group
100	Block
—	Housing unit, institutional living quarters, or street shelter

B. Supplementary Geographic Areas

400	Urbanized areas; urban/rural areas
300	Metropolitan areas (MAs), including freestanding metropolitan statistical areas (MSAs), consolidated MSAs (CMSAs), and their component primary MSAs (PMSAs)
200	American Indian and Alaska Native areas
500	Congressional districts
800	ZIP Codes
—	Traffic analysis zones
—	School districts
—	User-Defined Areas Program (UDAP)

nests 51 states (including the District of Columbia) within nine census divisions, those divisions within four regions, and regions within the whole of the United States.

The substate geography is most relevant to the purposes of this book. Each state is divided into counties and then the counties are further divided. For an illustration of the substate geographic units to be described, and how lower-level geographies nest within larger areas higher in the hierarchy, see Fig. 1 of Chapter 2.

Counties are the most basic subdivision of states, with 3,141 counties in the nation. In a few states, particular cities have special status as county equivalents (such as St. Louis in Missouri). County subdivisions are of three types that occupy the same position in the census hierarchy. The *minor civil division* (MCD) is the primary political and administrative subdivision of a county in 28 states. Often these are called townships or towns. In states where places are independent of any MCD the Census Bureau treats these as both place-equivalents and MCDs. The *census county division* (CCD) is a statistical subdivision defined in 21 states lacking well-defined MCDs. Finally, *unorganized territory* is defined as a residual area in MCD states that lack complete MCD coverage of the state.

The Census Bureau recognizes two types of *places*, those that are incorporated municipalities and those that are census designated places (CDP). The latter are areas that resemble incorporated places in that residents can identify the area with a name and local officials can define clear borders, but the CDP has not been incorporated as a municipality.

Although places are presented at a lower level in the hierarchy, as subsets of divisions of counties, note that places usually may occupy a higher level; they are divisions of the state. In practice, some places are quite large and may comprise whole counties (such as San Francisco) or even groups of counties (such as New York City with its five boroughs). Also, a place may overlap parts of two or more counties or county divisions, but they never cross a state line.

Census tracts are smaller areas than most places and subdivide counties, and most places, to a still lower level of the hierarchy. Although census tracts do not cross county boundaries, in many cases they cross the boundaries of places. *Block numbering areas* (BNA) are the equivalent of census tracts in counties where census tracts have not been established. The size of census tracts ranges from 2,500 to 8,000 persons, but average 4,000 persons, and BNAs may be smaller. The spatial area of either unit is smaller where the population density is greater.

The smallest unit tabulated from the census is the *block*, a subdivision of census tracts or BNAs that is defined by visible features of the landscape as well as nonvisible features such as legal boundaries. Blocks may contain zero population or a single institution, or they may contain a high-rise building with thousands of residents. Each block is identified by a 3-digit number, and sometimes with a character suffix(such as 234 or 117B) that is unique within its census tract or BNA. Blocks also are assigned to *block groups,* identified by the first digit of the block number, such as the 200 series, and these comprise a complete set of subdivisions of the census tract or BNA. Block groups are a substitute for the enumeration disricts used in reporting data from the 1980 census.

1.2. Pitfalls of Census Geography

Although the hierarchy of geographic levels is a relatively simple concept, two complications may trip the unwary. First, the same name may be applied to more than one spatial unit. For example, there is the Los Angeles *CMSA* (consolidated metropolitan statistical area), Los Angeles–Long Beach *PMSA* (primary metropolitan statistical area), Los Angeles *Urbanized Area*, Los Angeles *County*, and Los Angeles *City.* Clearly, it is important to pay close attention to the type of geographic unit or else the Los Angeles data being analyzed could be for the wrong area!

The second potential pitfall is that geographic areas at lower levels in the hierarchy may be subdivided by higher level geographic units. A common example is that of a census tract that lies partly inside a municipality and partly outside, creating a *split tract* with two portions. Some census tabulations will report data for this tract in three different places: once for the portion lying inside the municipality; once for the portion lying in another municipality or in the county remainder; and again for the whole of the tract in a section labeled *Totals for Split Tracts*. The analyst should carefully choose the data option that is preferred.[2]

---

[2]For analysis of change over time, the total for the split census tract is generally preferred because these borders are more stable. In some cases, analysts may need to analyze only that portion of the tract within the municipality. The simplest way to check whether a tract is split is to first see whether the tract is listed in the section on totals for split tracts.

1.3. Supplemental Units of Census Geography

In addition to this basic census geography, data are provided for a number of supplementary geographic areas (as listed in Table I). These overlap different levels of the basic hierarchy.

Two high-level categories are closely related and thus may be confused. One is the *urbanized area*, a central city (or cities) and surrounding closely settled territory ("urban fringe") that together have a minimum population of 50,000. The related concept is the *metropolitan area* (MA), termed *SMSA* in 1980, a highly populated and economically integrated area that is defined in one of two ways: It contains a city of at least 50,000 population or an urbanized area with a total metropolitan population of at least 100,000 (75,000 in New England). MAs are groups of one or more whole counties, except in New England where cities and MCDs are used. The key difference between the urbanized area and the MA is that urbanized areas take in only the densely developed portions of counties. In contrast, a suburban county added to an MA may often include not only the developed portion nearest the central city but also an extensive rural hinterland.

The MA geographic construct is used much more often than the urbanized area. MAs are classified into three types: a relatively freestanding MA is known as a *metropolitan statistical area* (MSA), whereas a very large MA that contains separately identifiable component areas is called a *consolidated metropolitan statistical area* (CMSA). In turn, the component MAs within the CMSA are termed *primary metropolitan statistical areas* (PMSAs). Thus, the Los Angeles CMSA contains four PMSAs: the Los Angeles–Long Beach PMSA (Los Angeles County); the Anaheim–Garden Grove–Santa Ana PMSA (Orange County); the Riverside–San Bernardino PMSA (Riverside and San Bernardino Counties); and the Oxnard–Ventura PMSA (Ventura County).

Additional special geographies consist of *American Indian reservations*, with boundaries defined by the Bureau of Indian Affairs and state governments, and *Alaska Native village statistical areas* (ANVSAs). Alaska Native villages are legal entities in Alaska, but their boundaries are not legally defined. These American Indian and Alaska Native areas often correspond to county divisions.

A very important special geographic entity is the *congressional district*, because the fundamental purpose of the census is for purposes of determining voting representation. These districts are defined by state officials or the courts for electing persons to the U.S. House of Representatives. These districts usually do not nest well within the census geography, cutting across geographic units at all levels of the census hierarchy below the state level, and often defined as groups of census blocks.

Traffic analysis zones are customized spatial units defined by local transportation or planning agencies. These small areas are reported only in a special transportation planning package issued by the Census Bureau to meet the needs of a particular user community. Emphasis is given to commuting behavior—place of work, mode of transit, etc. Given the detailed coding required, and their status as sample data, these data are among the slowest to be produced. For details of the 1980 package, see COMSIS Corporation (1983).

School districts are another special geography for which data are provided for the 1990 census. Special boundary files have been prepared to correspond to local school jurisdictions. Under the guidance of the U.S. Department of Education's National Center for Education Statistics, a set of tabulations has been prepared that resembles the standard format of the STF 3 summary tabulations. The content of the tabulations focuses on the demographic characteristics of school-age children and their social and economic living conditions.

Another special geography for which data are provided from the 1990 census is the five-digit *ZIP Code*. The coverage of ZIP Codes is defined by the U.S. Postal Service. They generally do not have firm physical boundaries, nor do they correspond to the basic census geography. Often several times larger than census tracts, ZIP Codes cut across the boundaries of states and counties, as well as lower-level geographic units such as census tracts and block groups. Businesses often require data in ZIP Code form for purposes of direct mail marketing programs, and data for ZIP Codes help them choose prospects from privately developed mailing lists.

Users should be aware that ZIP Codes are not defined as enclosed spatial areas but rather as sets of postal carrier routes emanating from a local post office. More than one set of routes may penetrate the same spatial area. For convenience, ZIP Codes may be mapped as spatial areas, but in some cases this may actually describe only the dominant ZIP Code for that area. Thus, the link between ZIP Codes and census data for a spatial area is not the exclusive match often assumed.

A Census Bureau program newly introduced with the 1990 data is the *User-Defined Areas Program* (UDAP). Users may specify any custom-defined geography and the Bureau will tabulate the data for those specific areas. The 1980 precursor to UDAP was the Neighborhood Statistics Program. Carried out for a reasonable user fee, the UDAP has several criteria for defining boundaries. Boundaries of custom areas must coincide with block boundaries (because these are the basic building blocks for higher level geographies) and contiguous areas may not overlap. Local parties are responsible for working out the boundaries. Finally, each user-defined area must meet a minimum population-size threshold.

1.4. TIGER Location System

The lowest level of census geography is the housing unit or structure located at an address. Some system is needed for simultaneously locating these addresses in physical space (defined by roads and significant geographic features), political space (defined by governmental boundaries), and statistical space (defined by census tract boundaries and the like). In the past these relationships have been established by manually drawing maps and plotting locations. Revisions were an extremely time-consuming process.

The TIGER (Topologically Integrated Geographic Encoding and Referencing) system was developed by the Census Bureau to automate the mapping and other geographic activities required to support its censuses and surveys. The system is based on topology, a branch of mathematics that describes the spatial relationships among points, lines, and areas in a two-dimensional plane. The TIGER database consists of these spatial relationships. Each line segment and point is coded with a series of geographic identifiers (name of street or place, address ranges of each street segment, latitude and longitude coordinates, political and statistical area codes, etc.). The detail in TIGER is so voluminous that the national TIGER/Line file is 25 Gbytes and occupies 125 high-density computer tapes. The average size for a single county is much more manageable, a median of 5.4 Mbytes, although in a very large county such as Los Angeles this runs to 135 Mbytes.

No census data are contained in TIGER; rather, TIGER is a geographic system that permits linking census data to specific locations. Its capacity for assigning street addresses to legal and statistical geographic areas makes it more precise and flexible than the zip code geography discussed above. For more details and a concise introduction to TIGER, see U.S. Bureau of the Census (1990a).

1.5. Comparability of Boundaries Over Time

One of the most important aspects of census geography concerns the comparability of areas over time. Any study of change from one census to the next must pay close attention to changes in boundaries.

States and counties have fixed boundaries that make them reliable units for trend analysis. MAs are made up of counties, and counties may be added as the region grows (or even subtracted). Therefore, it is wise to check which counties are included in an MA under investigation. If a county has been added from one census to the next, make necessary adjustments by adding county data to the earlier defined MA or by subtracting county data from the later defined MA until the area has comparable boundaries over time.

County subdivisions and places may change boundaries as governmental jurisdictions annex new territory. It is important to check for such changes because they are common in many states, particularly in rapidly growing areas and on the periphery of urban areas.

Census tracts provide the most stable small geographic units for measuring change over time. These statistical units are usually defined irrespective of municipal boundaries and are unaffected by such changes. As noted above under pitfalls of census geography, one problem is that tracts may be split by municipal boundaries, leaving one portion inside and the other outside. However, the Census Bureau provides a separate total tabulation for such split tracts, and that total should be used for studying changes over time.

A more difficult problem with census tracts is that their boundaries are changed occasionally. In growing areas, tracts are frequently subdivided into two or more new tracts. (The goal is to maintain the size of tracts at around 4,000 persons, although they may be as small as 1,500 or as large as 8,000.) At other times, tracts may be combined or their boundaries shifted to take into account new freeways or other major physical features. What is needed is a conversion table for matching up the parts of different tracts in order to define a consistent area over time. The front of the published census tract books lists two tables showing tract comparability over time; only tracts with boundary changes are listed. (For a complete file of 1980 to 1990 census tract comparability, the Census Bureau is preparing a machine-readable product called the TIGER/Comparability file.)

For example, a 1990 census tract numbered 2145.20 may comprise only part of a 1980 tract numbered 2145. If that old tract 2145 is looked up in the conversion table, we might find it listed as comprising two parts in 1990: tracts 2145.10 and 2145.20. With this information, we need only combine the 1990 census data for the two new tracts to achieve a comparable area for comparison with the 1980 census.

1.6. The Hierarchy of Data Availability

Roughly corresponding to the hierarchy of census geography is a hierarchy of data detail. Reports covering larger amounts of data detail (such as more categories per variable or more crosstabulation between variables) provide less geographic detail: Fewer areas are itemized and those are at higher levels of geography. Conversely, reports and computer files that provide information for larger numbers of places, including the most numerous smaller areas, usually compensate by providing less detailed data.

The hierarchy principle of data availability is that more detailed information is accessible for higher levels of geography. This principle is most clearly expressed in the printed publications from the census. More detailed tabulations are feasible for states and counties because they are less numerous than census tracts and blocks. In fact, block data are not even printed; instead, they are released only on microfiche or in computerized form. Even within the data on computer tape, a clear hierarchy principle is at work: The most detailed data—microdata—are released only for areas of at least 100,000 population. The controlling factor in computer files is less the volume of data and more the concern about confidentiality and sampling error in smaller areas.

1.6.1. Priorities for Detail

Space is limited in summary tabulations; therefore, choices must be made about which tables to produce and which variables to emphasize in crosstabulations. Table II illustrates the choices made for the 1990 printed reports, using two representative population and housing topics, marital status and tenure status (owner- or renter-occupied housing). As can be seen, much more detailed tabulations are provided at the state level than at lower levels of geography.

Marital status is crosstabbed by six race categories, 13 age categories, and two sex categories, for a total of 780 cells of data at the state level. That woud be a prohibitively large table to replicate for each place within the state, so compromises are made. By removing all categories except marrieds, and reducing the age categories, the crosstabulation by age is slimmed down to 84 cells for the lower-level geographies. At the lowest level, the table is reduced still further, by dropping age entirely (but adding back the other marital categories).

Tenure status is the most important housing characteristic, and the most fully crosstabbed housing variable in census reports, but tenure is reported with less detail than marital status. Sex is not part of the crosstabulation at even the state level, although a fairly large table is carried down to places as small as 10,000 population. At the lowest levels, the table is stripped to just race by tenure, requiring only 14 cells.

It is noteworthy how much of the space in these 1990 tables is devoted to race and Hispanic origin categories. Race is crosstabulated in every table, and in many more

Table II. Hierarchy of Detail in 1990 Published Reports Illustrated by Crosstabulations Available at Different Geographic Levels

Crosstab by: Topic	Marital Status			Owner or Renter Status	
	Race × Age × Sex: Marital Status	Race × Age × Sex: Married	Race × Sex: Marital Status	Race × Age: Tenure	Race: Tenure
State	Table 37 $6 \times 13 \times 2$ $\times 5$ *780* [a] [b]	Table 43 $7 \times 7 \times 2$ $\times 1$ *98* [a]	Table 43, 51–53 41×2 $\times 5$ *410* [a]	Tables 44, 46, 48 38×7 $\times 2$ *532* [a]	Tables 43, 45, 47 38 $\times 2$ *76* [a]
County	—	Table 57 $6 \times 7 \times 2$ $\times 1$ *84*	Table 57, 59, 60 28×2 $\times 5$ *280*	Tables 53, 55, 57 28×7 $\times 2$ *392*	Table 49 7 $\times 2$ *14*
Place with 10,000+	—	Table 64 $6 \times 7 \times 2$ $\times 1$ *84*	Table 64, 66, 67 28×2 $\times 5$ *280*	Tables 62, 64, 66 28×7 $\times 2$ *392*	Table 58 7 $\times 2$ *14*
Place with 2,500–9,999	—	—	Table 71–73 28×2 $\times 5$ *280* [c]	—	Table 67 7 $\times 2$ *14*
Census Tract or BNA	—	—	Table 2–7 5×2 $\times 5$ *50* [c]	—	Table 9 7 $\times 2$ *14* [c]
Sources: *(see Table III for more detailed citations)*	CP-1	CP-1	CP-1 CPH-3	CH-1	CH-1 CPH-3

Notes: "Race" includes Hispanic-origin tabulations as well.

[a] Signifies number of categories in each dimension of the crosstabulation: e.g., Race × Age × Sex × Marrieds with $7 \times 7 \times 2 \times 1$ indicates 7 race and age categories, 2 sexes, and only 1 marital category (marrieds), yielding 98 data items. Entries for total persons are not counted in number of categories, whether in rows, columns, or on pages for total population, except that subtotals categories are counted.

[b] Twice as many cells are published in this table as reported because raw data are then percentaged.

[c] In addition to total counts by race, tables in CPH-3 are repeated for six major race/Hispanic groups.

categories than in 1980, whereas age and sex are sacrificed at lower levels of geography. In fact, the housing tables omit sex of householder from the tabulation of age and tenure (so we cannot investigate female-headed households of different ages). Instead of the two-category sex variable, we find 28 or even 38 categories of race and Hispanic origin.

Choices must be made about what detail to report from the census in a limited amount of space. (For a discussion of federal legislative needs as a primary criterion, see Chapter 3.) It may be that race is a more politically significant variable than sex. Racial variations from place to place are likely to be much greater than variations based on gender. Also, specific racial categories have constituent members who lobbied vigorously for data on themselves—more so than members of age-sex categories. Nevertheless, age and sex are such fundamentally important variables for explaining behavior that it is disappointing they are less often reported in crosstabulations. Sex of householder, in particular, would require many fewer categories and data cells than those used for race and Hispanic origin. (For more explanation of the many race categories, and an appreciation of their use, see Chapter 10.)

1.6.2. Confidentiality

Regardless of decisions regarding content priorities, the Census Bureau does not release data if they will potentially identify individual households. Racial subgroups with small numbers also may not have data reported even for larger areas. For 1990 and 1980, details on racial subgroups are reported only in instances where the group, or its complement,[3] has at least 400 residents in the area. Thus, most of the table columns for detailed race categories will not be presented in smaller areas or in states with more homogeneous populations.

Confidentiality is the governing principle restricting access to microdata records as well. (Microdata are explained in a later section.) Access to these rich data at higher geographic levels permits much more detailed analysis than is possible using printed reports or computerized summary tabulations. The detailed information for individual households (minus names and addresses) is made available only in geographic areas of at least 100,000 population. Release of microdata in smaller areas might jeopardize privacy by permitting potential identification of unique individuals (such as the only elderly black woman living in a condominium).

1.6.3. Small Sample Sizes

In addition to restricting data detail to protect confidentiality, at the block level the Census Bureau only releases information asked on all questionnaires, not the more detailed information drawn from the sample, long-form questionnaire. There is too much unreliability to sample estimates in these very small areas because of the normal sampling variability on information items collected from only a fraction of a small group of households. The distinction between complete-count and sample information, and the problems of sampling error, are explained fully later in this chapter.

---

[3]The complement is the residual group defined by subtracting all identified groups from the total population.

☐ 2. The Two Forms of Census Data

Many census users do not recognize that the data are released in two very different formats. The most familiar data are published in summary tabulations for geographic areas, some as small as blocks. A second, very rich source of data consists of computerized records for individual households (without names or addresses) that are released for large geographic areas of 100,000 or more population. The summary tabulations are printed in books and also recorded in computerized form. Microdata, however, are only available on computer tape or CD-ROM (read-only computer disks).

In this section we overview the differences between the two forms of data, outlining their respective uses. Chapters 6 through 11 demonstrate detailed analyses of summary tabulations. Analysis of microdata, and their linkage to summary tabulations, is addressed in chapters 12, 13, and 14.

2.1. Summary Tabulations

Summary tabulations are tables that count the number of cases in different categories. This format yields a very condensed amount of information in printed tables or recorded in computer files.

2.1.1. Advantages of Summary Tabulations

Summary tabulations are prepared by the Census Bureau, counting the total cases for each variable in each local area. Not all possible combinations of variables can be tabulated, so only selected combinations are prepared.

The advantage of the summary tabulations is that they are prepared in standardized format for a great many places. The basic information from the census is thus made accessible with low cost and little effort by the user. The drawback is that the summary tabulations are prepared in fixed format with little flexibility. Custom tabulation cannot be carried out as it can with microdata.

Recognizing this constraint, in 1990 the Census Bureau began to offer two options for customized design. As mentioned above, the User-Defined Areas Program (UDAP) tabulates data to match specialized local geographies that differ from standard census geography. In addition, the Special Tabulations Program provides the opportunity for customized tabulations to be carried out by the Census Bureau from confidential internal record files. There is an added charge for both of these customized services.

2.1.2. Printed Reports Versus Computerized Files

Printed reports are the standard references for local census data. Even the users of computerized files make frequent reference to the published figures, if only to cross-check the accuracy of their computer data extractions.

Several advantages can be cited for using printed reports. Access to data in printed reports is easier and quicker than in computerized form because programs do not need to be written nor tapes mounted. (Simplified extract programs for use with CD-ROM greatly reduce the complexity of computerized access.) The printed volume is simply pulled from the shelf and the table of contents scanned. However, users need to exercise care in reading to which geographic area and subset of the population the data pertain. Tables on a subject may look identical except for those not-so-minor details (Batutis 1988). The ease of access also means that printed reports are much easier to browse than computerized data. Furthermore, the comparable format of published tables from one decade to the next makes investigation of changes over time fairly simple.

Published reports also have the advantage of requiring less technical skill for access and no computer equipment or specialized software. There is also less chance for error in retrieving data from a printed report than a computer file because each entry is specifically labeled in the printed report. Programming errors may also be detected by cross-checking computer-retrieved data against published sources.

Computerized files have three powerful advantages to compensate for their greater difficulty of access. The quicker access to printed reports only holds true if a small amount of data is being accessed. For large volumes of numbers, computer retrieval is much preferred. Also, computerized files may be read directly into analysis programs, thus speeding the analysis of data. This is especially true if data are analyzed for many places at a time. Finally, much more detailed data are accessible in the computerized files than in the printed reports. Although the printed reports have the most frequently needed data, specialized analysis may require data stored in the summary tape files as described in a later section.

2.1.3. Versions of Printed Reports

Table III provides an overview of printed reports issued from the 1990 and 1980 censuses. These are issued in three general groups. The first is a summary of population and housing characteristics, designated the CPH group of report series in 1990 (and PHC in 1980), that cover a variety of different geographies. The most often used reports are in series CPH-3, covering census tracts and block numbering areas in each MA or nonmetropolitan remainder of a state. These tract/BNA reports also provide data for places of 10,000 or more population, counties, and MSAs.

A second series covers population topics alone and is labeled CP (PC in 1980). The CP-1 reports include data items collected from the 100-percent count, while CP-2 covers social and economic data collected from the sample count. A third, parallel series addresses housing subjects: CH-1 and CH-2. A report is issued in each of these series for each state, as well as for the United States as a whole (see the notes to Table III). In addition, a number of special subject reports are planned for specific population and housing subject areas; however, these are not standardized in content for ease of comparison between decades.

The combined collection of all reports for all places in the United States fills a wall of shelves. Before seeking out data for a specific place, users should consider which report series is most useful for their immediate purpose. The titles of the reports should help to indicate their general content. It will also help to know the content of 100-percent count versus sample count data, as summarized previously in Table I of Chapter 3. Earlier in this chapter, we explained the different levels of geography to be found in the data, and these are referenced in Table III. In general, the most de-

Table III. Printed Reports Published from the 1990 Census

Report Series In 1990	Equivalent in 1980	Geography Covered	Data Type	Title in 1990
Population and Housing Reports				
CPH-1-xx	PHC80-P PHC80-V PHC80-3 (100% portion only)	State/Counties Local Gov'ts. Indian/Native Areas	100%	*Summary Population and Housing Characteristics*
CPH-2-xx	PC-80-1-A	State/Counties MCD/CCD's Places	100%	*Population and Housing Unit Counts*
CPH-3-xxx	PHC80-2	MSA/Counties Tract/BNA's Places>10,000	100% and Sample	*Population and Housing Characteristics for Census Tracts and Block Numbering Areas*
CPH-4-xx	PHC80-4	State/Counties Cong. Districts MCD/places>10,000	100% and Sample	*Population and Housing Characteristics for Congressional Districts of the 103rd Congress*
CPH-5-xx	PHC80-3 (sample portion only)	State/Counties Local Gov'ts. Indian/Native Areas	Sample	*Summary Social, Economic, and Housing Characteristics*
Population Reports				
CP-1-xx	PC-80-1-B	State/Counties MCD/CCD's Places, Indian/Native Areas	100%	*General Population Characteristics*
CP-2-xx	PC-80-1-C	State/Counties Places>2,500,MCD/CCD's Indian/Native Areas	Sample	*General Social and Economic Characteristics*
CP-3-x	—	U.S., Regions	100% and Sample	*Population Subject Reports*
Housing Reports				
CH-1-xx	HC-80-1-A	State/Counties MCD/CCD's Places, Indian/Native Areas	100%	*General Housing Characteristics*
CH-2-xx	HC-80-1-B	State/Counties Places>2,500,MCD/CCD's Indian/Native Areas	Sample	*Detailed Housing Characteristics*
CH-3-x	—	U.S., Regions	100% and Sample	*Housing Subject Reports*

Notes:

  x Designates one of 30 population or 10 housing special subject reports.

  xx Designates separately numbered report volumes:

      1 = United States

      1A = American Indian and Alaska Native Areas

      1B = Metropolitan Statistical Areas

      1C = Urbanized Areas

      2 through 52 = States: Alabama through Wyoming

  xxx Designates separately numbered report volumes for each MSA and non-MSA balance of each state.

tailed data published for counties, municipalities, or census designated places are found in CP-1 and CP-2 for population topics, and in CH-1 and CH-2 for housing topics. Analysis of census tracts and block numbering areas is only possible with CPH-3.

2.1.4. Versions of Summary Tape Files

Summary tabulations are released in several versions for the same geographic area, causing some potential confusion among users. Data published in books have just been described. We now turn to the computerized files known as summary tape files (STFs), of which there are four. These data are generally provided on computer tape, although a limited amount of the STF data is also available on microfiche and on CD-ROM (read-only compact laser disks).

Each of the STF data collections is released in three or four versions to cover different types of geography. Files A and B report data for the smallest areas, with a separate file issued for each state. File C is a national file covering higher levels in the census hierarchy down to the level of places with 10,000 or more population. File D covers congressional districts (only offered with STF 1 and 3). With the exception of the new CD-ROM medium, the content and availability of the different files are very similar to that available in 1980.[4] The following STF data sources are available for 1990 (Bureau of the Census 1989b). All files are issued on computer tape and in limited other media, as noted.

1. **STF 1** includes the 100-percent (full-count) population and housing items. Files A and B both cover counties, places, and census tracts; file A goes down to the block group, file B down to the block level.

    On microfiche, files A and B are available.

    On CD-ROM, files A, B, and C are available.

2. **STF 2** includes the same content as STF 1 but affords more detailed tabulations: Tables are repeated for each race and Hispanic origin. File A includes counties, places of 10,000 or more population, and census tracts. File B reaches down to places of 1,000 or more population, plus American Indian areas and Alaska Native areas. Block groups and blocks are not available in any version.

3. **STF 3** includes the sample count population, housing, and economic items. File A includes counties, all places, census tracts, and block groups (individual blocks not available). File B covers ZIP Codes.

    On microfiche, file A is available.

    On CD-ROM, files A and B are available.

4. **STF 4** covers content similar to STF 3, but adds additional detail and repeats tables for each race and Hispanic origin. File A includes census tracts, counties, and places of 10,000 or more population. File B includes counties and places of 2,500 or more population. Block groups and blocks are not available in any version. This is the most voluminous file among the STFs.

---

[4]For details of the 1980 STF products, consult the *Users' Guide: Part C. Index to Summary Tape Files 1 to 4* (U.S. Bureau of the Census 1983a). For 1990, the best current information on STF products is found in (U.S. Bureau of the Census 1991d: Appendix F).

The file structure of STFs 2 and 4 is more complex than the others because details are repeated for separate race and Hispanic-origin groups. Two sets of tabulations are reported for each geographic area in two different records. *Record A* is a series of tabulations covering the total population or total housing, and *record B* is a different series of tabulations repeated for the total population and each separate group. These A and B records are not to be confused with files A and B; they are separate, unrelated concepts. Details of the file structure are explained in U.S. Bureau of the Census (1983a) or in the technical documentation released with each file.

The STFs are the basis for all census publications. This is demonstrated in Table IV, which cross-references the published report series by their STF source files. Note that reports may derive from two different STFs if the report contains both 100-percent and sample data. Also, some volumes in a report series are for the nation or a collection of larger areas, such as MAs. Those volumes will be derived from a different version (file C) than the volumes that report data for small areas (usually file A or B).

2.1.5. File Structure Based on Census Geography

The summary tape files simply report the frequency counts for the number of cases falling in each cell of a table. A table made up of two sex categories and ten age categories would have 20 cells and would be reported in a series of 20 successive fields. Using the data dictionary included in the technical documentation for a file, users can tell which fields correspond to which cells in a table. A *record* for one location consists of the set of cells that describe the contents for a series of different tables. The same record layout is then repeated for each location in the file, with a separate record for each.[5] As explained above, STFs 2 and 4 have two separate records for

Table IV. Summary Tape File Sources for Data in Printed Reports

	100% Data		Sample Data	
	STF 1	STF 2	STF 3	STF 4
CPH-1	File A,C			
CPH-2-	File B,C			
CPH-3-		File A		File A
CPH-4-	File D		File D	
CPH-5-			File A, C	
CP-1-		File B,C		
CP-2-				File B,C
CH-1-		File B,C		
CH-2-				File B,C

Note: File C provides data for report volumes covering the whole of the United States, as well as MSAs, urbanized areas, and American Indian and Alaska Native Areas. The other files provide data for volumes covering places within individual states. (See Bureau of the Census 1989b for more information.)

[5]On some files, long records have been broken into *segments* to help users read the raw data into computer programs. This is particularly necessary with the CD-ROM data in dBase format.

each place, A and B, the latter of which covers the repetition of tabulations by race and Hispanic origin.

The complicated part of the data retrieval comes in locating the table for the correct geography. Knowledge of census geography is essential for working with STFs. As shown previously in Table I, the basic hierarchy of census geography and supplementary areas are described by *summary level codes.* For example, census tracts are identified in the files by records with summary level 080. (Note that trailing zeros are often filled by a more specific identifying code, such as 082, as explained in the documentation for each file.) See Appendix B for an example computer program.

STF data are reported in two alternative record sequences, even with two forms on the same tape. First, all geographic locations have records in *hierarchical geographic format,* with each subarea nested within the next higher level of a state. For selected types of geography, records are then repeated in *inventory-type files,* with all the records for places in the same geographic level listed in sequence. Some file versions may consist exclusively of inventory-type files. For details of each separate version see U.S. Bureau of the Census (1989b; 1991d: 1-1, 6-1, and A-7).

2.1.6. Selecting a Data Source

The combined published reports and summary tape files (tape, CD-ROM, or microfiche versions) present a wealth of opportunities. The seven decision rules listed in Table V may prove helpful when deciding which source of local data to use. These rules outline a conservative strategy that will minimize search effort and guide users to the most appropriate data for a given purpose.

2.2. Microdata (PUMS)

The microdata census format bears close resemblance to the raw data collected on the census questionnaires. With personal identifiers removed, these data are distributed on computer tape and in CD-ROM form as the Public Use Microdata Samples, or PUMS data. Technical documentation for these files is provided in U.S. Bureau of the Census (1983b).

2.2.1. Data Content and Advantage

PUMS data are taken from the sample, long-form questionnaires, and so they contain the full set of data obtained by the census. In addition to the raw census information, two additional groups of variables are added to the PUMS files. First, a number of variables are constructed from the raw census information. Examples of these derived variables are household income (summing income of individual household members) and a summary of household type that combines marital status and presence or absence of children.

In addition, a group of allocation *flags* is attached to each record in the PUMS file, indicating which variables have been imputed or corrected in accordance with the bureau's quality control data editing routines. Although the allocated values are carefully designed to sum to correct control totals, they may alter the true correlation of characteristics within an area's households. Therefore, when analyzing rela-

Table V. Summary Guidelines for Choosing Among Data Sources

1) Rely on books, microfiche, or custom reports before computerized data.

- If only a few places are being analyzed, the published books, or microfiche, may prove more accessible than the computerized data.
- If the particular data are only available in computerized form, ask a state data center or another source to provide a custom printed report. (See the list of state data centers in Appendix C.)

2) Access the least detailed data necessary.

- Totals or simple tabulations should not be searched out of more complex reports. Those simple data can be found in three places:
  - reports titled "Summary" or "Counts" in Table III;
  - similarly titled tables at the front of more detailed reports; or
  - reports for lower-level geography, especially CPH-3.

3) When detail is desired, seek out the most detailed published reports that include a given level of geography.

- Lower level reports often include summary data for higher level geographies as well, but in less detail than found in higher level reports:
  - Fuller national and regional data on a subject are provided in special subject reports (CP-3 and CH-3) than in lower level reports.
  - Within each state, the CP-1 and 2, or CH-1 and 2, reports provide more detailed data on states, counties, and places than found reported for those geographies in the census tract/BNA report series (CPH-3).
  - At the census tract/BNA level, the CPH-3 series provides the only data.

4) When using computerized summary tabulations, make use of CD-ROM or other files stored "on-line" on computer disk before turning to tape versions.

- Data can be accessed for individual places much more quickly and easily from computer disks than tapes. Extract programs packaged with CD-ROM disks make access more user friendly than with tapes. Turn to tapes when the data are not available on disk or when large numbers of places are being analyzed.

5) When using computerized summary tabulations, rely on STF 1 and 2, which are 100-percent count data, before STF 3 and 4.

- STF 1 and 2 are smaller sized and more manageable, and they also are more accurate. Their data have no sampling error, whereas the sample count data in STF 3 and 4 are subject to sampling variability. Even the 100 percent variables included on STF 3 and 4 have sampling error because they also are taken off the long-form questionnaires. Therefore, use the STF 1 and 2 files unless the only place the desired data are recorded is on STF 3 and 4.

6) Within the pairs of STF 1 and 2, or STF 3 and 4, preference should be given to STF 1 and 3, respectively.

- Those files are less bulky than their counterparts. The main reason to use STF 2 or STF 4 is that they provide detailed tabulations by race and Hispanic origin. A number of other important variables also are reported only on those files. Seek out more detailed data in STF 2 or 4 only if that is the only place it can be found.

7) Turn to microdata for customized tabulations and analysis only if no suitable, existing tabulation can be found or if the user needs to analyze statistical relationships between variables at the individual or household level.

tionships among variables for a set of households, analysts may choose to select only cases where complete data on those variables were obtained directly from the households. The allocation flags serve that purpose. (See the discussion of allocation and substitution in a later section.)

The great advantage of the PUMS data is that custom tabulation or other analysis can be carried out. Variables can be related to one another for any subset of cases and in any combination imaginable. The drawbacks of PUMS are two-fold: 1) the data are voluminous and only available in computerized form and 2) the data are not reported for small geographic areas of less than 100,000 population.

2.2.2. Alternative Series of PUMS

The PUMS data are sampled from the the Census Bureau's master sample detail file (drawn from the long-form questionnaire) according to three sample rates: 1-in-1000, 1-in-100, and 5-in-100. The smaller sample rate is used for files covering the national population, whereas the larger 5-percent sample is most useful for analysis of local areas.[6] That large sample size is afforded only on the A sample files. A version B also has some local uses despite its limitation to a 1-percent sample, but that file is intended to identify metropolitan areas and non-metro remainders of states. A version C, offered in 1980 only, identifies urbanized areas and urban/rural portions of states (see U.S. Bureau of the Census 1983b: 5–6).

The PUMS files are samples that require weighting to reflect the complete population and housing counts. In 1980, the files were self-weighting, with weights defined as the inverse of the sample fraction: e.g., counts in the five-percent sample are multiplied by 20 to represent the complete count. In 1990, a specific weight is provided on each population and housing record, closely reflecting the inverse of the sample fraction. The more specific weights are used to ensure that PUMS data will weight up to totals published in summary tabulations.

2.2.3. Subarea Identification

Large subareas may be identified in PUMS data if they contain at least 100,000 population and if the remaining portion—the complement—is also of at least 100,000 population. In 1980, these subareas were known as *county groups*; in 1990, they are termed *Public Use Microdata Areas* or *PUMAs*. These subareas are defined according to boundaries specified by state data centers or local planning agencies. The subareas may consist of whole counties or, in less populated areas, of groups of counties that are aggregated to meet the 100,000 population threshold. Within large counties, subareas may be classified as aggregations of municipalities and other census designated places, or, in 1990 only, the subareas may be defined as groups of census tracts.

Three types of locations are recorded in PUMS files: place of residence on census day, place of residence five years earlier, and place of work for employees in the week

---

[6]In 1980, a 2.5-in-100 sampling rate was employed for a subset of variables in the 5-in-100 sample. For budgetary reasons, the migration and place of work data were coded for only half the cases, and left missing on the other half, effectively decreasing the sampling rate for those variables.

preceding the census. Each of these locations is identified by the same system of subarea codes, so it is possible to trace migrants or commuters (see Chapter 14). However, note that subareas based on census tract groupings can be used to identify only the current place of residence, not work place or prior residence.

Different sets of subareas may be defined in the A or B file versions, with more areas of smaller size identified in one version or the other. The availability of small-area detail differs widely from state to state because the subareas are locally defined. For example, in 1980, three subareas of Los Angeles County are defined in file A and 28 subareas in file B. Dade County (Miami), Florida, has the same 11 subareas identified on both files, while the entire Atlanta metropolitan area has only three subareas shown on both files. In general, more numerous, smaller sized areas are of greater use in PUMS-A with its larger, 5-percent sample than in PUMS-B.

2.2.4. Hierarchical File Structure

The key to understanding how to use PUMS is the hierarchical structure of the data records. Each housing record, designated by a file type H, is followed by one or more person records, P, depending on the household size. Persons living in group quarters, not in housing units, receive a blank housing record before their person record. Thus, the sequence of records in the data file resembles this:

$$H - P - P - H - P - H - P - P - P - H \ldots$$

An illustration of how these data look, and their difference from summary tabulations, is provided in Fig. 1. The advantage of the hierarchical file structure is that it preserves information about who lives with whom, and what kind of housing unit they share. The records can be processed in different ways to accommodate different purposes. Four types of files are often created as extracts from the raw PUMS data:

1. persons only file: read only the P records into a data file (type P extract) ;
2. housing only file: read only the H records into a data file (type H extract) ;
3. all persons plus their housing: combine the H record with each P record, outputing a combined record for each person (type PH extract); and
4. householder person plus housing: combine the H record plus the first P record, pertaining to the householder—a subset of the type PH extract (type PHH).

Appendix B outlines computer programs for reading the hierarchical raw data records into the four types of files usable for statistical analysis. Each of these types of data files can be used to address particular questions.

2.2.5. Applications in Analysis

A type P file can be used for detailed demographic or economic analysis, such as explaining fertility or salary on the basis of a person's sex, age, race, or educational attainment.

A type H file can be used to explore housing characteristics in depth. An example question might be how the value of a house relates to the number of rooms, year built, and recency of purchase.

SUMMARY DATA

- Basic unit is an identified geographic area
- Data summarized on people and housing in areas
- Available for small areas

Illustrative Summary Data

City	Total Pop.	Occupied Housing Units	Number of Persons Per Unit	Renter Occupied Units	Gross Rent		
					Under $80	$80 – 99	$100 – 149
Weston City	110,938	49,426	2.2	31 447	858	3,967	13,282
Smithville	21,970	7,261	3.1	2,492	37	190	1,766
Junction	17,152	5,494	2.7	822	11	29	238

PUBLIC-USE MICRODATA

- Basic unit is an unidentified housing unit and its occupants
- Unaggregated data to be summarized by the user
- Allows detailed study of relationships among characteristics
- Not available for small areas

Illustrative Microdata*

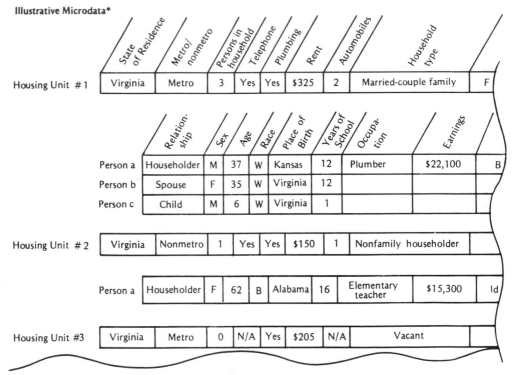

*Public-use microdata samples do not actually contain alphabetic information. Such information is converted to numeric codes; for example, the State of Virginia has a numeric code of 51.

Figure 1. Comparison of summary tabulation data with information on microdata files. [*Source:* U.S. Bureau of the Census (1983b: 2).]

Type PH files combine characteristics of the person with housing unit information. With these data we can estimate such things as the percentage of children living in overcrowded houses or the number of school children per bedroom in different types of newly built houses.

Type PHH files are more commonly used than type H because they combine characteristics of the householder with the housing unit information. With these data we might inquire how race or age of householder also relates to house value. As another example, we could calculate how homeownership rates rise with age of householder, tabulating these separately for different education levels.

Other, more specialized combinations of the data can be crafted to address very specific purposes. PUMS is the most flexible form of data yielded from the census. The uses of microdata are developed more fully in later chapters, beginning with Chapter 12.

☐ 3. Sampling and Accuracy in the Census

Most users think of the census as a (nearly) complete count. They would be surprised to learn that the majority of decennial census data are collected from a sample and has a certain amount of sampling error. In addition, questionnaires are edited for internal consistency and missing values, and certain imputations may be made to improve the quality of the data. Finally, there is the possibility of a more fundamental alteration in the data—an adjustment of the total for estimated undercount. All these topics pertain to users' acceptance of accuracy in the census.

3.1. 100-Percent Versus Sample Count Questions

As discussed in Chapter 3, the data collection process has two parts. The main part is a collection of a small amount of basic information from all of the people. These 100-percent data are supplemented by a larger amount of detailed information drawn from only a sample of the total respondents. (See Table I in Chapter 3 for a list of the 100-percent and sample variables.) Those sample data are then *weighted up* to reflect the numbers we would expect to find for the entire population. As a result, they *look* like a complete count even though based only on a sample.

Census users need to understand this distinction because the sample data are subject to the variability common in sample surveys. As a result, the sample data are less accurate than complete count data for neighborhoods and other small areas. Sampling error pertains to some key information that is widely relied on, such as income and education.

3.1.1. Sampling Rate

The fraction of respondents that received the long-form questionnaire differs by size of tabulation area. In most areas, 1-in-6 of all households (16 to 17 percent) received the long form. Because sampling variability grows more serious if samples are smaller sized, the Census Bureau has sampled small places at a higher rate to compensate. In jurisdictions such as counties or incorporated places with an estimated population of fewer than 2,500 in 1988, one-half the households received the long form. In contrast, in some very populous census tracts or block numbering areas, the sample rate was reduced to 1-in-8. This sampling plan was essentially the same as in 1980, except for the greater economy sought by the sparser sampling in larger areas.

3.1.2. Reporting the Sample Count Information

The sampled data are compiled in two different ways: All of it is used to tabulate characteristics for areas and some of it is used in microdata files. A subsample of the sample count records is taken to make up the microdata files. Of course, this sample of records from the long-form questionnaires includes the 100-percent variables as well as the additional sample variables.

Many users are unaware that variables are from a sample count, despite clear labels on those data, because sample numbers have been *weighted up* in summary tabulations to match the totals derived from the full count. For example, if the full count shows 2,000 households, but income is only collected from a sample of 300 households, the income data are scaled up to the full count total by multiplying the number of sample count households in each income category by a 2,000/300 ratio.[7] As a result, the published tables of sample data (both on paper and in computer files) reflect an estimate of the true numbers that would have been collected from the entire population.

3.2. Sampling Variance and Statistical Significance

Most users ignore the sampling error in the data. Usually, that is not a problem, but an informed census user should know when to be concerned about sampling error. There is no sampling error associated with the 100-percent data. In practice, we accept those numbers as *truth*.[8] However, conclusions drawn from the sample data have less accuracy. Therefore, users need to know the level of uncertainty surrounding a given number or difference. Not all numbers from the sample data need be checked for accuracy, but the greater the importance placed on a given conclusion, the more important it is to verify the accuracy of the underlying data.

Sampling variability is variation that occurs by chance because a sample is surveyed instead of the entire population. This error is randomly distributed and a factor statisticians accept as normal when small numbers of persons are selected from a population.

The importance of sampling error is that it introduces uncertainty about the true value derived from sample data. This uncertainty involves a range that can be calculated with some assurance, yielding a confidence interval around the estimate. The notion of *statistical significance* addresses the impact the size of this confidence interval has on the validity of calculated estimates. For example, measurement of a particular change over time, or of differences between two groups or two places, may

---

[7]The actual procedure used is more complex because the data are scaled up to match detailed control totals defined by type of household, age, sex, race, tenure, and other factors. The procedure for 1980 is described in Appendix D to the published census tract books.

[8]If not all persons have been completely counted, that is error attributed to underenumeration, an issue addressed in the next section.

be insignificant in the face of the rather large confidence interval that could surround calculations with the sample data. Therefore, we need to know how large the confidence interval is and whether it invalidates conclusions we may draw.

Each published report from the Census Bureau contains an appendix explaining exactly how to calculate the standard error and confidence interval for estimates in the report. Those reports are concise and clear, but it may help to provide a more general explanation.

3.2.1. Calculating Standard Errors and Confidence Intervals

One primary factor determines sampling error and the size of the confidence interval: The range of uncertainty increases as the sample grows smaller. Figure 2 illustrates how the confidence interval for a sample estimate of 50 percent widens as the reported base becomes smaller. With a reported population base of 2,000, the estimate has a confidence interval of + or −5 percentage points, which means we can be statistically confident that the confidence interval from 45 to 55 percent contains the true population value.[9] Once the reported size of the base falls below 600, the confidence interval widens rapidly, approaching + or −10 percentage points or more.

A special feature of significance testing with census data is that calculations are carried out with the size of the *reported base* in the tabulation, i.e., after weighting up, not the size of the underlying sample. Thus, the reported base of 2,000 really represents only 332 observations if the area was sampled at 1-in-6. Note also that the base for an estimate varies by the particular selection in the tabulation, not by the total size of the area. For example, if the estimate applies to the total population, such as the percent born in their present state of residence, then the total population is the base. However, if the estimate pertains to only a subset of the population, such

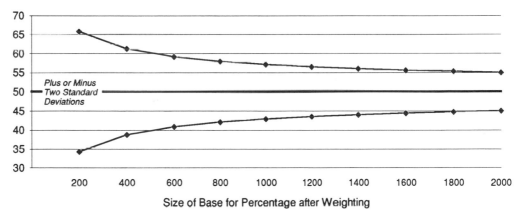

Figure 2. Confidence intervals widen with smaller reported sample bases. This example illustrates the case of a percentage estimated at 50 percent, with weighted bases as shown, and with a sampling fraction of 1-in-6.

[9]The example assumes a 95-percent confidence interval, defined by the range of two standard deviations on either side of the estimate.

as the percent of females that are natives of the state, then that subset (females) is the size of the base.

Formulas published by the Census Bureau take the sampling rate into account in two ways. First, they alter the standard error calculation by multiplying the numerator by 5 (the denominator of the sampling fraction minus 1), effectively adjusting the base back to the underlying sample size derived from a 1-in-6 sample. Second, they adjust the resulting standard error for the rate of sampling in each locality. Sampling rates on the order of 1-in-2 lead to standard errors that are about 60 percent as great as those calculated for the more typical 1-in-6 sampling fraction. The specific adjustments vary for each variable in the census, but most fall into a very similar range.[10] Confidence intervals can then be constructed by multiplying standard errors by 2 (or 1.96), yielding a range in which 95 percent of estimates from all possible samples would fall.

The necessary standard error for a census estimate can be retrieved from a table in the back of every published volume, or it can be calculated from a formula, one for estimated totals and one for percentages.

Estimated totals:

$$Se(\hat{Y}) = \sqrt{5\hat{Y}(1 - \frac{\hat{Y}}{N})} \,,$$

where

Se = Standard error;
N = size of area (total number of cases); and
$\hat{Y}$ = estimate of characteristic total.

Estimated percentages:

$$Se(\hat{p}) = \sqrt{\frac{5}{B} \hat{p}(100 - \hat{p})} \,,$$

where

Se = standard error;
B = base of estimated percentage (total number of cases); and
$\hat{p}$ = estimated percentage.

3.2.2. Statistical Significance

The statistical significance of a difference between two estimated totals or percentages is calculated from the confidence interval. The difference being tested might

---

[10]Calculation of standard errors with summary data according to the Census Bureau procedure is a three-step process. First, one applies the basic standard error calculation to the estimated total or percentage. Second, one determines what is the sampling rate that applies for the specific locality by looking this up in a table in the back of the published book (Table D for 1980 census tracts). Third, one finds in Table C an adjustment factor pertaining to the specific variable under the actual sampling rate. This adjustment factor accounts for the clustering of persons in households as well as the sampling rate. The basic standard error should be multiplied by this adjustment factor.

measure change between two censuses, or a difference between two places or two groups. Is the difference random, due to sampling error, or is it real? Differences larger than the confidence interval are said to be statistically significant.

The confidence interval for a difference between two estimates is found by first squaring the standard error of each separate estimate, then summing the two and taking the square root of that sum. When multiplied by 2 (or 1.96) this yields an estimate of the range holding 95 percent of all possible sample estimates, a range within which a difference could occur by chance.

Table VI provides a summary guide for judging the likely statistical significance of a difference drawn from summary tabulations. The top portion shows the required size threshold for an absolute difference to be significant, depending on the size of the reported base and the magnitude of the estimates that are being differenced. The bottom portion reports thresholds for differences between two percentages. Col-

Table VI Required Size of Difference for Statistical Significance in Data from Summary Tabulations*

Difference in Absolute Values

Size of Most Variable Estimate	Difference Derived from Two Reported Bases of Size:				
	6,000	4,000	2,000	1,000	600
2,000	254	220			
1,000	201	191	156		
800	183	176	152		
600	162	157	143		
400	134	132	124	108	
200	97	96	93	88	80

Difference Between Two Percentages

Size of Most Variable Estimate	Difference Derived from Two Bases of Size:				
	6,000	4,000	2,000	1,000	600
50	4.5	5.5	7.8	11.0	14.2
40	4.4	5.4	7.6	10.8	13.9
30	4.1	5.0	7.1	10.1	13.0
20	3.6	4.4	6.2	8.8	11.4
10	2.7	3.3	4.7	6.6	8.5

Note: Calculations assume common 1-in-6 sampling rate. Areas with 1-in-2 sampling will achieve significance at values approximately 60% of those listed. Areas with 1-in-8 sampling will achieve significance at values around 15% greater than listed.

*Calculations assume standard error adjustments recommended by the Census Bureau for typical variables in the sample count. Based on experience of the 1980 census, estimated thresholds for significance must be increased by a sizable amount for some variables:

migration	71%
poverty status of persons	28%
ancestry	24%
nativity and place of birth	13%
school enrollment	9%

umns in the table define the reported size of the base that contains the estimate, whereas rows define the size of estimate in question.

Note that a difference requires two estimates derived from two bases. Conservative use of the table would be to identify the entry based on the smaller of the two bases and the more variable of the two estimates. The *more variable estimate* is that which is closest to half way in the range of possibilities: 50 percent in the case of percentages, or half the size of the base in the case of absolute values. (Estimates nearer to the limits of possibility are less variable because they are confined by the limit.)

As an example of how the table can be used, assume that we have found the number of two-bedroom housing units has increased by 100, increasing from 700 out of a total stock of 2,000 housing units to 800 units out of a total of 2,200. Number of bedrooms is a sample count variable and is estimated with some variability, so we cannot be certain that an increase of 100 is large enough to be significant. Looking down the 2000 column (for the size of the smaller base) and across the 800 row (for the size of the most variable estimate), we find that a difference of 152 units would be required for statistical significance.

A second example compares percentages between two places. One town has 40 percent of its population with a college degree, versus 35 percent in the other. Can the first town claim its citizens are better educated? If both towns have 3,000 population, we need to interpolate between the 4,000 and 2,000 columns. (A simple average is close enough.) Based on the more variable estimate of 40 percent, a difference of 5.4 percentage points is required at 4,000 population and 7.6 points at 2,000 population. Averaged, this suggests a difference of 6.5 percentage points is needed for signficance. Therefore, we conclude that the first town's apparent advantage could be due to sampling variability and is not large enough to be reliable. There is no statistical significance to the difference between the education level of the two towns.

For more precise assessment of smaller differences, the user can interpolate between categories of the table to find thresholds that correspond to specific size bases and specific values for the two estimates being differenced. For detailed calculation, recourse to the basic statistical equation is recommended. This is well explained in the appendix to published census reports and data file documentation. However, Table VI should serve most census users well as a general guide to the likely significance of a difference.

3.2.3. Significance in Microdata

Microdata require somewhat different calculations of significance than shown here, and specific formulas are given in the technical documentation to the PUMS data (U.S. Bureau of the Census 1983b: 17–30). The difference comes in the sampling rate and the weighting of data. In the summary tabulations, weighting is carried out for users, but in microdata users must weight the data themselves. In 1980, users simply applied the inverse of the sampling fraction, e.g., a data from a 5-percent file were weighted by 20. In 1990, explicit weights were added to the PUMS files and users should apply these. For significance tests, users must first weight the microdata observations and then adjust the standard error formula differently from the summary tabulations. Where the formula shown above for summary data has a 5 in the numerator under the square root (for a 1-in 6 sampling fraction), in microdata we replace that as shown in Table VII.

Table VII. Adjustments for Microdata Sampling Fractions

Sampling Fraction	Standard Error Sample Factor
5% (1-in-20)	19
2.5% (1-in-40)	39
1% (1-in-100)	99
0.1% (1-in-1000)	999

3.2.4. Summary Guidelines on Sampling and Significance

Users may ignore sampling error in summary tabulations for variables that are derived from the complete count data. The issue of significance arises only for data derived from the sample count or that are microdata (because all microdata are a sample of records).

Analysts should focus attention on sampling error to an increasing degree if the re ported summary tabulation base for a calculation falls below 2,000 or if the microdata base falls below 300. This small base may occur either in very small areas or if calculations are performed for very small subsets of the population. In particular, remember it is the base for the specific percentage that counts, not the size of the total sample.

It is most necessary to address sampling error if the analysis emphasizes differences—either changes over time or differences between two places or two groups. Small differences may not be significant if the sample bases are small. Table VI provides a guide for determining whether more detailed investigation of significance may be required. If the observed differences are greater than the example confidence intervals reported there, the differences are surely significant. Analysts should calculate their own significance tests of marginally small differences if they are deemed crucial to the analysis.

3.3. Editing and Imputation of Basic Records

To maximize the value of census data, the Census Bureau exercises editing procedures to improve the data quality. After the questionnaires have been collected and coded, they are screened for erroneous or missing entries. Efforts are made to catch these problems before submitting the questionnaires to central processing. Two general procedures, *allocation* and *substitution*, are used to generate as complete and accurate a count as possible. These procedures are applied to the basic records counted by the census, and thus this information is fed into all the printed tabulations, summary tape files, and microdata.

3.3.1. Allocation

Allocations are made in central processing to fill missing entries or to replace unacceptable entries. The assignment of acceptable codes is needed most often when the questionnaire item was left blank or when an inconsistent answer was given. An example of inconsistency is when the householder states a length of occupancy in the housing unit that is longer than his or her own age. The former must be changed to conform to the latter.

When an item is left blank, the procedure is to replace the blank with information borrowed from a different household. The replacement information is drawn from the last complete questionnaire that had characteristics similar to those known for the present case. This statistical allocation is believed to yield overall results that are more accurate than the alternative: leaving the information missing (and the person uncounted for that topic).

3.3.2. Substitution

In some cases, the entire record may be missing for a housing unit or person. This can occur either because of failure to obtain an interview or because of a mechanical processing error with the completed questionnaire. In such cases, if the Census Bureau believes that the person or housing unit exists in fact, a full set of characteristics is *substituted for the missing record.* The replacement data are drawn from a previously processed record that is simply duplicated. Substitution is applied to only the 100-percent data items, and thus it does not affect the content of sample data.

3.3.3. Accounting for Imputed Information

In view of the seriousness of allocation, and especially substitution, the Census Bureau publishes tables counting the number of imputed cases in the back of the major report series for each state (CP-1, CP-2, CH-1, and CH-2, and their 1980 equivalents). The census tract/BNA reports also identify the tracts or BNAs for which 20 percent or more of persons or housing units were substituted.

Summary tape files include tables counting the number of allocated or substituted entries for most data items. Similarly, microdata include allocation flags marking individual cases that have been adjusted on particular items.

3.4. Adjusting for Undercount

No other feature about the census has garnered as much attention in the past few years as the issue of a census undercount.[11] Population counts collected by the census have taken on such political and economic importance, as discussed in the preceding chapter, that states and localities are fearful that their full population size may not be represented. Previously, knowledge about undercount was too imprecise to support an accurate adjustment. However, as our technical knowledge increases, it becomes feasible to adjust the census to achieve a total count that is more true. In turn, that capability makes the matter more controversial.

---

[11]Net undercount is the difference between underenumeration (failing to count persons) and overenumeration (counting persons more than once).

Sheer undercount has no significance in these debates. Rather, it is *differential* undercount that is important, because only an undercount that affects one location or one population group more than others will influence the distribution of political power or funds. Accordingly, any adjustment system must be sensitive enough to distribute accurately extra population to specific locations and specific population groups. Effectively, the Census Bureau would be asked to generate substitution records (see above) for persons never *directly observed or known* to live at an address.

During the decade of the 1980s, the Census Bureau engaged in a large-scale research effort aimed at solving the theoretical and practical problems of measuring census coverage and adjusting census counts. Two major aspects of this research—demographic analysis and the post-enumeration survey—are described below. Leading statisticians at the Census Bureau (and many outside experts) became convinced that these new capabilities were strong enough to be applied to census adjustment. Other outside experts and political leaders were more skeptical. Technical development and debate will likely continue through the 1990s as well.

3.4.1. Demographic Analysis of National Undercount

In conjunction with the 1990 census, the Census Bureau fielded an unprecedented research effort to estimate the size of the undercount.[12] That effort takes the form of two main projects: demographic analysis and the post-enumeration survey. The former is most useful at the national level and is discussed first.

Demographic analysis of undercount uses records of births, deaths, emigration, and immigration to estimate how many persons of each age, sex, and race should be counted in the census. Any difference from the census defines an under (or over) count. With good administrative records for the nation, this should be an accurate method. However, several problems need to be solved. Emigration is not well recorded; in addition, there is the matter of illegal, and undocumented, immigration. Demographic analysis must make some well-researched assumptions regarding these issues.

A surprising source of other difficulty has been birth records, particularly for black males born in the south before 1940. The Census Bureau's earlier estimates of those incomplete records have proven erroneously high, causing a larger expected count than found. This error became apparent once the undercount was graphed by age for more than one census. Black males born in the 1930s were heavily undercounted, much more than others, at age 25 in 1960, age 35 in 1970, and age 45 in 1980 (see Freedman 1991: Fig. 1). Census Bureau staff recognized the error and began to correct it going into the 1990 census (Robinson 1988).

The advantage of the demographic method is that an internally consistent series of estimates have developed over time. Recently revised results of that analysis are displayed in Fig. 3 for six decades.[13] The evidence indicates a shrinking undercount each decade, falling from 5.4 percent undercount in 1940 to 1.2 percent in 1980, but

---

[12]For a very useful brief review that is also easily attainable, see Wolter (1991).

[13]These data are taken from a press release by the U.S. Bureau of the Census announcing revised demographic estimates of undercount, June 13, 1991, Table 3. These revisions update those in U.S. Bureau of the Census (1988). The estimates for earlier decades have been revised slightly downward for consistency with current assumptions.

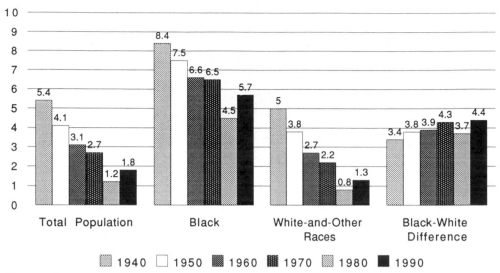

Figure 3. Percent net undercount by race: 1940, 1950, 1960, 1970, 1980, and 1990. (*Source:* see text.)

rising to 1.8 percent in 1990. Blacks have been undercounted at a much higher rate than all others (whites and other races), and the differential has remained just about constant over the decades. The Census Bureau could be very pleased at reducing the undercount from earlier levels, except that the undercount is greater in 1990 than in 1980.

3.4.2. Post-Enumeration Survey for State and Local Adjustment

Although the nationwide undercount error was estimated to be very small in 1990, that error was substantially greater in some places than others. The problem is how to assign this undercount to specific locations. An entirely different procedure is required.

A *post-enumeration survey* (PES) was fielded after the 1990 census to cross-check a sample of 170,000 housing units in approximately 5,400 block clusters (Hogan 1990). Through a complex matching scheme, information on *extra* persons captured in the survey can be used to infer the degree of undercount by type of person and type of geographic area (Hogan 1990). The method uses a *dual system estimator* for calculating the number of persons missed by the census.[14] The best description is the capture-recapture method of estimating fish stocks in a lake. By tagging a number of fish, releasing them, and then recapturing a sample, we can estimate from the number of fish found tagged in the second sample how many total fish are in the lake. It is not so easy with the PES: People can't be tagged like fish, and there are a number of ambiguities. Nevertheless, after substantial technical development and testing, the system appears workable, although a larger sample would have helped reduce the sample error from the estimates (see Wolter 1991).

---

[14]The basic method has been long known to demographers. See Shyrock and Siegel (1976: 503–505).

The purpose of the PES is to allocate adjusted counts to small areas in the nation. For this reason, a total of 1,392 *post-strata* were defined, those being specific types of persons defined by four race groups, six age groups, two sexes, region of the country, type of location, and type of housing (rented or owned). Overall, this method estimated an undercount rate of 2.1 percent, very close to that estimated by demographic analysis. However, the specific undercount estimated in each of the strata would be used to adjust the numbers of persons of each type in each local area.

Some of the detailed results from this process are illustrated in Figs. 4 and 5, covering males of different ages and races in Los Angeles–Long Beach PMSA,[15] depending on housing type.[16] Blacks in owner-occupied housing would need to be adjusted by a much larger factor than the other race groups (Fig. 4). The adjustment for children and young adults would be well over 10 percent among blacks. Elderly adjustment would be relatively small. Among black renters (Fig. 5), the adjustments would be as great as among owners. However, for other race groups, the adjustments are much higher among renters than owners. An increase of more than 15 percent would be made to Hispanic renters and nearly as much to Asian renters. The *other* category (largely white, non-Hispanic persons) also shows much greater undercount among renters than owners. As a final point, observe among the elderly that black and other renters have negative adjustments, signifying that those persons have been *over-counted*. If these adjustments are employed, the numbers of persons in those categories would be *subtracted* by a few percentage points.

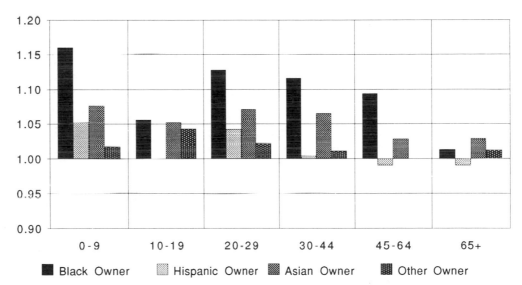

Figure 4. PES adjustment factors for owner-occupants in Los Angeles City. The factors are multipliers required to increase the enumerated population in each category to reflect the estimated true count. (*Source:* see text.)

[15]Los Angeles is one of four cities so large that it is separately identified as a stratum. Other locations are grouped together by size class or type in a region.

[16]These data are the final, smoothed adjustment factors and were made available to the author as part of a special packet sent to commentators on whether the census should be adjusted (per a May 24, 1991, notice in the *Federal Register* posted by the U.S. Department of Commerce).

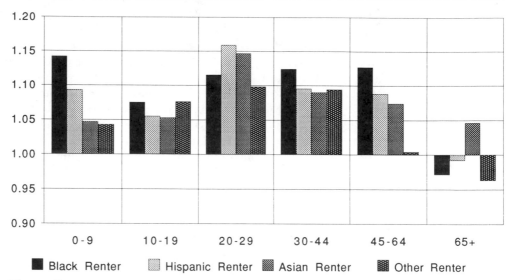

Figure 5. PES adjustment factors for renter-occupants in Los Angeles City. The factors are multipliers required to increase the enumerated population in each category to reflect the estimated true count. (*Source:* see text.)

These Los Angeles data show a degree of undercount that is virtually unmatched in all the post-strata. A few scattered places have high undercount rates that greatly exceed the national average of 2.1 percent, but California stands out with a 3.7 percent undercount. Only New Mexico (4.5) has a higher estimated percentage state undercount. The combination of California's large minority population and the high rates of undercount for those groups has produced by far the largest absolute undercount of any state in the nation. Most of the other states are near the national average undercount rate and would either be harmed or not benefit much from adjustment.

If the PES were to be used for adjustment, the factors relevant to each location and population group would be used to weight up (or down) the population in each block and census tract. The best available solution is to allocate the estimated missing persons to neighborhoods according to the age, sex, and race profiles of the undercounted persons. Although the PES is a housing unit survey, these missing persons cannot be assigned to households and housing units. Instead, they would be recorded in a special category called *Count Adjustment* like persons in group quarters, part of the population, but not living in the residential population. This solution would add to the size of local populations and alter the age, sex, and race numbers but little else.

3.4.3. Decision Not to Adjust the Census

In 1980, the Director of the Bureau of the Census made the decision not to adjust the census counts. In 1990, that decision was reserved for the Secretary of the Department of Commerce, which supervises the Census Bureau. The Secretary decided not to adjust, announcing his decision on the predetermined deadline of July 15, 1991, despite a recommendation for adjustment by the Director of the Bureau and by a majority of the Bureau's own experts (Mosbacher 1991). Secretary Mosbacher based his decision primarily on the grounds of accuracy, but also out of concern for the tradition of head counting rather than statistical inference.[17]

As the Secretary stated, "There is general agreement that at the national level, the adjusted counts are better. . . Below the national level, however, the experts disagree with respect to the accuracy of the shares measured from an adjustment" (Mosbacher 1991: 4). The problem is that the adjustments would be made based on the results of a sample survey (to be described) and the sampling error in that survey creates some uncertainty. The result is that in many cases, particularly small states or localities, or those with a smaller degree of undercount, there is the risk that the adjusted figures would be *less accurate* than the unadjusted count.

Although we might be statistically confident that certain major cities or entire states had suffered sizable undercounts, any adjustment implies a redistribution of the *relative* count (and, hence, political power and funds) from other locations. Simply stated, the technical problem is that the undercount may be highly concentrated, and therefore knowable with certainty in some places, but the compensating adjustments are so thinly distributed across many other locations that they are statistically uncertain in those other locations. The smaller the area and the smaller the relative undercount in an area, the less certain the adjustments become. (A larger sample survey would be needed to reduce the uncertainty, but that was not politically supported, as discussed in Section 1.5.4 of Chapter 3.)

Ethical–political grounds also were cited by the Secretary for not supporting adjustment. For one, "If the scientists cannot agree on these issues, how can we expect the losing cities and states as well as the American public to accept this change?" (Mosbacher 1991: 5). The Secretary also expressed great concern that so much judgment was required in the statistical inferences for determining the adjustment, and that these judgments had different political consequences:

> "What is unsettling, however, is that the choice of the adjustment method selected by Bureau officials can make a difference in apportionment, and the political outcome of that choice can be known in advance. I am confident that political considerations played no role in the Census Bureau's choice of an adjustment model for the 1990 census. I am deeply concerned, however, that adjustment would open the door to political tampering with the census in the future" (Mosbacher 1991: 5-6).

Finally, the Secretary expressed concern that adjusting the census would erode the incentive for localities to strive for full enumeration. Why would local officials expend scarce resources to improve the census count when they could bank on an adjustment? That would only increase the need for a heavier reliance on adjustment in the next census.

A useful and appropriate analogy for debating the ethical merits of adjustment is that of voting. Not all persons who are eligible become registered voters, and not all

---

[17]For a detailed account of the technical issues and politics surrounding the 1990 decision, see Choldin (1991); for details of the 1980 decision process, see Mitroff, *et al.* (1983).

registered voters decide to vote. Although it would be more *accurate*, in the sense of representation, to decide an election by a large sample poll, that is contrary to the assumptions of a democracy. (That also would open the opportunity for statistical tampering and control by computer.) Although a census may distribute political power like an election, there is an important difference. The persons not counted still must be provided with police, roadways, and other essential public services, part of which is paid for by state and federal aid that is distributed on the basis of a census count. Thus, the individual not counted is giving up more than his or her own political weight; he or she is placing greater financial burden on the public. Given this distinction, an argument could be made for adjusting the census count as part of revenue sharing formulas, even if the census count is not adjusted for political apportionment purposes.

Adjusting for undercount will continue to be an issue confronting census data users. For example, given what we know about the Los Angeles area's large undercount, on what population numbers should business planners rely? The Census Bureau does not plan to release its adjusted figures for local areas, even though a complete set of adjusted tabulations are potentially available for use. The number of lawsuits that have been filed to force release of the adjusted figures makes it clear that this matter will continue to be debated for some time.

☐ 4. Conclusion

There is a lot to know about census data. This chapter has surveyed some of the most essential issues: census geography, alternative sources of data, and issues of accuracy (sampling, imputation, and undercount adjustment). Additional details about specific issues of data availability are offered in the context of specific analyses in the chapters that follow.

5 □□□

Strategies of Presentation

A thorough knowledge of what is in the census and how to analyze the data effectively will not, unfortunately, guarantee accurate and understandable communication of results. Effective presentation of findings in tables, graphs, maps, and diagrams is an essential part of research methodology. These exhibits present the viewer with the most important facts the analyst has distilled from a mountain of raw data. This chapter focuses on tables and graphs, the two most widely utilized exhibits for presenting statistical information.

In the past, little attention has been given these presentation skills in most books on methodology or in university education. Design of tables and graphs was assumed to be self-evident. However, there is growing appreciation among statisticians and policy analysts for the difficulty of conveying the meaning of numbers. Attention to some basic principles will help analysts prepare much more effective reports and presentations with census data.

□ 1. General Principles of Presentation

1.1. The Art of Presenting Numbers

Judgment within the realm of art is required for both the design of exhibits and their integration with the rest of the presentation. In the past this has been a difficult concept for scientists to accept. They tend to assume that laws of science, not art, should dictate the presentation. Of course, astute observers have learned that this separation is impossible.

> Since numbers do not speak for themselves, analytic tables require careful planning and oblige the table maker to steer a course between art and artifice. Without art, he may fail to convey his evidence to the reader, but technique can also be used to deceive. Thus, *the most important rule of table making is this*: Arrange the table so that the reader may both *see* and *test* the inferences drawn in the text. (Davis and Jacobs, 1968:497; first emphasis added)

A leading marketing statistician beseeches his peers to respect the art of tabular communication: "The presentation of data of course involves judgement. But that is true of any form of analysis. . . . Judgement is largely what distinguishes a good analyst from a lesser one" (Ehrenberg 1977: 297).

Similarly, an eminant urban economist and policy analyst concludes:

> Those who like clear rules and universal principles may be distressed by the amount of judgment entailed in designing a table. There *are* fundamental purposes from which all tactics derive; they can be summarized as follows:
>
> - To persuade the reader to look at your table.
> - To enable him to see in it the message or conclusion that you draw from the data.
> - To convince him that the data are trustworthy. (Lowry 1983: 23; emphasis in the original).

If tables require artful judgment, what about graphs? Those are more purely visual than tables and place greater emphasis on art. Yet Edward Tufte mounts a scathing attack on artistic treatment of graphs in his now-classic work, *The Visual Display of Quantitative Information*:

> "Graphical competence demands three quite different skills: the substantive, statistical, and artistic. . . . Allowing artist-illustrators to control the design and content of statistical graphics is almost like allowing typographers to control the content, style, and editing of prose." (Tufte 1983: 87)

Tufte's point is that artistic decoration is not sufficient. If statistical integrity and substantive relevance are to be achieved, a thoughtful union of skills is required. Art and science must be combined.

1.2. Four Principles to Guide Judgment

Four principles stand out in the contemporary philosophy of how evidence should be presented in reports. Foremost is the requirement for *integration*, namely, that the evidence should play a central role meshed with the words of analysis and not be mere decoration or placed in a divorced position such as in an appendix. The second principle emphasizes *speed* or efficiency of communication: The viewer should absorb the ideas quickly and with a minimum of confusion. Third, each exhibit should attract the viewer and then draw attention deeper into successive layers of content. This principle we term *engagement in depth*. Finally, the exhibits should persuade the viewer of their accuracy, that they are *trustworthy*. Each of these principles deserves elaboration.

1.2.1. Integration

Experienced communicators of analysis view tables and graphics as part of a seamless information flow taken in by the reader/audience. As much as possible, these exhibits should be integrated with the stream of ideas from the text. This *integration* principle leads to:

- writing text that leads into or follows from the exhibit, rather than literally repeating in words what is shown in the exhibit;
- arranging the numbers, labels, and line art to convey easily the idea expressed in the text; and
- placing the exhibit as closely as possible to the paragraph where it is referenced (best is on the same page).

1.2.2. Speed or Efficiency of Communication

Modern communications take place in a hurried environment. Ideas must be conveyed quickly. Analysts cannot expect the audience to labor for even two minutes over a table or graph. This seems unfair when the analyst herself often has devoted hours to its preparation, but if the exhibit doesn't "click" within 10 seconds or less, the reader will not bother to study it further. Even a thorough inspection may well last fewer than two or three minutes.

"The criterion for a good table is that the patterns and exceptions should be obvious at a glance. . . ," according to Ehrenberg (1977: 277). He stresses maximizing the ease with which the reader can see and absorb the data. Similarly, Lowry makes frequent reference in his design considerations to the "hurrying eye" of the reader who is scanning the table's labels and data entries. Even greater emphasis is placed on speed of comprehension in graphics because the eye races over the visual displays.

Our emphasis is on efficiency of communication—speed being a primary measure of efficiency. The bottom line is that speed tells. A well-designed and smoothly integrated exhibit will meet the speed test. A clunky one won't.

1.2.3. Engagement in Depth

Ideally, each exhibit should attract viewer attention. At one level this attraction is superficial: a title may state the topic very simply or the lines of a graph may provoke curiosity about content. However, for the exhibit to engage the viewer, he or she must quickly see beyond the initial attraction to deeper layers of content.

Speed of communication is required for engagement, not to dispense with the exhibit quickly, but to draw the reader in deeper and stimulate reflection on the meaning in the exhibit. Any initial difficulty the viewer has with making sense of the graphic may discourage that deeper inspection.

The exhibit will continue to hold the viewer until one of three events intervenes to break the engagement: (1) the viewer becomes confused and disoriented, disrupting the flow of information absorption, and breaking concentration; (2) the viewer finds the content being portrayed of little interest in light of the report's overall theme; or (3) the viewer "bottoms out," having extracted all the available information from the exhibit. Obviously, the designer of the exhibit wishes viewer attention to continue to this end.

Exhibits with greater content may engage viewers for even longer intervals before they bottom out. A frequent expression of Tufte's is that graphics should "repay careful study," i.e. longer study should reward the viewer by uncovering new levels of meaning. However, the danger is that an overly complex graphic will confuse the viewer and cut short the attention span. The challenge is how to engage the viewer in depth with a large quantity of smooth flowing information.

To avoid drowning the viewer in a sea of complex information, the data should be structured hierarchically so that successive layers can be absorbed. Engagement is achieved by offering the viewer's eye manageable chunks of information, one leading to another.

The goal is to create a richly informative exhibit that is dense with information but open and accessible to the eye. The well-designed table or graph achieves engagement by attracting the viewer's attention, provoking curiosity, and then encouraging exploration of the multiple viewing angles and layers of information.

1.2.4. Trustworthiness

Exhibits are not mere decoration. They make factual statements in support of the written argument of a report. Exhibits are persuasive when they reinforce that argument and appear to be constructed truthfully.

One of Lowry's three fundamental purposes in table design is to convince the reader that the data are trustworthy. He stresses detailed explanation of the data's source and notes to explain various definitions or concepts. All this information is placed at the bottom of the table, lending an aura of weighty substantiation.

An additional strategy involves the very structure of the data and labels in the table. Above we have cited Davis and Jacobs's primary rule of table making: "Arrange the table so that the reader may both *see* and *test* the inferences drawn in the text" (Davis and Jacobs, 1968:497; emphasis in the original). The reader should be able to check through the numbers to verify conclusions drawn.

Data are less directly accessible in graphs, and deceit is more commonplace. Tufte devotes an entire chapter to graphical integrity where he exposes the misrepresentation of numbers in graphs that are more decorative than factual.

Scholars and professionals who exercise judgment in producing their work to meet these four overarching criteria—integration, speed, engagement, and trustworthiness—will find more receptive audiences. Their work will be read more fully; their ideas absorbed; and their recommendations understood and respected. Analysts whose work exemplifies this contemporary philosophy also will be well recommended themselves.

2. Presentation of Tabular Data

Since the majority of data results are presented as tables, attention to some basic rules of tabular presentation can improve greatly the quality of written reports.[1] Following this we turn to the related issue of graphic presentation.

---

[1]Ira S. Lowry provides a major source of counsel. For his final work product at the Rand Corporation, he summarized his accumulated experience with the formatting of tables in a generous working paper, "Designing Readable and Persuasive Tables" (Lowry 1983). Lowry's advice on tabular presentation seems particularly well suited to the reporting of analysis with census data.

2.1. Types of Tables

Two broad types of tables are commonly found: raw data tables and analytic tables. Raw data tables serve a library function, storing data for others to analyze. Published census data tables are of this form, as are some of the appendices to specialized reports. Sometimes such tables are called statistical tables (Kaplan and Van Valey 1980: Appendix E). Such raw storage tables are very dense and require great care in reading (Batutis 1988).

Analytic tables, in contrast, seek to make a point from the raw data. If raw data tables are passive holders of information, analytic tables seek active communication.

We can further subdivide the analytic tables into two types: analytic *method* tables and tables for *presentation* of results. Method tables seek to depict the analytical logic, the sequence by which numbers are processed. Presentation tables may carry less responsibility for showing how the results were calculated, but the viewer will still wish to compare the numbers to the totals and to one another.

2.1.1. Text Tables Versus Fully Formatted Tables

Presentation of the three table types may take either of two formats: text tables and fully formatted tables. A text table is briefly constructed to present a result and is inserted right into the text. This can be much more effective than including the numbers as parts of a paragraph. Consider this example paragraph:

The 1990 census revealed a higher overall percentage of minority population in Home City (34.8 percent) than in Neighbor City (30.8 percent). However, Neighbor City had a higher percentage of population who were Hispanics (16.4 percent) than in Home City (14.6 percent). The County had a lower overall minority percentage (25.7 percent) than either city, but this was due primarily to the county's much lower percentage Hispanic (10.3 percent). The 15.4 percent black population in the county was actually higher than in Neighbor City, although still less than in Home City.

Despite our best efforts to structure the comparisons in text form, the reader still has trouble grasping the overall pattern of the numbers. The problem is even worse when the labels describing the numbers are more lengthy than simply "black" or "County." Alternatively, we may report these numbers in a text table (Table I).

Table I. Example of a Brief Table Inserted in the Text

	Total Percent Minorities	Percent Blacks	Percent Hispanics
Home City	34.8	20.2	14.6
Neighbor City	30.8	14.4	16.4
County	25.7	15.4	10.3

The text table provides a much better view of the numbers than embedding them in sentence structure in a paragraph. The reader is free to make his or her own comparisons, and to test whatever conclusion is drawn by the author.

Fully formatted tables, by contrast, would add a title and full column and row headings, as well as a source and notes. Lowry observes that a fully formatted table contains "three main elements: the data, its identifying labels, and its credentials" (1983: 9). Each of these elements has certain governing principles and we address them in turn in later sections.

2.1.2. Choice of Table Type

It should be obvious that only small versions of presentation tables are suitable candidates for positioning as text tables. Fully formatted tables are required for presenting (1) raw data tables, (2) analytic method tables, and (3) larger presentation tables.

There is no firm boundary dividing the three. The exact construction of the table depends on the table's purpose. The table maker may wish to balance three guidelines. A text table may be most effective for persuading the reader to *look at the table*. However, a fully-formatted presentation or method table may be more effective for helping the reader to *see the intended message*, particularly if more than a few numbers are involved. The presentation or method tables must also be fully formatted if the reader is to be convinced that the *data are trustworthy*.

Fully formatted tables must be especially well designed if they are to attract the reader as effectively as the smaller text tables. (We will consider those design principles shortly.) But their greater size also allows for greater depth of information. With this depth and the credentials displayed at the bottom, fully formatted tables project a more satisfying and trustworthy image.

One rule is clear: Raw tables are to be avoided in analytical reports. Novice data users often copy raw data verbatim and report it as important information. However, such raw tables fail to communicate the point quickly. Instead, they force the reader to perform calculations and rearrangements of the data. Analysts must ask: How can I extract meaning from raw data and then display it to make my point quickly?

2.2. Arranging the Data

Fully formatted tables have two major design tasks: arranging and then labeling the data. We will address each major task in turn.

The emphasis in data arrangement should be on "seeing the data." The key design considerations for achieving speed and engagement center on the ease of absorbing the data and making comparisons. Two particular issues are emphasized here: how to focus attention on comparisons and the use of spacing or boxing to direct attention. A number of supporting technical matters are addressed after that.

2.2.1. Focusing Attention on Comparisons

The heart of every table is a comparison: how one variable rises or falls across categories of another variable. The simplest example is the time series, such as population growth over time. We might also make comparisons between places, or compare different population subgroups to one another and also between places. A relevant comparison is the substantive link that integrates the exhibit with the report and engages the viewer's interest.

The primary rule of data arrangement is that numbers to be compared must be adjacent. This is often the controlling principle for laying out data. Substantive decisions are required because the analyst first must know which comparison is the point.

For example, a table showing educational attainment by different racial-ethnic groups might report data separately for each sex and for each of several different places. At least three comparisons could be emphasized: across racial-ethnic groups, between sexes, or across places. Given the row-and-column format of tables, it is possible to make only two comparisons at a time: one along the rows and one down the columns.

Table II shows three different configurations for these data. Alternative A places the sexes within each race side-by-side, thus emphasizing comparison of educational attainment by males and females of the same race. Scanning down the columns we can also study the changes in educational attainment for each group across places. Place A has the highest percentages achieving high school graduation for every

Table II. Alternative Data Groupings Stress Different Comparisons

	Percent High School Graduates					
Alternative A	Blacks		Hispanics		White Non-Hispanics	
	Male	Female	Male	Female	Male	Female
Place A	55	60	51	59	72	75
Place B	49	54	47	55	62	65
Place C	47	52	49	57	68	71
Place D	52	57	49	51	68	71
Place E	50	55	44	52	52	59

	Percent High School Graduates					
Alternative B	Male			Female		
	Blacks	Hispanics	Non-H Whites	Blacks	Hispanics	Non-H Whites
Place A	55	51	72	60	59	75
Place B	49	47	62	54	55	57
Place C	47	49	88	62	57	55
Place D	52	49	68	57	51	60
Place E	50	44	52	55	52	62

	High School Graduates Index[a]					
Alternative C	Male			Female		
	Blacks	Hispanics	Non-H Whites	Blacks	Hispanics	Non-H Whites
Place A	.76	.71	1.00	.83	.82	1.00
Place B	.68	.65	.86	.75	.76	.76
Place C	.65	.68	.94	.72	.79	.73
Place D	.72	.68	.94	.79	.71	.80
Place E	.69	.61	.72	.76	.72	.83

Note: [a]Ratio of the percent high school graduates within each group to the percent graduates for a reference group:
   Male reference group: Non-Hispanic white males in Place A (72% high school graduates)
   Female reference group: Non-Hispanic white females in Place A (75% high school graduates)

group. With the male and female columns adjacent, we can also compare them across places.

If we wished to compare races instead, we must rearrange the data, as shown in Alternative B of Table II. There we place the races side-by-side, comparing them first among males and then among females.

Alternative C affords an easier means of comparison. Each percentage is expressed as a ratio to a common reference group: the percent high school graduates among non-Hispanic whites in Place A. Thus, black males in Place A have an achievement level 0.76 of the reference males; and black females in Place A have an achievement level 0.83 of the reference females, a smaller gap.

These index numbers make the comparisons even clearer, but at a price. The comparison between races, not sexes, is firmly embedded in the index numbers by choosing separate male and female reference groups. If we wanted to stress comparison between sexes, as in Alternative A, a separate set of index numbers must be constructed. This highlights the importance of choosing the comparison that is most relevant to the analysis.

2.2.2. Use of Spacing and Boxing

A critical element of table layouts is the spacing used to direct attention. Note how spacing is used in Table II to group columns of data for comparison. In Alternative A, the spacing is much closer between columns for the two sexes, and wider between racial groups, helping to channel the viewer's attention to a comparison between sexes. In Alternative B, the emphasis is reversed, with appropriate changes in spacing. Certainly the viewer can still compare males and females in Alternative B, but that is portrayed as a secondary objective.

Spacing can be supplemented by boxing, as in Table III. A few lines ruled onto a table, or formatted as borders around cells of a spreadsheet, can bring added clarity to a table. However, the spacing often has to be more generous in order to make room for these lines. Fewer lines are preferable to a cluttered table.

The boxes serve to demarcate major groupings of information in the table. While the row and column headings are marked off by lines, along with the totals, the numbers at the heart of the table are separated only by carefully designed white space.

Table III. Alternative Data Groupings Stress Different Comparisons

Alternative A

| Location | Percent High School Graduates | | | | | |
| | Blacks | | Hispanics | | White Non-Hispanics | |
	Male	Female	Male	Female	Male	Female
Place A	55	60	51	59	72	75
Place B	49	54	47	55	62	65
Place C	47	52	49	57	68	71
Place D	52	57	49	51	68	71
Place E	50	55	44	52	52	59
Total Area	51	56	48	55	64	68

Vertical lines should be avoided between columns of numbers to be compared, because these erect barriers to free eye movement across columns, creating something Tufte (1983) terms a "gridprison." Emphasis on seeing the data leads us to value the open window of data created by the central box of Table III.

2.2.3. Additional Technical Considerations to Facilitate Comparisons

A number of additional technical considerations are necessary for designing data arrangements to facilitate comparisons. Some of the more important factors are discussed here.

2.2.3.1. Comparisons by Rows or Columns.
Which is more effective, comparison down columns or across rows? There is some disagreement among experts on this question. In the preceding example we did both, but the primary comparison was along rows, comparing groups of two or three. Lowry believes such horizontal comparisons are easiest for up to six numbers, but Ehrenberg (1977) is emphatic that numbers are best compared down columns, not along rows.

Ehrenberg's lengthy argument is compelling, because he provides the reader with examples for self-testing. One of his key strategies is to group the numbers close together to minimize eye movement and enhance speed of absorption. Close spacing is easiest to achieve between rows (i.e. for scanning down columns). Given the standard dimensions of paper (in vertical or portrait position), and the need for horizontally typed labels, each additional column takes more room and spaces the numbers further apart. Overall, fewer columns than rows can be fit on a page, and therefore larger series of numbers require comparison down columns.

2.2.3.2. Ordering Rows and Columns for Comprehension.
Tables are not always designed with well-ordered rows and columns. Sometimes categories are laid out alphabetically, not following a substantive relationship; sometimes they are even more haphazard. A poor ordering hampers the speed of absorption and may disengage the viewer's attention.

Time is always well-ordered, either from the past to the present or from the present to the future. Other interval variables such as income, age, or number of rooms in the house also follow an internal order from smaller to larger.

In the case of variables that lack categories with any necessary order, such as race, sex, house type, or place, two alternative rules apply. The first ordering rule is to rank the categories in relation to the magnitude of a key variable of relevance to the analysis. For example, places might be ordered by total population size, or by the size of their unemployment rate or level of income, whatever is most relevant to the substance of the analysis.

A second, conflicting principle states that the categories should be ordered in the same fashion in each neighboring table. Once the reader is familiar with a certain order, it is best to reuse it across successive tables. The choice between the two principles requires judgment: The analyst should make a deliberate choice in ordering categories within the tables and maintain comparability across tables.

2.2.3.3. Number of Digits to Show.
The matter of rounding and significant digits is often a dilemma. On the one hand, analysts wish to convey the data as precisely as

possible, for example showing percentages of 27.72 percent instead of 27.7 percent or even 28 percent. This precision also minimizes rounding error and permits the numbers to add to 100. On the other hand, the greater number of digits makes the numbers harder to absorb and compare. Also, the degree of precision printed out by the computer is often a false pretense: Measurement and sampling error usually dwarf any differences that exist between 27.72 and 28 percent.

Ehrenberg argues that the extra digits are not merely unnecessary; they actually subtract from the information likely to be gained from the table. That is because two digits (e.g., 28 percent) are all that can be usefully absorbed and the surplus digits are a *barrier* to communication. Similarly, Lowry acknowledges that "the more numbers to be compared, the fewer digits each should have. . . . If your message depends on the exact values of lower-order digits, it's probably too subtle to be persuasive anyway. If you want your reader to look at the higher-order digits, show him those only" (Lowry 1983: 9).

2.2.3.4. Transformation to Ratios. The best way to show the relationship between two sets of numbers is to establish a ratio or index number. "If you save the reader work, he'll reward you by greater comprehension" (Lowry 1983: 9). For example, if the issue is the relationship between educational attainment of different groups, try forming a ratio of one to the other. That has been demonstrated already in Alternative C of Table II. Note that in the example we could have divided all groups by an external reference group, one drawn from outside the table. A useful candidate would be the county-wide (or national) average level of high school achievement. Index numbers focus attention on comparisons in very specific ways, sparing the reader the task of performing mental calculations.

2.2.3.5. Avoid an Overload of Alternatives. Some analysts overdo the helpfulness by reporting several alternative transformations in the same table. It taxes the reader to show both the absolute numbers and percentages in the same table. He or she must skip columns to make comparisons and otherwise dodge the unwanted information. That leads to losses of both speed and engagement.

The crosstabulation procedures in major computing packages usually violate the overload principle. As standard procedure, they print out four numbers for every cell in the table, such as:

$$278$$
$$6.88$$
$$24.93$$
$$37.61$$

These entries correspond to the absolute number, the percent of the table total, the percent of the row total, and the percent of the column total. A table with six rows and five columns would consist of 30 sets of these numbers, for a total of 120 numbers. Certainly, this raw computer printout should not be copied into a report! Most readers are uncertain which number is the appropriate one to focus on and compare across cells. If nothing else, the analyst should make that choice for the reader. (We address the problem of choosing appropriate percentage bases in Chapter 12.) If more than one set of percentages is useful, perhaps a second table is warranted.

Tables that focus on analytic method, as opposed to presentation of results, may require a greater mix of data and transformations. In this case, the point of the table is to show the alternative formats through which the data can be processed. Naturally, the row and column headings must exhibit even greater care to guide the reader through these alternatives.

2.3. Labeling the Data

If the goal in data arrangement is *seeing the data*, the goal in labeling is *understanding what the data mean*. Without adequate text on the table, the numbers may mean nothing to the reader. Perhaps the most difficult task in preparing tables is designing the title and other labels that describe the data.

It helps to think of a hierarchy of information, with the parts all working together to engage the viewer's attention in depth. The title, row and column headings, and footnotes should be used:

> ... selectively to support each other and to guide readers to the depth of information they need. As you move from one to the next, get increasingly specific rather than just repeating. (Lowry 1983: 11)

At the top level is the table's title, which may be characterized as like a newspaper headline. Its purpose is to "... contain enough information so that the reader can decide whether to linger or pass on" (Lowry 1983: 11). After this initial attraction, control of the viewer's attention passes to the row and column headings and then to the data themselves where the reader's eye is encouraged to make comparisons between different categories. The most detailed information is contained in notes at the bottom of the table so as not to obstruct the rest of the table.

In this section, we first address the kinds of information that labeling must provide and the potential locations for the information. Then we turn to a more detailed consideration of how each of the table parts should be designed.

2.3.1. Necessary Information and Its Location

Four kinds of information are usually needed to interpret an item of data in a table, according to Lowry:

- the attribute (variable or characteristic) that is measured: age, sex, income, fuel consumption, etc.;
- the unit of account: number of cases, percentages, dollars, gallons, etc.;
- the population or set to which the attribute pertains: residents of California in 1990, all job holders, the stock of all housing units, etc.; and
- the subset to which the entry pertains: e.g., racial groups or sexes (each a subset of the population); e.g., new housing units (a subset of the housing stock), etc.

Lowry neatly summarizes the challenge facing the table designer: "The problem is to reduce all this information to a minimum of words so that it will fit neatly into the space available and leave the reader with time to ponder the entries rather than the labels" (Lowry 1983: 10).

Three general places in the table may hold information describing the data: the title, the row and column headings, and the notes at the bottom. Information should be reported with increasing detail as we move from the title to the row and column headings and finally to the notes at the bottom.

Table IV shows an example of how the different types of information may be positioned throughout a table. The *attribute* described in the table is the origins of recent movers to Los Angeles County and is identified in four locations: the title ("Origins of Recent Movers"), the row dimension heading ("Residence 5 Years Ago"), the row headings ("Same MSA," etc.), and in the last part of the note at the bottom of the table. From the first to last location, the description of the attribute becomes progressively more detailed.

The *unit of account* in Table IV is a percentage—specifically, a percentage distribution that sums to 100 for each population subset. This is identified in the *spanner* that extends over the column labels. The *population* covered by the table is addressed in two places. The title identifies the geographic location of the population and its nature: "Recent Movers to Los Angeles County." The first part of the Note at the bottom then clarifies that the population consists of not all persons, but only those at least five years old and who had changed residence in the five years preceding the census. Finally, *subsets* are specified by the column headings as blacks and Hispanics. The total population also is treated as a subset in terms of the table's structure.

As in the example, the attribute and the population often require the most description. The former is the variable of interest and important for engaging the viewer. The latter is the sample base on which the attribute is measured. Failure to specify the unit of account or relevant subsets may lead to confusion when reading the table. Trustworthiness is enhanced when a table is fully labeled.

2.3.2. Designing the Table Parts

Each of the parts to a table plays an important role. Because each is different, the technical details are best considered separately for each.

2.3.2.1. Table Titles. A short title is generally preferable. This should be enough to draw the reader's attention and distinguish one table from another. To illustrate increasing brevity, Lowry gives the example of these three alternatives:

Option A:

Brown County Renters in 1974, Classified by Household Income in 1980 Dollars, Age of Head, Marital Status, and Presence of Children, and Monthly Contract Rent and Utility Expenses

Option B:

Housing Expenses, By Income and Life-Cycle Stage: Renter Households in Brown County, Wisconsin, 1974

Table IV. Annotation of Label Contents

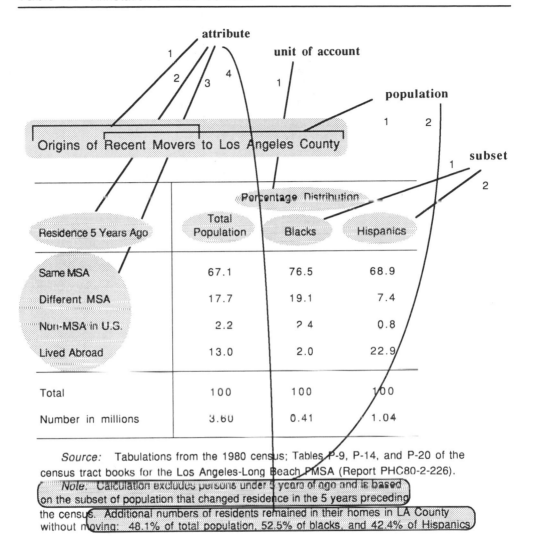

Source: Tabulations from the 1980 census; Tables P-9, P-14, and P-20 of the census tract books for the Los Angeles-Long Beach PMSA (Report PHC80-2-226).

Note: Calculation excludes persons under 5 years of age and is based on the subset of population that changed residence in the 5 years preceding the census. Additional numbers of residents remained in their homes in LA County without moving: 48.1% of total population, 52.5% of blacks, and 42.4% of Hispanics.

Option C:

Renter's Housing Expenses, By Income and Life-Cycle Stage

The first option is clearly too wordy. The reader needs to study it at length, thus violating the speed commandment. Remember that the title is merely intended to persuade the hurrying reader to look at the table, giving some idea of its substance.

Option B is more streamlined, condensing age, marital status, and children into the overall concept of life-cycle stage, and eliminating the unnecessary information that income in expressed in 1980 dollars. (That can be given in the row and column headings or, better yet, in the notes at the bottom of the table.) Finally, Option C reduces the word count still further by dropping the place and date reference. That is justified if the report is about a single place at a single point in time and the reader can be assumed to know that the data pertain to Brown County in 1974.

How short the title can be also depends on the optional use of subtitles. Examination of popular journals such as *American Demographics* will reveal numerous examples of this type of format. We can extend the example above to a fourth alternative that makes use of a subtitle in different size type:

Option D:

Renters' Housing Expenses
Average Monthly Gross Rent, by Income and Stage in Life Cycle

Alternatively, the subtitle might highlight the place and date of the data. In that case the information shown in the subtitle of Option D would need to be moved to the row and column headings.

A basic decision when designing a title is whether the title will describe the *data* within the table, as do the examples above, or whether the title will describe the *interpretation* to be drawn from the table. In fact, stronger viewer engagement may be gained by highlighting the point of the table and not just its topic, as in this example:

Housing Expenses Rise for Middle-Aged Renters
Average Monthly Gross Rent, by Income and Stage in Life Cycle

Or, placing the story point in the subtitle:

Renters' Housing Expenses in Brown County, 1974
Rents Rise in the Middle Life-Cycle Stages for all Income Groups

The point-emphasis format is useful for popular communication of research findings. This is considered generally unacceptable for scientific communication, but it is very effective in executive summaries of reports, public presentations with visuals, or in reports to citizens and local officials.

Finally, we must consider the size and treatment of type in the title. A series of words set in all capital letters is difficult to read and should be avoided. Similarly, extensive underlining is unnecessarily cluttered and less legible. Computer-based word processing has opened up a world of options, including a mixture of fonts and sizes, and bold face and italics. The basic rule is to use these options judiciously, keeping the variety of type faces and/or treatments to a maximum of two or three on a page.

2.3.2.2. Row and Column Headings. The title identifies what the table is about, but the major burden of communicating the meaning of the data falls on the row captions and column headings. The row captions are often referred to as the *stub* of a table and the column headings may be called the *boxhead*. However, the more commonplace terminology of rows and columns is usually sufficient.

Space is most constrained in this part of the table, and so considerable art is required. Lowry cautions that the row captions and column heads should enable the reader ". . . to interpret every entry *precisely enough for your purposes.* . . Don't over-burden labels with information that could be relegated to notes at the bottom of the table" (Lowry 1983: 13; emphasis in the original).

A simple table layout, showing changes over time in marital status, is given in Table V. The table title is accompanied by a subtitle designating the geographic area covered by the table. This is a much more visible location than embedding the geographic area in the title and is a useful strategy if tables are being prepared for a series of geographic areas.

The row headings give the marital status categories, introduced by the row dimension title (stubhead): Marital Status. The column headings give the year and gender group (Males or Females). These column headings are introduced by a spanner proclaiming the unit of account reported in the tables cells: Percent of Adult Population in Each Status.

Percentages can be confusing because they can be formed in more than one way with the same data (as discussed in Chapter 12). The design of Table V helps to orient the viewer to the direction of percentaging in two ways. First, a % sign is placed by the number at the head of the column, and then the column is anchored with 100 to show where the percentages sum to a total.

The example makes use of both careful spacing and boxing to organize the table. The addition of lines or boxing is especially important for organizing column categories that form a complex hierarchy (as in the migration examples in Chapter 9) or when the units of account differ between columns of the table.

2.3.2.3. Table Notes. Detailed explanation of table entries can be provided with notes at the bottom of the table. These notes are used for at least three types of information: (1) fuller expression of row and column labels that were condensed for reasons of space; (2) detailed definitions of important concepts; and (3) notation of

Table V. Marital Status Changes, 1970–1980

Census Tract 4602

| | Percent of Adult[a] Population in Each Status | | | | | |
| | Males | | Females | | Change 1970–80[c] | |
Marital Status	1970	1980	1970	1980	Male	Female
Married, spouse present	68.2%	56.0%	61.1%	51.2%	–12.2	–9.9
Formerly married[b]	5.2	10.3	19.4	23.0	5.0	3.6
Never married (single)	26.6	33.7	19.5	25.8	7.2	6.3
Total	100	100	100	100	NA	NA
Number of Adults[a]	2158	2162	2422	2377		

Source: Tabulations from Table P-1 (1970) and Table P-1 (1980) of the published census tract books for the Los Angeles SMSA.
Notes:
[a]Adults are defined for marital status purposes as all persons aged 15 and older in 1980 and all persons aged 14 and older in 1970.
[b]Includes widowed, divorced, and separated persons.
[c]Difference between the 1980 and 1970 percent.
NA—not applicable

exceptions from the stated conditions in the table. Some of the notes are keyed with symbols (a, b, etc.) to specific entries in the table. It is also possible to precede the keyed notes with a general note describing the data more generally.

The notes at the bottom of the table are important for establishing the credibility of the data. Imagine what questions the reader might have about the table: What is meant by "income"? Is that for families or households, or is it per capita income? Or, what exactly is the population base for these numbers? Column heads are frequently too brief to spell these out.

Lowry emphasizes the importance of establishing what he calls "the credentials" for the data. His view from the trenches is worth attention:

> If you've persuaded the reader to look at your table with a specific issue in mind and if you have formatted and labeled the data so that he understands the comparisons you want him to make, you should get his assent to your conclusions. But you can still lose a reader who decides that your data are untrustworthy. Especially if he disapproves of your conclusion, he's likely to seek this way out. So it's important to supply your data with convincing credentials. (Lowry 1983: 20)

At a minimum, the notes at the bottom must account for the source of the data. "The most important credential is a clear account of where the numbers came from and what happened to them along the way" (Lowry 1983: 20).

The accounting used by Lowry first gives a short "Source" description and then a longer "Note" of explanation written in text form. The following example shows both features (Lowry 1983: 18, 21):

> *Source:* Tabulations by HASE Staff of records of the screening survey for Site II.
> *Note:* Estimates are based on data from a sample of 2,541 housing units, including 371 for which market value was not reported; the latter are distributed by value in the pattern of other units of the same sizes. Entries exclude an estimated 1,630 owner-occupied mobile homes. Distributions may not add exactly to totals because of rounding.

Of the 48 tables offered by Lowry in an appendix of examples, 83 percent contained a general note, such as above. On average the examples also contained two keyed notes, referencing specific parts of the table, and the median total of keyed notes and general notes ran seven lines.

The marital status example above (Table V) has four keyed notes. One defines the ages of adult population, noting the change in definition between censuses. Another explains that formerly married persons include widowed, divorced, and separated persons. That information might help the reader to understand why there are so many more females than males who are formerly married (widows). A third note explains the manner in which change between 1970 and 1980 was calculated. Finally, NA is defined as meaning *not applicable*. This abbreviation sometimes also stands for *not available* or *no answer*.

Placing these details at the bottom of the table, out of the way, is part of the hierarchical strategy for organizing complexity. The reader can access the data in the table quickly and review the notes if desired. Even if unread, the notes make a visual statement that the table is carefully crafted and that the analysis is well substantiated. Or, as Lowry summarizes: "Altogether, the note should leave the reader with the sense that the author is careful with numbers and their interpretation. . . ." (Lowry 1983: 21).

2.3.2.4. Citing the Census. Census data users should note the source of their data if they wish to gain credibility. A bare reference, such as "Source: U.S. Bureau of the Census," is laughable. The knowledgable reader will wonder whether or not the author realizes how much data—and how many different reports—there are from the census. Should we trust this neophyte's judgment?

The more credible approach is to cite the report series, the table number, and any alterations to, or selections from, the raw data. (This information will prove invaluable to the analyst as well, once a week or two has passed and memory is growing dim.) An example might read:

> *Source:* U.S. Bureau of the Census, CP-1-20, Tables 18, 24; Hispanic portion of the data excluded. [or, STF 2, Tables PB2, PB6; California state file]
>
> *Note:* Hispanics were excluded to eliminate overlap with the race categories and to maximize the sample of Americans Indians and blacks. Five-year age groups were interpolated from the ten-year age groups given in the census data using the Karup-King method described in Myers (1992). Distributions may not add exactly to totals because of rounding.

These few lines will bring greater confidence to the reader by making the analyst's choices appear less casual. The notes may help the analyst to remember as well.

☐ 3. Graphic Presentation of Analysis

Most of the rules for designing effective tables apply to graphs as well. The major difference is that the data pattern is more prominent in a graph. The viewer can access the pattern initially with very little reading of any captions. Nevertheless, the viewer must still scan all the captions to establish the meaning of the visual pattern.

The visual aspect of graphs raises some unique technical considerations that must be addressed. First, however, it is helpful to understand the revolution that has occurred in statistical graphics in recent years.

3.1. Graphic Revolution

Graphic presentation of data experienced two revolutions in the 1980s, one working against the other. The spread of microcomputers brought computerized graphing programs to millions of desktops. Graphs could now be drawn almost instantly, with a minimum of effort, and without any requirement for art or judgment. The results of such automatically produced graphs, although cleanly drawn and neatly labeled, were rarely commendable.

The second revolution was ushered in largely by one man, Edward R. Tufte, through his landmark book, *The Visual Display of Quantitative Information* (1983). A statistician with a penchant for graphic design, Tufte was appalled by the degradation of statistical graphics in the twentieth century. The computer revolution was now permitting a vast increase in the volume of nonsensical or wasteful graphics. Tufte sought to restore statistical graphics to higher intellectual and artistic standards by linking graphic practice to modern philosophies of statistical analysis and visual communications theory.

The two revolutions—one of technology and one of thought—may coalesce in the 1990s. Tufte's ideas are being quoted in computer industry magazines, and he reports on work he has carried out for IBM (Tufte 1990: 60). Further improvements in computing technology make Tufte's ideas ever more practical. The spread of the graphical user interface, popularized by the Apple Macintosh and later extended to IBM-compatible equipment, is leading to new software that emphasizes statistical graphics. Analysis of data from the 1990 census will benefit immensely from the merging revolutions.

3.2. Selected Principles from Tufte

We will not attempt to repeat all of Tufte's lessons. By now, those lessons are much better known, and more accessible, than the craft of producing persuasive tables. Instead, we will highlight a few of the main principles derived from Tufte, and then apply these to the task of graphing data from the census.

Among other attributes, Tufte believes graphic displays should:

- encourage the eye to compare different pieces of data. . . .
- be closely integrated with the statistical and verbal descriptions of a data set (Tufte 1983: 13).

As an overall summary principle, this seems most succinct:

> Graphical excellence is that which gives to the viewer the greatest number of ideas in the shortest time with the least ink in the smallest space (Tufte 1983: 51).

Although it is impossible to maximize all four of these dimensions simultaneously, Tufte's intent to promote efficient graphic displays is clear.

3.2.1. Focus on Comparisons

The goal of every statistical graph is to make comparisons: changes over time; differences between places; differences between groups; or a combination such as differences between the rate of change of different groups in different places. Better designed graphs help the viewer to see more complex comparisons more easily.

"Information consists of *differences that make a difference*" (Tufte 1990: 65; emphasis in the original). Elements of a graph that do not help the viewer see differences are noninformation or disinformation. These wasteful elements distract the viewer and detract from the communication.

3.2.2. Architecture of Complexity

A central task in designing a graph is how to organize visually a complex set of data. Tufte urges that graphs should be analyzed for their "viewing architecture" in order to create and evaluate designs that organize complex information. Hierarchical organization is central to maintaining clarity in the face of complexity:

> The information shown is both integrated and separated: integrated through its connected content, separated in that the eye follows several different and uncluttered paths in looking over the data (Tufte 1983: 159).

The viewing architecture includes not only the data pattern but also the various captions and text on the graph. Later we will describe a model of how the "cycling eye" scans over a graph, taking in new levels of information on each sweep.

3.2.3. Data-Ink Ratio

A simple rule for judging the communication efficiency of a graph is to maximize the *data-ink ratio*. Tufte describes that ratio as the "proportion of a graphic's ink devoted to the non-redundant display of data-information" (Tufte 1983: 93). Extraneous ink often is used for embellishments that merely detract from the communication power of a graph. Such graphs often have only a few data points and represent only trivial relationships. So their designers compensate by decoration. Instead, Tufte would like to see more data and less decoration.

The perfect data representation under this rule is the dot—not the bar or the pie wedge. Borders around the graph, grids, and captions should also be minimized, either with thinner lines or by elimination.

Computer-generated graphs typically violate the rule about maximizing the data-ink ratio. Since they are not manually prepared, the analyst perceives no advantage to economizing on the amount of ink and number of lines drawn. Pie charts, which Tufte abhors, are too difficult to attempt by hand, but they are a snap with a computer. Similarly, bar graphs require far more line work than placement of a dot, and they have proliferated under computerization. One reason is that most computer equipment cannot draw a very clean line on a diagonal. With their rectangular shape, bar (and column) graphs are well suited to those equipment limitations, so designers of graphing software have emphasized bar and column charts.

3.2.4. Chart Junk

Computers have also opened the door to Tufte's hated *chart junk*, consisting of extraneous nondata-ink: overbusy grid lines, excess tick marks, heavy borders, and eye-straining cross-hatching. The latter is a special category that Tufte labels *vibrating chart junk* for its unintentional optical art. This moire vibration is probably the most common form of graphical clutter, often generating great noise to cloud the flow of information from technical and scientific studies.

Figure 1 presents three variations of a graph. The top is an example burdened with chart junk. The bars are distinguished by diagonal and horizontal cross-hatching, providing enough differentiation to separate the three data series (blacks, Hispanics, and whites). However, the viewer has to work to remember which cross-hatching represents each group, and all three bars vibrate to varying degrees. The chart is also embellished with a legend possessing a deep, black shadow for extra effect. All of

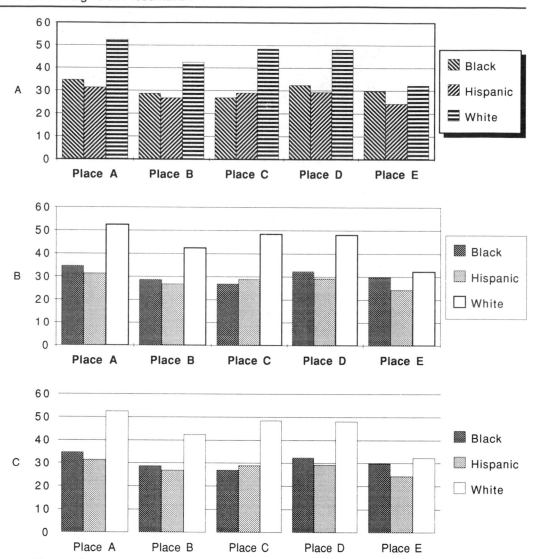

Figure 1. Three variations on a graph. Each uses progressively less ink to depict the percentage high school graduates among males, by race and location.

these treatments detract from the true data being depicted—a comparison of educational attainment among three groups in five places.

The middle version is much calmer and projects the data values for comparison more easily. Following Tufte's recommendation, we have replaced the crosshatching with screens of varying density or shades of gray. The white bars may not contrast greatly with the background, but this choice is warranted for two reasons: (1) the bars protrude so far beyond the others that less emphasis is needed to distinguish them; and (2) the color white corresponds well with the label given that group.

The example also increases the data-ink ratio of the graph by reducing the line weight on the border around the plot area and by entirely removing the border—and shadow—around the legend. A simple comparison of the legends on the two exhibits reveals how much more effective is the reduced ink version. Plainly, less is more.

The bottom alternative attempts to reduce the nondata-ink even further. The gridlines are reduced in weight, and all axes are removed. The percent labels are sufficient to represent the vertical scale, and the bottom axis is indicated by the bottom of the bars. Nor do we need the tick marks that separated the groups of bars for each place in the earlier version. Each cluster is sufficiently demarcated by white space alone. When compared to the middle alternative, the tick marks and bottom axis seem rather aimless. To remove them we need only select the option to make the horizontal axis and tick marks invisible.

3.2.5. A Comment on Grids

A special form of chart junk is the grid. Tufte recommends that the grid should usually be toned down so that it is only faintly visible in the background and doesn't compete with the data (Tufte 1983: 112). He reminds us that the grid is for convenience in plotting and not for putting into a presentation.

Computer-generated graphs need no grid to help with plotting. Grid lines should be shown only if they help the viewer and do not detract in other ways. Very thin horizontal lines may be acceptable; even better, print them in a soft gray—or halftone—instead of black. The examples shown in this chapter (and throughout the book) were produced with commonly available microcomputer software.

When plotting by hand, grids are essential, but the necessary graph paper comes in two common forms: one has a dark green grid and the other has a pale blue grid. Pale blue is much preferred. To see the reason why, photocopy a graph prepared on each type. (The photocopy test is significant, because any analysis will be copied if it is important and will be shown to others.) On the green-grid paper, the grid turns black and the graph is overwhelmed by a *grid prison*. In contrast, the pale blue lines disappear when photocopied. With the pale blue graph paper, the analyst must draw in the axes and grid lines the viewer actually needs. Far better to be in a situation of deliberate choice than to suffer the burden of an unwanted grid.

3.3. Practical Guidelines for Better Data Graphs

Specific rules for designing better graphs can be summarized in a compressed format. These rules are derived from a decade's experience teaching census data users how to draw sensible graphs for their reports. The guidelines have been condensed into five broad categories, *The Five C's of Good Graphics*. As displayed in Table VI, each of the guidelines draws attention to a different set of practical considerations.

3.3.1. Centering and Scaling

The most basic problem is how to center the graph on the page and then scale it properly on the vertical and horizontal axes. When drawn by hand, graphs may not have enough room left at the margins for captions. Alternatively, the plot may be mis-scaled so that it only occupies a small portion of the plot area; or the plot may surge to the very top of the page, invading the title region like a mountain peak slicing the clouds.

Table VI. The Five C's of Good Graphics

1. Centering and Scaling

Do not let computer technology dictate; avoid default options:
Deliberately set the high and low values for the vertical scale.
Move the legend out of the way to gain more room on either axis.

Hand-drawn is often more intentional and elegant:
Define axes 1 inch from left paper margin and 2 inches from bottom.
No plotting should extend within 2 inches of the top margin.
Choose a scale that accommodates your data within this space.

2. Choice of Format

Select a format for the graph that communicates the data concept most clearly:
Bar and column charts emphasize side-by-side comparisons.
Line charts show trends, slopes, and handle many data points.
Area charts shade beneath a line for effect, but cannot compare many data series.
Scatterplots correlate two data series, often with very many data points.
Pie charts have limited use—representing budget breakdowns as slices of a coin.

3. Correspondence of Lines or Patterns with Substance

Choose a line or pattern order to correspond with a substantive order in the analysis.
When lines represent multiple data series:
Order line weights from thin to heavy, or dotted to solid.
Order line symbols from open to closed, or square to X to circle.
Order line color along a spectrum; beware how it will photocopy.

When patterned bars represent multiple data series:
Order pattern densities from white to gray to black (avoid cross-hatching).
Order pattern color along a spectrum; beware how it will photocopy.

4. Captions: Communication with the Cycling Eye

Communicate quickly with the viewer in words:
Write a short title reflecting the content of the graph.
Use a subtitle—in smaller size type—to spell out details.
Label each axis clearly.
Write additional messages to the viewer in the plot area.
Footnote the source and other technical comments.

Imagine how the viewer's eye will cycle through the graphic:
Arrange the written information in a hierarchy of detail.
Give the viewer quick initial access to the graph's content
　　—backed up by more detailed explanation.

5. Content is Most Important

A good idea deserves an engaging, well-designed graph.

But the fanciest graph cannot compensate for lack of an interesting, underlying idea.

Computerized graphing software is programmed to scale a graph within certain spatial boundaries automatically, eliminating the crudest problems. However, computer-generated graphs introduce more subtle errors of scaling. Unlike the analyst, the computer does not know what the graph is about, and so identical scaling procedures are applied to all graphs whether appropriate or not. Consider the examples shown in Fig. 2.

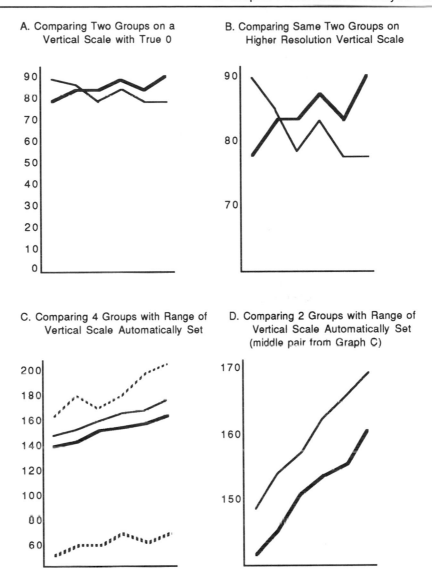

Figure 2. Scaling variation can mislead the eye about data variation.

The top two graphs represent the same data graphed at two different vertical scales. Option A plots the data on a scale with a true zero at the bottom, but leaves a lot of empty white space. Option B on the right uses more of the plot area to offer a higher resolution picture of the data variation, thereby amplifyng the line movements. Often it is desirable to accentuate subtle differences in this manner, but the eye is potentially deceived as to the true magnitude of the effects. Analysts must exercise judgment in this decision.

The bottom two graphs of Fig. 2 explore the scaling problem with an example showing the effects of computerized default procedures. In both figures, the computer has automatically scaled the vertical axis to accommodate both the lowest and highest values found in any data series to be graphed. Only two of the same data series are graphed in the right-hand option, and the computer has rescaled the axis accordingly. The slopes of the two groups appear much steeper in Option D than the middle pair in Option C.

In this example, the eye is attracted to the line pattern, especially their slopes, but the apparent variation is actually controlled by the vertical scale. (Note that stretching out the horizontal scale would have a comparable effect on the slopes.) To compare between graphs, the viewer must study the changes in the scale closely and make some difficult calculations.

To avoid misleading the viewer or creating a barrier to comparisons between graphs, the analyst should intervene and deliberately set the low and high values for each graph. The scale should be standardized with the same maximum and minimum scale values for graphs that potentially will be compared. This might lead to awkward or inefficient use of space and require that some groups be plotted separately from others. However, depending on the objective, excess white space may simply indicate a relationship more truthfully. In any event, the analyst's judgment is better about these matters than the computer's.

Deliberate control of scaling must also be accompanied by self-discipline. So powerful is the visual effect on the viewer's sense of magnitude and slope, that Lowry has suggested analysts must resist temptation to *manipulate the scale for argumentative purposes*. To that end he recommends that every chart contain "some normal reference value . . . zero, the mean, the long-term trend, etc. . . . If a zero point is not shown on the scale, the axis should be conspicuously broken—a zig-zag is the usual device" (Lowry 1991).

3.3.2. Choice of Formats

Most graphing software will automatically format a graph from data that are entered, usually offering a column (vertical bar) graph. Software programmers believe that this format looks best within the limitations of computer videoscreens and printers. The rectangular structure of graphic elements avoids the *jaggies* frequently observed when lines trace a diagonal path. Also, contrary to Tufte's data-ink maxim, the software programmers may feel that the greater amount of ink consumed in bar and column graphs is more impressive than the sparse lines of a line graph.

Analysts should let the data concept suggest a graphic format and then choose that option. Bar and column charts are better suited for some purposes; different formats may work better in other contexts. A variety of graphic formats are employed in the analyses of the following chapters; here we only briefly summarize the advantages of the alternatives. The second panel of *The Five C's of Good Graphics* (Table VI) suggests the main differences in advantage for each format.

Some types of graphs are better for helping the viewer to make comparisons. Bar graphs are best for side-by-side comparisons. Line graphs are best for comparing slopes or changes across the entire width of the graph. According to Tufte, pie charts never should be used: ". . . the only worse design than a pie chart is several of them, for then the viewer is asked to compare quantities located in spatial disarray both within and between pies. . ." (Tufte 1983: 178).

Line charts are superior to other formats on many accounts. They can handle more data points and maximize the data-ink ratio more than all other formats except the scatterplot, and they are drawn much more easily than other formats. But line charts have an important limitation: they require that the categories on the horizontal axis be arranged on an interval (or at least ordinal) scale. For example, we can plot the average earnings of males and females (separate lines) across age groups (from 15 to 75 or more) on the horizontal axis, or across education levels (from less than high school to college graduate), or across time (from 1960 to 1990). But we cannot plot a line graph across categories that have no logical connection or hierarchical order, such as race, marital status, or place.

Demographers almost always use line graphs, because their horizontal axis is often devoted to age, an interval variable, and also because more data can be displayed that way. Hopefully, improvements in computer equipment will make this option more attractive to a wider range of analysts because it is much preferred by the viewer.

To encourage greater use of line graphs, consider the following strategy: Plot a line graph with different symbols for each data series. Then suppress (make invisible or color white) the lines connecting the symbols of each series. Instead, draw those lines in by hand using a straight edge and a pen. This procedure can yield very impressive results because it uses the computer for lettering and plotting but adds a manual touch.

The scatterplot is even more efficient at representing data than the line graph. This is a widely used diagnostic tool for evaluating bivariate relationships in a data set. We can plot each case as a separate dot, showing its location on two dimensions, one for each variable. A number of examples are presented in subsequent chapters. For example, one effective use we will make of the scatter format is to plot the average household size in 1990 (vertical axis) against the average household size for the same place in 1980 (horizontal axis). The diagonal line represents equal household size in both years. Instead, most places are plotted well below the line, indicating decrease in household size, but some are plotted above the line, indicating an increase (see Chapter 7, Fig. 3). This is a very efficient design for comparing change over time in each place (in the microviewing perspective) and for comparing changes between many places (in the macroviewing structure).

A final graphic format we can use is the area graph. This is essentially a line graph with the area shaded beneath it. Despite its greater use of ink, Tufte acknowledges that this format resembles a view of a horizon (the border between sky and hilly terrain) and he suggests that a shaded, high-contrast display might occasionally be better than the unshaded line graph floating in white space. Area graphs give emphasis to the cumulative area enclosed by the line and can be used with great effect.

3.3.3. Correspondence of Lines and Patterns to Substance

Effective use of lines and patterns requires some strategic thought. Too often the choice is haphazard: The analyst draws the first data series with one line or pattern and then, without thought, selects a different one for the next series. This decision process is the same process the computer follows. The same series of lines and patterns is applied in the same order to all graphs regardless of content. We can do better with very little effort.

The basic rule is that the variation of lines or patterns should correspond in some intuitive manner to a natural order based on the substantive differences between the data groups. For example, the two better variations of the graph in Fig. 1 shaded the bars to reflect the racial concepts of white, black, and Hispanic. Not only did this shading make it easier for the viewer to track the three data series, but there was no sensible alternative. Had we reversed the order from the intuitive to something different (e.g., white population represented by bold black bars) the viewer would need to consult the legend repeatedly.

A more general guideline may prove useful. Unlike race, sex, and a few other variables, a great many concepts can be represented by an internal ordering useful for comparisons. Variable dimensions that can be labeled poor, low, small, or old have an intuitive image of being weak. The contrasting dimensions that are rich, high, large, and young or new have an intuitive image of being strong, full, or bold. This polarity of weak versus strong can be represented visually, as shown in Fig. 3.

For example, a graph with multiple data series, one for each of three education groups, could be made intuitively accessible to the viewer if the darker, bolder lines

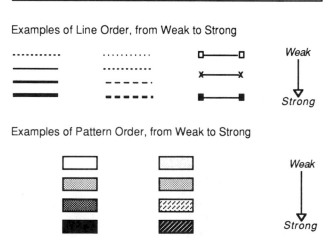

Characteristics with a Natural Order for Comparison

	Weak Image	Strong or Full Image
Income	Poor	Rich
Education	Low	High
Age	Old	Young or New
Size	Small	Large
Date	Future, Past	Present

Examples of Line Order, from Weak to Strong

Examples of Pattern Order, from Weak to Strong

Figure 3.   Line and pattern variation should correspond to the substantive order of comparison.

and patterns were assigned to the higher education group. Similarly, if we were graphing the percent homeowners of three age groups across income on the horizontal axis, it would work best if the younger persons were assigned the bolder line or pattern. Using the dotted line for elderly persons represents them as less vital, graying, or fading away, while the bold line for young adults represents them as vigorous and active. In a related concept, we may assign the boldest line or pattern to the present, showing the earlier dates as fading into the past, or showing future projections as fading into the future because they are more speculative and less certain.

A competing principle says that the most important group in the graph should receive the boldest line or pattern. A study of the elderly, for example, might compare them with other groups and emphasize their importance by a bold line. The emphasis rule may often be the best choice, but the problem of how to order the other groups remains. The principle of ordered comparisons is a better general rule to follow, with important exceptions warranted for emphasis. However, never assume that the computer-selected order makes any intuitive sense. Analysts must supply their judgment.

3.3.4. Captions Must Communicate

The principles for labeling graphs are very similar in spirit to those for tables. However, one exception is usually that less specific detail is expected. The graph will often refer to a table as the source of the data, and the details are given there. Alternatively, the graph may depend on the report text for greater explanation. A graph without any associated documentation may suffer loss of trustworthiness.

Another difference with tables is that the viewer's eye is drawn to the data much more quickly, bouncing back to the labels for only the briefest orientation. The written communication must be very quick, heightening the importance of the speed principle. The viewer's eye scan might follow this rapid sequence:

data picture;
  title;
  caption for vertical axis;
  categories on horizontal axis;
data picture;
  legend (if there are multiple series to sort out);
data picture;
  rescan legend (perhaps once for each series);
data picture;
  rescan captions on horizontal axis;
data picture;
  rescan subtitle;
  rescan caption for vertical axis;
data picture;
  explanatory notes;
data picture.

The captions on the two axes are only glanced at as the viewer races from the title back to the data picture in the graphic. The simpler that labels can be stated the better. The viewer will likely return to them later for a closer look.

Similarly, the viewer hastily scans the legend and tries to match these symbols with the data picture. The eye may dart back and forth from legend symbol to each data series. Difficulties encountered here often are a source of frustration in reading multiple series data graphs, hence, the importance of the intuitive correspondence discussed above. Tufte goes so far as to recommend against any legend. Better, he says, to write messages directly on the data picture with arrows indicating which line is which.

Explanatory notes (describing the source or offering technical comments) are desirable but should be placed out of the way, and in smaller sized type, so as not to interfere with the earlier stages of information flow. Essential concepts that must be understood sooner are best stated in a subtitle to the graph and may be as many as three lines long. See the examples of multiple-layered explanations in graphs published in *American Demographics*. Their designers appear to expect the viewer's eye to cycle through the graphic, taking in a new level of documentation on each sweep. (In 1989, *American Demographics* won the American Statistical Association Award for Outstanding Statistical Presentation in the Media.)

3.3.5. Content Above All

Not the least concern is the content of the graph. A good idea deserves to be communicated clearly with a well-designed graph; but the fanciest graph cannot rescue a bad or nonexistent idea.

The central theme of Tufte's work is that graphical excellence must mesh with substantive content. The reader will have a much easier time making sense of a graph if it is logically related to ideas just presented in the text. The more interesting the idea in the text, the better motivated the graphic viewer. However, if the graph presents ideas that are different from what the text leads the viewer to expect, then comprehension or viewer interest will fail. The emphasis on content is thus a restatement of the integration commandment: text, numbers, and graphic must work together to be effective. Content is the theme that ties together the three media.

4. Conclusion

The overarching principles for presenting effective exhibits have permeated all of our specific advice on tables and graphs. It is especially interesting how much the different experts agree with one another, whether discussing tables or graphs. Judgment takes different forms in tables and graphs, but the underlying principles are the same. Numbers don't speak for themselves, and computers don't think for humans. Analysts must focus on the requirements for effective communication.

Integration is at the heart of report writing. When should the author switch from text to a table or from a table to a graph? As a rule, numbers should be stated as part of the text until the argument exceeds three or four numbers. Beyond that point it is too difficult for the reader to compare the numbers. The author might then switch to a text table or a fully formatted table. When a great many numbers are to be compared, graphs are superior. With income and age distributions we often have a great many numbers. Graphs are also useful as a change of pace, to break the tedium in a long string of tables. (The reverse is also true.) And graphs also may help to draw attention strategically to a point that could be easily described by a table.

The speed principle emphasizes more efficient communication. Ideas "that are there" may not count if they cannot be absorbed quickly. Analysts can profit by paying close attention to their own reactions as they read tables and graphs. Where are the stumbling blocks? What seems to speed the communication? With careful reading, one's discrimination improves. This resource can then be brought to bear on one's own work.

Engagement is the goal of all communicators. If deep and important content is being presented, flowing from integration with the text, and if it is packaged with artful judgment to enhance speed of absorption, then the viewer will likely be well engaged. Census data often tell powerful stories of great interest to persons who live in the area described. Analysts can use exhibits to engage their audience in deep reflection.

Finally, trustworthiness is of prime importance for persuading the audience. Exhibits that are eye catching but don't appear well substantiated may lead to a loss of credibility. Exhibits must be designed with integrity from the beginning and then documented with sources at the bottom. Engagement plus trust leads to a convincing presentation.

Part 2

Analyzing Local Changes

6

□□□

Local Housing Conditions

Housing provides the bedrock for local census data. Virtually all persons live in housing units, and different types of units attract different types of people. Housing data are closely interwoven with population data for this and other reasons. Indeed, the way in which the census or other population surveys contact people is through their housing units.

The housing stock provides a relatively firm foundation for local analysis for several reasons. Housing units are highly visible, relatively permanent, and largely immobile. In contrast, people come and go, and households change their membership drastically. Housing units also have unique identifiers that help locate them in space. They have addresses and they have parcel numbers on tax assessors' rolls.[1] The only comparable identifiers for people are the social security number, the passport, or the drivers license number. But these personal identifiers may not accurately reflect the current location of the individual. Finally, we have better administrative data in local areas on the numbers of housing units than on the number of people. Building permits record the construction of new housing at fixed locations. In contrast, we may record births and deaths of people, but in the United States, unlike many other nations, we do not register their location in between the two events.

---

[1]William Diemer (1987), a longtime official with the City of Los Angeles, has presented a fascinating account of how tax assessor data, and other administrative records can be used to cross-check and even improve on the housing stock base that underlies the census. See also the discussion by William Butz (1987).

For these reasons, and because of the growing problem of gathering information from personal questionnaires, census planners are debating whether future censuses in the United States should become even more heavily based on housing unit counts, supplemented by a sample survey of person characteristics. It is becoming more important that we understand how the housing stock changes and what its relation is to the population.

Housing is a vital characteristic of communities in its own right. Homeownership is widely regarded as part of the *American dream*, the belief that you can get ahead by hard work, of which one of the chief fruits is a home you own. By this view, a declining homeownership rate in a local area would signal declining well-being. Housing also deserves close study in communities because it varies so much from neighborhood to neighborhood. According to White (1987), housing types are distributed far more unevenly than any other census characteristic except race. For a rich, graphic view of spatial patterns in housing characteristics, see Adams (1987). How these housing differences interact with local population differences recently has been recognized as a rich field for inquiry (Myers 1990a).

1. Data for Describing Housing Conditions

Recall from Chapter 3 how the *household* differs from its *housing unit*. The household is the person or group of people occupying a housing unit. (Households can be either the members of a family who live together, a person living alone, or a group of unrelated persons.) The distinction between the residents (the household) and the dwelling (the housing unit) is important to keep clear because both concepts are used in analyzing housing.

Housing conditions are measured in two broad sets: One is the physical and economic nature of the housing stock, while the other set is made up of characteristics of the household's *fit* to the housing unit. Housing stock characteristics include the type of housing (whether it is owned or rented and number of units in the structure), its size (number of rooms or bedrooms), age or year built, adequacy of plumbing, and its cost. Household fit characteristics include the level of crowding (persons per room), the level of affordability (percentage of household income spent on the rent or mortgage), and the length of time the household has occupied its dwelling.

1.1. Measuring Housing Problems

Housing problems are measured by a mix of these *physical stock* characteristics and *fit* characteristics. For example, federal, state, and local housing programs measure housing needs based on a combination of the number of physically deficient units and the number of households that are too large or too poor for their units.

Only two measures of physical quality are recorded by the census: (1) the presence or absence of complete plumbing and (2) the age of the dwelling unit. Plumbing deficiencies have nearly disappeared as a housing problem in the United States, falling from more than 40 percent of units after World War II to less than one percent by 1980 (Clemmer and Simonson 1983). The number of housing units built before

1940 is sometimes used as an indicator of poorer quality housing, especially in community development block grant or revenue sharing formulas, but this is only approximate. Many older houses are of even better quality than those built in the 1940s or 1950s.

As physical problems have diminished, the problems of fit—particularly affordability—have grown more prominent. Overcrowding is usually defined as more than one person per room, although that standard was as high as 1.5 persons per room earlier in the century (Baer 1976). Over time, crowding levels have declined markedly, largely due to falling fertility and shrinking household size, but also due to the construction of larger housing units. After 1980 the levels of overcrowding began to increase again for the first time in decades, not because of smaller housing units, but instead because households doubled up to afford rising rents and because of growing numbers of families with larger numbers of children.

Affordability has emerged as the primary source of housing problems in most cities. This is measured as the percentage of household income devoted to montly housing expenses. As reported below, detailed tables of these levels of payment burden are reported separately for renters and owners for areas as small as census tracts.

1.2. The Format of Housing Data

Housing data reported in census tract books are published in three basic tables, with only minor differences among the 1970, 1980, and 1990 reports. The large number of data items compressed into these tables is potentially confusing to users. Table I summarizes the content of the three tables. Observe that the first table summarizes the 100-percent count portion of the data, while the other two report sample data. The first table addresses issues of occupancy and utilization, plus value and rent, while the second reports on structural characteristics and equipment, and the third reports on costs of housing. Each of the tables is repeated for a number of different racial groups (including Hispanics).

An important improvement achieved in the 1990 data is that information on structure type (*Units in Structure*) was moved from the sample data of Table H-7 to the 100-percent count data of Table 9. (See Table I in Chapter 3 for a listing of other, minor data changes between 1980 and 1990). The 100-percent count housing information has the additional advantage of measuring socioeconomic status in small areas more accurately. Because they are free of sampling error, measures such as percent owner occupancy and median house value or rent provide excellent indicators of socioeconomic status. In contrast, education or income data are only collected on a sample basis and are subject to more error, as discussed in Chapter 4.

Two different *universes* are reported in the census tract tables. As discussed in Chapter 3, some questions in the census pertain to all housing units and others only pertain to housing units that are occupied, or to the households that occupy the units. For example, tenure (owner or renter status) is only asked of households, whereas structure type can be observed even if the unit is vacant. Thus, in two of the tables, the top half of the data pertains to all housing units and the bottom half pertains only to occupied units. It is important to recognize these two different totals and choose appropriately.

Table I. Content of Housing Tables in Published Census Tract Books

Table Number 1990	Table Number 1980	Data Type	General Topic	Universe and Content in 1990

9 (subsequent tables contain data for separate races and Hispanic origin)	H-1	100%	Occupancy, Utilization, and Financial	Housing Units Universe
				  Vacancy Status
				  Units in Structure (in Table H-7 in 1980)
				  Number of Rooms
				Occupied Units (Household) Universe
				  Tenure by Race and Hispanic Origin
				  Persons in Unit/per Room
				  Value and Rent

32 (subsequent tables contain data for separate races and Hispanic origin)	H-7	Sample	Structural Characteristics	Housing Units Universe
				  Year Structure Built
				  Condominium Status
				  Number of Bedrooms
				  Plumbing, Heating, Telephone, etc.
				Occupied Units (Household) Universe
				  Tenure by Year Moved Into Unit
				  Heating and Telephone
				  Vehicles Available
				  Tenure by Mean Income and Poverty

33 (subsequent tables contain data for separate races and Hispanic origin)	H-8	Sample	Financial Characteristics	Occupied Units (Household) Universe
				  Selected Monthly Owner Costs
				  Income by % of Income for Owner Costs
				  Gross Rent
				  Income by % of Income for Renter Costs

Local analysts in resort or farming and logging areas should be alert that the definition of the housing universe changed slightly between the 1980 and 1990 censuses. Prior to 1990, most housing tables were reported for a universe of *year-round units.* Excluded from the universe were seasonal units used for recreation (beach cottages, hunting cabins, ski condominiums, and the like) or temporary lodging used by migratory farm workers, loggers, and others. In 1990, these seasonal units were added into the housing universe. The definitional change has no practical effect in most urban areas, but the change may be major in resort areas or some rural areas. The magnitude of the change can be assessed for each local area by reviewing the number of seasonal and migratory units recorded in 100-percent count housing tables for 1980.

1.2.1. Sources of More Detailed Data

Other published sources provide more detail than shown here but only covering *places* with 10,000 or more population. The reports on *General Housing Characteristics* (Series CH-1) cover the 100-percent count data, while the reports on *Detailed Housing Characteristics* (CH-2) cover the sample data. The principal advantages afforded by those two series are twofold. First, they offer most variables crosstabbed by tenure. In the published census tract housing tables, that detail is only afforded for race, year moved into unit, and financial characteristics. The other advantage of CH-1 and CH-2 is that they report data crosstabbed by many more categories of race than the five used in the census tract books (CPH-3).[2]

Even more detailed tabulations may be found in the summary tape files. We will cite specific tables from those sources when it adds significantly to analyses that follow. On the whole, for most analyses the published sources serve very well. The summary tape files are especially valuable for bringing the detail of the CH-1 and CH-2 report series down to the census tract level. In addition, when analyses must be conducted for many places at a time, the machine-readable summary tape files greatly facilitate the research.

□ 2. Measuring Changes in the Housing Stock

Analysis with housing data is illustrated with Tract 1392 close by the Ventura freeway in the San Fernando Valley. This tract started as a 1940s and 1950s single-family development, but since 1960 it has added development of large apartment complexes.

2.1. Tenure

The most basic item of housing data is *tenure*—whether the unit is rented or owned by its occupants. In 1970, Tract 1392 had about an even mix of owners and renters—46.7 percent owners—but this declined to 43.1 percent in 1980. Owners accounted for only 34.8 percent of the household growth over the decade.

Table II presents a basic accounting of changes in this area's housing stock. Note how the indented row headings define the key subsets used for describing the housing stock. Occupied units plus vacant units sum to all housing units, and occupied units (equaling the number of households) are further subdivided into owners and renters. The percent owners is also known as the homeownership rate and is defined as the number of owner-occupied units divided by all occupied units (or the sum of the owner- and renter-occupied housing units). A frequent error is to use all housing

Table II. Basic Accounting of Housing Change

	1970	1980	Change	% change
All housing units (or year round units)	1,366	2,032	666	48.8
Occupied units	1,325	1,909	584	44.1
owners	619	822	203	32.8
renters	706	1,087	381	54.0
Vacant	41	123	82	200.0
% owner occupied	46.7	43.1	−3.6	
% vacant	3.0	6.1	+3.1	

[2]For details on the race and Hispanic origin categories in use, see Chapter 10, particularly Table I.

units as the denominator, in which case the percent owners and percent renters will not sum to 100.

2.1.1. Homeownership by Age

Homeownership attainment by different age groups is of major public concern after a decade of acute affordability problems (Apgar *et al.* 1991). Are young families still able to purchase a home? Are elderly persons being forced to move into rental housing? Answers to such questions require tenure data by age. A major improvement in local census data in 1990 is the publication of tenure data by age of householder. In 1980, such data were available in *Metropolitan Housing Characteristics* (Series HC80-2), but only for places of at least 100,000 population. In 1990, tenure tabulations are published in CH-1 by seven age categories[3] (and by an extensive list of race groups) for places as small as 10,000 population. With these data, analysts can now calculate homeownership rates for specific age groups, comparing these across different communities.[4]

For analysis of changes over time, census users will need to restrict themseleves to the published data available for 1980 in HC80-2 for larger areas, or else access the computerized records on STF 2, Table A15. Unfortunately, what is gained in 1990 for smaller places is lost for larger places. *Metropolitan Housing Characteristics* is no longer being published. That source provided much more detailed, published crosstabulations than any other at the subnational level. Of special value to analysis of homeownership by age, households were further specified by race and by household type (marrried couple, other female head, and other male head). With those data, one could distinguish the housing circumstances of married couples in an age group from the generally less favorable situation of female-headed households, separating the experience of one race group from the entire population. At the level of counties or large cities, opportunities for homeownership analysis are much weaker in 1990 than in 1980.[5]

2.2. Structure Type

The most basic physical description of the housing stock is *structure type*. This is defined in the census as the *number of units in structure*, a term that is potentially confusing. The concept seeks to distinguish dwelling units that are part of multifamily buildings from those that are in duplexes or single-unit structures. For example, the category *5 to 9* counts the number of units in an area that are part of buildings with a total of five to nine units. This is a count of units, not of the number of buildings. However, for the one-unit category, these counts are the same.

---

[3]The published age categories are under 25, 75 and older, and then every ten years in between.

[4]Computerized data are available for 1990 for the entire population on the easier-to-access STF 1, Table H12, or by detailed race group on STF 2, Table HB10. The latter source also adds one more age category— 75 to 84—and moves the open-ended top category to 85 or older.

[5]For analysts who still value that breakdown by sex and marital status, controlling for age, the best 1990 source is STF 2, Table HA6; however, the best breakdown by race (STF 2, Table HB10) excludes the sex and marital status dimension. For those data, in 1990, analysts will need to go all the way to microdata, constructing custom tabulations for areas of at least 100,000 population.

One source of confusion derives from the fact that we often find tracts where only a few units are listed in categories that call for 10 to 49 or 50 or more units. In theory, this is impossible, because all the units in the structure should be recorded in that category. Inexperienced analysts may interpret these small counts to mean the number of structures, instead of units. There are two explanations for these incomplete counts. Residents may overestimate the number of units in their structures, such as when they count all apartments in their complex instead of just those in their separate building. Sampling error also plays a role, because the units in structure data are obtained from the long form sent to only a sample of residents (except in 1990).

The meaning of the one-unit category is also potentially confusing. We often refer to these structures as *single-family*, but the structure type has nothing to do with family status at all. These are actually single *household* structures. A second problem concerns the notion of *detached and attached* units. The single-family detached unit is easily recognized, but the attached unit may be confused with multifamily. Attached dwellings are row houses or town houses that sit on individual foundations but which have walls that adjoin one another. This form of housing has traditionally been concentrated in the Middle Atlantic states, ranging from Washington, D.C., up to Philadelphia (but also including New York and Boston). In recent years, town house developments have found their way into suburban regions around most large cities, because they offer more affordable single-family housing in an era of rising land prices.

A final difficulty attending the one-unit structure category is that traditional single-family housing often overlaps with mobile homes. In 1970, the two categories were published as one. More recently, they have been separated, but mobile homes have come to look increasingly like traditional single-family construction.

Analysts will need to combine the structure type data into a reduced set of categories that reflects the choices predominant in their region. In rural areas, mobile homes may assume a major role, whereas in New York City, apartment structures of 50 or more units may be very important. Elsewhere these housing types may not warrant individual analysis.

Table III combines the structure type data into four categories that are generally most useful. The largest size structure is five or more units, as this seems to capture the essence of an apartment building, distinguishing it from the smaller scale duplexes or fourplexes, while at the same time avoiding some of the respondent error noted above with regard to the largest structure categories. Mobile home is retained as a separate category, even though none are present in the example area.

The exhibit shows a rising proportion of apartment housing, from 35.2 percent in 1970 to 47.4 percent in 1980. Over the decade 483 such units were added, doubling the number existing in 1970. Single-unit construction also continued over the decade, adding 188 units in that segment of the stock, but this slower rate of construction led to a lower proportion of single-family houses in 1980 than in 1970. These new one-unit structures were very likely attached units built at higher densities, but we cannot measure this from the published data.

Published data for 1990 afford better quality data about structure type. For the first time, structure type has been moved to the complete count questionnaire, eliminating the sampling error in previous censuses that reduced its accuracy in small ar-

Table III. Change in Housing Units by Structure Type
San Fernando Valley Tract 1392

| | -1970- | | -1980- | | Change | |
	Number	% of Total	Number	% of Total	in Number	in Percentage
1 unit	881	64.5	1069	52.6	188 .05	−11.9 .05
2 to 4 units	4	0.3	0	0.0	−4 n.s.	−0.3 n.s.
5+ units	481	35.2	964	47.4	483 .05	12.2 .05
Mobile Home	—a	—	0	0.0	—	—
Total	1366	100	2033	100	667	—

Notes:
aMobile homes are combined with 1-unit structures in 1970 tract books; therefore, they must also be combined for 1980 prior to computing changes over the decade.
 Sample data; statistical significance tests are indicated:
　.05　　95 percent confidence that change is different from 0.
　.10　　90 percent confidence that change is different from 0.
　n.s.　　Not statistically significant.
Source: PHC(1)–117, Table H-Z, and PHC 80-2–226, Table H-7.

eas. With this greater accuracy, the published tract reports for 1990 now distinguish beween attached and detached single-unit structures.

2.3. Size of Units and Households

The census places strong emphasis on the size of units and the number of persons residing per unit. These questions have always been asked of all households, not of a subsample as with structure type.

2.3.1. Rooms per Unit

The size of units is reported by exact number of rooms up to eight or more. This is the most accurate way to have respondents judge the size of their homes.[6] The census reports median number of rooms for four categories: all year-round units, occupied units, owner-occupied units, and renter-occupied units. These are sometimes useful summaries, but the distribution by number of units contains more information. Tract 1392 has relatively few units at either the largest or smallest size categories.

Table IV describes changes in the stock for four size categories of units. The smallest units have only one or two rooms (studios and one-bedrooms). These comprised 13.9 percent of the stock in 1970 and their number kept pace with overall construction over the decade. At the opposite end, large units of 7 or more rooms comprised only 2.1 percent of the stock in 1970, but their number grew so much that they made up 10.1 percent of the stock in 1980. The five to six room category experienced the greatest absolute change, growing by 318 units, accounting for nearly half of all the housing increase.

---

[6]Square footage is not known to many residents, especially renters. Instead, among apartment renters the number of bedrooms is the best known. In cases other than a standard one- or two-bedroom apartment, however, the count of bedrooms may be biased. This is because larger households tend to convert other rooms to bedroom use, so that the number of bedrooms in a unit depends on who lives there. For all these reasons, the number of rooms seems the most reliable measure of unit size.

Table IV. Change in Housing Units by Number of Rooms
San Fernando Valley Tract 1392

	-1970-		-1980-		Change	
	Number	% of Total	Number	% of Total	in Number	in Percentage
1 or 2 rooms	190	13.9	282	13.9	92	0.0
3 or 4 rooms	783	57.3	858	42.2	75	−15.1
5 or 6 rooms	364	26.6	682	33.6	318	6.9
7+ rooms	29	2.1	210	10.3	181	8.2
Total	1366	100	2032	100	666	—

Source: PHC(1)–117, Table H-1, and PHC 80-2-226, Table H-1.

These changes may be due to more than new construction alone. It bears emphasis that the increase in large units could be accounted for by remodeling and room additions. The one-unit structures present in 1970 were fairly small: Less than 400 of all units had five rooms or more, implying that the majority of the 881 one-unit structures must have been below this size. (Among owners, excluding rented single-family houses, the median size of unit was 5.1 rooms.) This small size is not uncommon for early postwar single-family developments. Over time, successive occupants of small houses tend to enlarge their units by converting garages into rooms and by adding master bedroom suites.

Analysts must apply a substantial amount of judgment, *reading between the lines* of the numbers. Granted that significant remodeling may have occurred, we also note that 188 single-family units were added to the stock. Many of these likely were in the largest size category. No data are available in the census to tell us exactly how much each process contributed to the change in dwelling sizes. Analysts must supply their own interpretation.

2.3.2. Persons per Unit

The number of persons living in each unit parallels the data structure for number of rooms in the unit. In 1980, exact numbers of occupants are published up to eight or more, and medians are again given as summary measures. The issue of household size is vital for many analyses and will be given more attention in subsequent chapters. We address here the portion of the data published as part of the housing tables. The match between household sizes and size of unit is clearly of importance.

Table V presents four categories for analyzing household size: one person living alone, two, three or four, and five or more persons. These categories are selected on the basis of their social significance and their frequency of occurrence. The exhibit shows that more than one-quarter of all households were occupied by persons living alone in 1970. This category grew faster than average, accounting for 31.2 percent of all households by 1980. The number of two-person households increased nearly as fast and was the largest category in 1980, accounting for 36.4 percent of all households. Together these two small categories amount to two-thirds of all occupied units.

A growing concentration of small households is commonly found throughout the Los Angeles region (and the United States). Areas with more apartments have even

Table V. Change in Occupied Units by Number of Persons
San Fernando Valley Tract 1392

| | -1970- | | -1980- | | Change | |
	Number	% of Total	Number	% of Total	in Number	in Percentage
1 person	352	26.6	595	31.2	243	4.6
2 persons	431	32.5	694	36.4	263	3.8
3 or 4 persons	421	31.8	523	27.4	102	-4.4
5+ persons	121	9.1	97	5.1	-24	-4.1
Total	1325	100	1909	100	584	—

Source: PHC(1)–117, Table H-1, and PHC 80-2, Table H-1.

smaller household sizes than found in this San Fernando Valley tract, but even in areas with 100 percent single-family units, occupancy by one- and two-person households has grown to be quite substantial. Comparison with data on marital status and age, as discussed in following chapters, may lead to a more detailed explanation in each local area of who makes up these small households.

2.3.3. Overcrowding

Overcrowding occurs when larger households are squeezed into smaller units and is usually defined as a ratio greater than 1.0 persons per room. Given the small size of the households in Tract 1392, it should not be surprising that relatively few households (4.5 percent) are overcrowded.

Overcrowding of housing units in the United States has generally declined to a low level in concert with shrinking household sizes. We might expect to find the highest incidence in urban areas that have expensive housing and large numbers of immigrants. In Los Angeles County, 11.2 percent of households were overcrowded in 1980, increasing from 8.5 percent in 1970. This increase reflects the rapid growth of the Hispanic population, which has a much higher likelihood than average of being crowded—35.2 percent in 1980—because of their substantially larger household sizes. (By comparison, the incidence of overcrowding among blacks is only 11.0 percent, about average for the county.) In fact, Hispanics accounted for 62.2 percent of all overcrowded households in Los Angeles County in 1980 even though they made up only 19.8 percent of all households. As shown in Chapter 10, in the case of Pasadena Tract 4620, levels of overcrowding can be especially high in neighborhoods of recent Hispanic immigrants. Attention to crowding conditions is especially important in such communities.

2.3.4. Graphing the Match of Unit Size and Household Size

Despite the recent increase in overcrowding in some locations, underutilization of the housing stock has emerged as an important issue. Stories are often told of persons living alone in oversized housing units. How common is this in a particular area?

One way to compare the growth of units by number of rooms and size of household is to overlay their trends for corresponding categories. This has the effect of dramatizing differences or similarities in two different trends. The critical step is to decide the correspondence between categories of the different variables. To place them on the same graph we must match the size categories according to some rule. In this case, the best rule is that households should have between 0.5 and 1.0 persons per room. Following this rough guideline, we will define a *very small* category as one person and one to two rooms; a *small* category as two persons and three to four rooms; a *medium* category as three to four persons and five to six rooms; and a *large* category as five or more persons and seven or more rooms.

Figure 1 portrays the absolute growth of units and households according to these size categories. It is immediately obvious that the growth in households in this area has been almost all in the small and very small categories, whereas the growth in units has been concentrated in the medium and large size categories. This contrast was also indicated in the summary statistics given in Chapter 2: The median household size in this neighborhood fell from 2.20 to 2.02, but the median number of rooms in units rose from 4.0 to 4.3. Those seem like relatively weak opposite trends, but they are dramatized by the graph in Fig. 1. Here we can see also which categories account for the movement in the medians. Such trends are likely to be found in many neighborhoods across the nation.

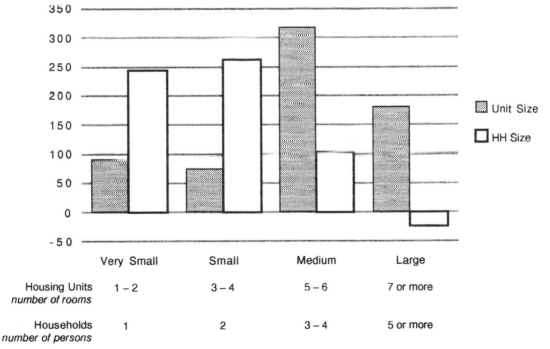

Figure 1. Change in number of households and housing units of each size in San Fernando Valley Tract 1392 from 1970 to 1980.

2.4. Price and Affordability of Housing

Like overcrowding, housing affordability is a matter of the fit between the characteristics of the housing unit (in this case its price) and the characteristics of the household (its income). Although price and affordability are related, they are distinctly different concepts. Data on prices over time can be used to measure the socioeconomic level of a local area, provided that it is properly adjusted for inflation. Affordability, on the other hand, measures the burden of monthly housing expenditures relative to residents' incomes. The problem of overpayment for housing (more than 30 or 35 percent of income) has become the dominant source of housing problems since the 1970s.

2.4.1. Price Trends

The simplest and most concise representation of prices is the median house value for owners and the median contract rent for renters.[7] These prices are reported in Table VI for San Fernando Valley Tract 1392.

Inflation increased prices substantially over the decade, raising median house values fourfold and rents by a factor of 2.5. As discussed more fully in Chapter 11, the easiest way to measure the relative price level over time is to express the tract median as a percentage of the median price for the region. When seen in this perspective, it is clear that this neighborhood's prices increased faster over time than did average prices in the region.

In 1970, house values were only 80 percent of the regional average, but over the decade they increased to almost equal the region's. In contrast, rents were already well above the region's but increased still further until they exceeded the regional average by 51 percent. Overall, the price picture suggests that the rental units are more upscale than the owner-occupied units.

These data provide an easily attainable approximation of the local area's relative expensiveness. What is left out of the calculation is any detail on prices by size category or age of unit. Such detail is obtainable from the census only in the most detailed summary tape files.[8] In particular, it would be useful to separate out the value changes for existing housing units from that contributed by new construction.

2.4.2. Measuring Affordability

Given the growing importance of affordability problems, the 1980 census added an entire table devoted to measuring affordability. Because the 1990 version is virtually identical, with dollar categories updated for inflation, we reproduce the 1980 table, H-8, in Table VII. The first portion of that table is devoted to owners' affordability and the rest to renters (gross rent). For both owners and renters, a tabulation is

---

[7]Gross rent may be preferred to contract rent. That concept adds in the costs of basic utilities (electric, heat, and water). However, contract rent has the advantage that it is a complete count variable and is not subject to the sampling error of gross rent. Thus, contract rent may be preferable for analysis in small areas.

[8]For 1980 data, detailed house value crosstabulations are reported in tables STF 4: HA26-28 and STF 2: A28. Detailed contract rent tabulations are given in STF 4: HA 77-79 and STF 2: A29. In 1990, these detailed crosstabulations are not available in any STF series.

Table VI. Standardizing Housing Prices to a Regional Basis for Comparison over Time

	Current Dollars		% of Regional Average	
Median Prices	1970	1980	1970	1980
House Value	$19,500	$83,300	80	95
Contract Rent	$141	$368	128	151

made, first, of households paying different levels of housing expenses, followed by a tabulation of percentage of income spent on housing. Those percentage expenditures are grouped by the income level of the household.

This complex table is the source of our best data on housing affordability. Better data are not reported in the CH-2 reports for larger places.[9] The key entries in the tables are the number of households paying a given percentage of their income for housing expenses. The limit used to define affordability varies from purpose to purpose. Sometimes the old rule of thumb of 25 percent is used. Other times the standard is set that households should pay no more than 30 percent of their income for housing. Still a third option is to set the limit at 35 percent or more. Given the variety of alternatives, it is fortunate that the census provides a series of categories. Data can be aggregated to measure those paying a higher percentage of their income than each of the alternative standards.

In 1970, the published data combined the 25 to 29 and 30 to 34 percent payment brackets into one large category of 25 to 34. Therefore, comparisons to 1970 data must make use of the higher standard of overpayment—35 percent of income or more. According to that standard, 29.3 percent of the renters in our case example neighborhood faced an overpayment problem in 1970. By 1980, this percentage had grown to 46.0, nearly half of all renters in the area. By way of contrast, in Los Angeles County as a whole, the percentage of renters suffering an affordability problem increased from 28.0 percent in 1970 to 35.0 percent in 1980. The greater increase in affordability problems in this San Fernando Valley neighborhood is surely related to the greater increase in rents observed there, as discussed previously.

The level of affordability problems depends not only on housing prices in the area, but also on the standard chosen for comparison. Table VIII shows how the percentage of households suffering affordability problems decreases as our standard of affordability is raised to a higher percentage of income acceptable for housing payments. For example, if we set the standard at less than 25 percent of income, rental affordability problems affect 54.3 percent of county renters. At the higher standard of 35 percent of income, affordability problems afflict only 35 percent of renters. Similarly with homeowners, a smaller share suffers an affordability problem if we measure it with a higher standard.

[9]Even the 1990 data in STF 3, Tables H50 and H59, are little better. Their sole advantage in detail is to add one more income category to the upper end of renters and, for owners, one more category to the lower end.

Table VII.

Census Tracts

	Los Angeles city, Los Angeles County—Con.		
	Tract 1392	Tract 1393	Tract 1394
Specified owner-occupied housing units	569	577	1 004
MORTGAGE STATUS AND SELECTED MONTHLY OWNER COSTS			
With a mortgage	435	457	838
Less than $100	—	—	—
$100 to $199	75	22	42
$200 to $299	79	40	90
$300 to $399	38	113	70
$400 to $599	118	98	137
$600 or more	125	184	499
Median	$432	$485	$724
Not mortgaged	134	120	166
Less than $100	83	26	14
$100 to $199	51	82	96
$200 or more	—	12	56
Median	$94	$119	$180
HOUSEHOLD INCOME IN 1979 BY SELECTED MONTHLY OWNER COSTS AS PERCENTAGE OF INCOME			
Less than $10,000	67	79	107
Less than 15 percent	8	5	—
15 to 24 percent	21	14	28
25 to 29 percent	—	—	7
30 percent or more	38	48	54
Not computed	—	12	18
Median	33.8	50+	43.6
$10,000 to $19,999	141	116	174
Less than 15 percent	61	39	55
15 to 24 percent	19	36	61
25 to 29 percent	13	6	21
30 percent or more	48	35	37
Not computed	—	—	—
Median	17.5	19.1	19.4
$20,000 or more	361	382	723
Less than 15 percent	212	209	335
15 to 24 percent	81	75	215
25 to 29 percent	27	15	49
30 percent or more	41	83	124
Not computed	—	—	—
Median	13.5	14.0	16.0

Census Tracts

	Los Angeles city, Los Angeles County—Con.		
	Tract 1392	Tract 1393	Tract 1394
Specified renter-occupied housing units	1 081	3 216	780
GROSS RENT			
Less than $80	14	7	—
$80 to $99	7	—	—
$100 to $149	7	6	11
$150 to $199	99	35	101
$200 to $249	60	172	170
$250 to $299	139	456	192
$300 to $349	95	510	124
$350 to $399	80	659	169
$400 or more	580	1 358	13
No cash rent	—	13	
Median	$414	$382	$326
One-family house, detached or attached	304	165	171
Median gross rent	$500+	$500+	$490
HOUSEHOLD INCOME IN 1979 BY GROSS RENT AS PERCENTAGE OF INCOME			
Less than $10,000	354	822	279
Less than 15 percent	14	—	—
15 to 19 percent	22	—	—
20 to 24 percent	—	—	13
25 to 29 percent	8	6	14
30 to 34 percent	306	727	226
35 percent or more	4	89	26
Not computed	—	—	—
Median	50+	50+	50+
$10,000 to $19,999	381	1 280	315
Less than 15 percent	8	12	11
15 to 19 percent	41	40	24
20 to 24 percent	40	268	90
25 to 29 percent	80	361	63
30 to 34 percent	34	221	63
35 percent or more	178	373	58
Not computed	—	5	—
Median	33.2	29.4	27.3
$20,000 or more	346	1 114	186
Less than 15 percent	83	301	83
15 to 19 percent	99	322	34
20 to 24 percent	59	318	48
25 to 29 percent	57	112	10
30 to 34 percent	37	44	11
35 percent or more	11	9	—
Not computed	—	8	—
Median	19.5	18.9	16.5

Table VIII. Percent of 1980 Households with Monthly Housing Expenses in Excess of a Certain Percentage of Income

	% of Income Standard	Tract 1392	LA County
Renters	25% or more	66.0	54.3
	30% or more	53.3	43.2
	35% or more	46.0	35.0
Owners	25% or more	29.3	28.6
	30% or more	22.3	20.7

Overpayment is much more likely among lower income households. Accordingly, the census breaks out the data by income bracket. It is useful to see how high the affordability problem is at each income level, depending on the standard selected for measurement.

Figure 2 graphs the percentage of an income group (separately for owners and renters) that spends more for housing than the designated standards. Renters in the lowest income category have a very high incidence of affordability problem, no matter what the standard. Conversely, renters in the higher income bracket have a very low incidence of affordability problems, no matter what the standard. In the middle bracket, however, there are a lot more *borderline* cases that will be added to the affordability caseload if a lower standard is employed. The level of affordability problems in this income bracket drops markedly if a higher payment standard (35 percent or more of income) is employed.

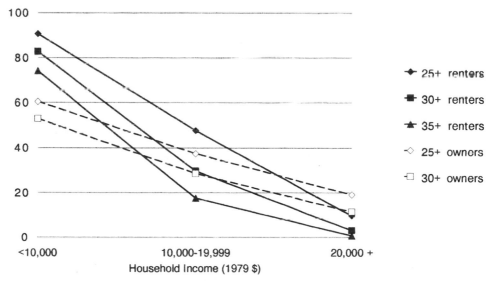

Figure 2. Incidence of affordability problems, by income level, for Los Angeles County in 1980.

The affordability problem incidence of homeowners looks considerably different from that of renters. The top category published for this group is 30 percent or more, preventing analysis of the 35 percent standard.[10] Nevertheless, the basic pattern is clear: Homeowners in the lowest income category have a much lower incidence of affordability problems than renters in the same bracket. (Many of these lower income homeowners are elderly persons with mortgages paid off.) The incidence for homeowners does not decline as sharply as for renters, so that by the $20,000-plus-income category affordability problems for homeowners exceed those for renters. It bears emphasis, however, that homeowners get some of their housing payments returned to them through income tax deductions for mortgage interest payments. Therefore, the calculation of affordability problems from census data should be adjusted in some manner to account for this unequal tax treatment of homeowners' incomes.

2.5. Year Structure Built or Age of Housing

Age of housing is often used as a measure of housing quality. The presumption is that areas with housing built before 1940 are likely to suffer from greater dilapidation or at least outdated housing facilities such as plumbing and heating. On balance, the older housing stock is inferior; however, the gentrification movement of the past two decades has exposed the greater quality found in some parts of the older stock.

Demographers recognize other analytical uses for data on age of housing, given the parallel to population age. Year built is equivalent to year of birth. In fact, the decade of construction defines a dwelling vintage that can be analyzed like a cohort that ages over time. Two alternative perspectives on these data are elaborated below.

The ages of housing units in local areas are published by the census under the heading of *Year Structure Built* (see the middle panel of Table I). These data are drawn from the sample questionnaire and so are subject to sampling variability. There also is a small degree of respondent error that is nonrandom, particularly on the part of renters. Respondents in more recent censuses tend to assume that their units are a little newer than they really are, so that the older categories seem to shrink and the newer categories grow (see Tippett 1987; Baer 1990). Nevertheless, the data for year structure built seem approximately accurate and they usefully reflect the development history of the locality.

Table IX, part A, reports the ages of dwelling units for our example neighborhood in the San Fernando Valley. The proportion of new units (built within the preceding 10 years) is about the same in 1970 and 1980, at 33 to 36 percent. However, the proportion of older units, at least 30 years old, has risen markedly from less than 1 percent to nearly 20 percent.

The other way of looking at these age data is to trace vintages over time as these *cohorts* of housing grow older.[11] Part B of Table IX shows how many units built in each

---

[10]In 1980, more detailed categories for percentage of income spent on housing are reported in the summary tape files: STF 4: HB50 and HB52. In 1990, STF 4 also reports the most detailed expenditure categories (Tables HB16 and HB17), with a top category of 50 percent or more.

[11]The cohort concept is explained in detail in Chapter 8.

Table IX. Year Structure Built and Age of Housing
San Fernando Valley Tract 1392

Part A—Change in Housing Units of Each Age

	-1970-		-1980-		Change	
	Number	% of Total	Number	% of Total	in Number	in Percentage
Under 10 years	456	33.4	720	35.4	264 .05	2.0 n.s.
10–19 years	374	27.4	412	20.3	38 n.s.	–7.1 .05
20–29 years	527	38.6	520	25.6	–7 n.s.	–13.0 .05
30+ years	9	0.7	381	18.7	372 .05	18.1 .05
Total	1366	100	2033	100	667	—

Part B—Survival of Housing Units from Each Vintage (Decade of Construction)

	-1970-		-1980-		Change	
	Number	% of Total	Number	% of Total	Loss/Gain	% Remaining
1970s	—	—	720	35.4	720 n.s.	100
1960s	456	33.4	412	20.3	–44 n.s.	90.4
1950s	374	27.4	520	25.6	146 .05	139.0
1940s	527	38.6	374	18.4	–153 .05	71.0
pre-1940	9	0.7	7	0.3	–2 n.s.	77.8
Total	1366	100	2033	100	667	NA

Notes:
NA—not applicable

Sample data; statistical significance tests are indicated:
.05　　95 percent confidence that change is different from 0.
.10　　90 percent confidence that change is different from 0.
n.s.　　Not statistically significant.

Source: PHC(1)–117, Table H-2, and PHC 80-2, Table H-7.

decade, or vintage, are recorded in two different censuses. The large infusion of new, 1970s housing was recorded in the 1980 census as 720 units. The 1960s vintage is measured with about the same number of units each decade; the shortfall of 44 units in 1980 is not statistically significant. More discrepancy is seen in the 1940s and 1950s vintages. Fully 153 units were lost from the 1940s vintage between 1970 and 1980 (with consequent gains by the 1950s vintage). Only 50 to 60 units of this loss/gain can be attributed to sampling error, and so the changes to the vintages are *significant.* Unless we are to believe that over 100 homes built in the 1940s were demolished in the 1970s, it is apparent that some of this vintage has passed into the 1950s category due to respondent error. Research has shown that the distinction between 1940s and 1950s construction is the most difficult for respondents to recognize accurately (Baer 1990; Tippett 1987).

　　The San Fernando Valley example illustrates a rather extreme amount of respondent error. Most tracts and larger areas show greater stability over time in vintage estimates. For example, Fig. 3 graphs the relevant data pertaining to Los Angeles County. Proportionally little change is evident in most vintages, although with this extremely large sample size, the differences are statistically significant. The only vintage losing units is the oldest. Most of the loss is likely due to demolitions, but a

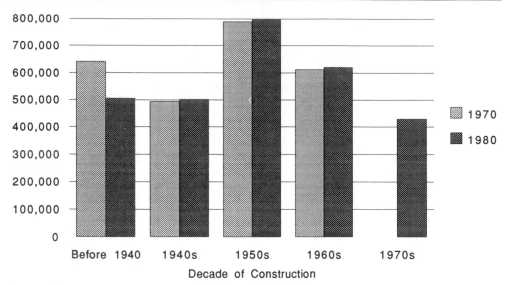

Figure 3. Number of housing units of each vintage, as observed at each census, in Los Angeles County. Units built in the first three months of the decade—between January and April of 1970 or 1980—are classified with the decade that precedes the census year.

small amount is probably attributed to respondent error that has underestimated the age of structure in 1980.

The accuracy of year-built analysis can be greatly improved in two ways. One is to resort to older censuses, closer to the date of construction, to estimate the size of older vintages. Barring the unlikely event of substantial demolitions, or mergers, these earlier estimates should be good for several decades. Second, the analyst may wish to consult the local planning department. Any substantial land clearance for highways or redevelopment should be well known to local authorities. Local administrative records also can provide the numbers of demolitions and building permits recorded in different decades.

2.6. Duration of Occupancy

A key link between the history of the housing development and the history of the population is provided by the duration of occupancy. Dwellings that are newer cannot have held residents for long; in contrast, older units may shelter longtime residents. With the passing of time, those longtime residents will have aged in place, thereby linking the history of development to the age structure of the local area (Myers 1978).

The duration of occupancy is published by the census under the heading of *Year Householder Moved into Unit* (see the middle panel of Table I). Like the year-built data, occupancy duration data are drawn from the sample questionnaire and so are subject to sampling variability. Beginning in 1980, occupancy durations of owners and renters were tabulated separately, permitting more detailed analysis of the homeowners who are generally longer residing. However, analysis of changes since 1970 must revert to combining the two tenure groups together. In general, the housing data on occupancy duration provide more categories of duration than the population data on mobility, which has only two categories—less than five years, or longer (see the discussion on migration and local mobility in Chapter 9).

2.6.1. Theory of Residential Mobility

Residential mobility theory can help with the interpretation of occupancy duration data. In addition to the effect of new construction, two key variables are known to be most important in explaining mobility (see Clark 1986). First, younger people are much more likely to move than older persons. Second, controlling for age, renters are much more likely to move than owners. Thus an apartment neighborhood that is attracting young adults would be expected to have the shortest occupancy duration. Conversely, a neighborhood of single-family homes sheltering older or late middle-aged households would have the longest occupancy duration. Most neighborhoods would fall in between these two extremes, depending on their mix of owners and renters, their population age profile, and the age of their housing.

Two common misconceptions about occupancy duration deserve recognition. One popular, but often misguided, perception is that a high degree of recent arrivals reflects transience in the neighborhood. With new construction, such a condition usually prevails in any subdivision or apartment building newly occupied for the first time. By examining the year-built data, analysts could subtract the new units from the recently occupied category, thereby correcting this error of interpretation.

Similarly, it may also be erroneous to interpret the normal turnover process in apartment areas as transience or instability. The neighborhoods around universities can be remarkably stable, with the same type of persons persisting year after year even though different people move in each year. In fact, it may prove that the opposite is true: Neighborhoods without inmovers become very unstable and change dramatically. Existing residents simply grow older, all children leave home, and the neighborhood character is transformed eventually to a senior citizens retirement center.

Occupancy duration can be looked at in two ways: either how many households have recently arrived or how many households have resided a long time. For analysis of changes over time, the latter construction is more meaningful. Longtime occupants indicate the likelihood of demographic transition in a neighborhood due to aging. In addition, longtime occupants may indicate a sustained attractiveness by the neighborhood. Finally, longtime occupants also indicate how many homes have been held off the market, preventing the opportunity for new persons to move in.

2.6.2. Change in Occupancy Duration

Table X, part A, compares the 1970 and 1980 data on occupancy duration. In 1980, over 70 percent of the households had resided less than five years in their unit. The proportion of short-term residents rose slightly over the decade, by about seven percentage points. We assume that virtually all of these had resided outside the neighborhood as well, but the census provides no data on prior neighborhood of residence (although see Chapter 9 for discussion of migrants' origins by city, state, or nation).

Conversely, the proportion of households that occupied their homes for 20 or more years rose from 4.9 to 9.3 percent, reflecting the growing number of older

Table X. Year Moved In and Duration of Occupancy
San Fernando Valley Tract 1392

	-1970-		-1980-		Change			
	Number	% of Total	Number	% of Total	in Number		in Percentage	
Under 5 years	849	64.1	1362	71.3	513	.05	7.3	.10
5–9 years	208	15.7	205	10.7	–3	n.s.	–5.0	.10
10–19 years	203	15.3	164	8.6	–39	n.s.	–6.7	.05
20+ years	65	4.9	178	9.3	113	.05	4.4	.05
Since previous census	268	20.2	342	17.9	74	n.s.	–2.3	n.s.
Total	1325	100	1909	100	584		NA	

Part B—Survival of Households by Decade of Initial Occupancy

	-1970-		-1980-		Change			
	Number	% of Total	Number	% of Total	Loss/Gain		% Remaining	
1970s	—	—	1567	82.1	1567	n.s.	100	.10
1960s	1057	79.8	164	8.6	–893	.05	15.5	.10
pre-1960	268	20.2	178	9.3	–90	.10	66.4	.10
Total	1325	100	1909	100	584		NA	.10

Notes:
NA—not applicable

Sample data; statistical significance tests are indicated:
.05   95 percent confidence that change is different from 0.
.10   90 percent confidence that change is different from 0.
n.s.   Not statistically significant.

Published tract data for 1970 combine occupancy durations of owners and renters, preventing the separate analyses for each tenure possible with 1980 and 1990 data.

Source: PHC(1)–117, Table H-2, and PHC 80-2–226, Table H-7.

housing units. Virtually all of these very longtime residents are likely homeowners. Data published separately for owners and renters in 1980 permits construction of a graph such as Fig. 4 for San Fernando Valley Tract 1392. Here we see that at least 95 percent of renters have occupied their homes less than 10 years, while this is true of less than 65 percent of owners. Indeed, over 20 percent of homeowners in this neighborhood have lived in their units for 20 years or more. This percentage is not as high as may be observed some places. If all the owner-occupied homes in a neighborhood were at least 20 years old, we would expect 30 to 40 percent to have been occupied by their current residents for that long. Neighborhoods more desirable than average could hold even higher proportions of longterm occupants.

2.6.3. Retention of Early Occupants

A complementary way of structuring the occupancy duration data is presented in part B of Table X. Here the data are grouped by decade of initial occupancy instead of by length of occupancy. The last column of the table reports the percentage remaining in 1980 from the occupancy group observed in 1970. It is striking that only

15.5 percent of the occupants of the 1960s are still resident in 1980. This contrasts sharply with the 66.4 percent of early occupants (from before 1960) that are still resident. The decline of 90 households in this early residents category is marginally significant (at the 90 percent confidence level).

The group of remaining early residents is certainly made up almost exclusively of homeowners, whereas the 1960s occupants may contain many renters. (With the more detailed published tabulations of 1990 data, analysts easily will be able to track occupancy groups separately for owners and renters between 1980 and 1990.) The greater persistence of these early occupants is also reflected in Fig. 4: The percentage of owners who have resided 20 years or more exceeds that residing 10 to 19 years.

2.6.4. Interpretation

This neighborhood in the San Fernando Valley may actually consist of several neighborhoods or layers of residents. Many of the early arrivals live on in their 1940s and 1950s tract homes. These are relatively permanent residents. At the other extreme are the renters who occupy apartments on a relatively short-term basis. In between these two groups is a third group of homeowners who are more recent arrivals and who are unlikely to remain to the next census.

For all but the longest time occupants, this tract may be a *stepping stone* neighborhood. As noted above, its houses are modestly sized, and large sections have experienced construction of higher density apartment housing. Homeowners with growing families, and who are able, may have left for other locations. In addition, this neighborhood was caught up in the overall price inflation that gripped southern California in the late 1970s. As some have argued, this inflation encouraged homeowners to cash in their equity and move up to bigger homes more rapidly than they otherwise might have (Masnick *et al.* 1990). These factors explain, at least in part, the apparently low degree of very longtime occupancy in this neighborhood.

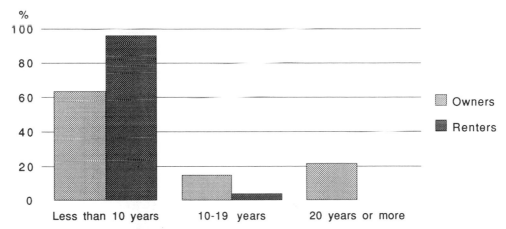

Figure 4. Length of time households have occupied their homes in 1980, by tenure, in San Fernando Valley Tract 1392.

2.7. Neighborhood Life Cycles and Filtering

Closely related to age of housing and occupancy duration are the concepts of *neighborhood life cycle* and *filtering*, both of which are widely recognized, but not well understood. The general idea is that a neighborhood passes through a series of stages from when it is newly built to when it is older and run down. Changes may occur on two dimensions, both in terms of the age of residents and in terms of the economic status of the neighborhood.

A new neighborhood often attracts residents who are young adults. Many of these adults remain in their homes for a long time, aging in place, especially if they are homeowners. The result is that the population age profile may grow older in concert with the housing age (Myers 1978). A further consequence is that household size may drop sharply in certain neighborhoods (Gober 1990). Eventually, turnover will occur in the neighborhoods and a new wave of residents will begin the cycle again. These changes are easily tracked with census data on housing, household size, and population age structure, as demonstrated in following chapters. The spatial patterns in urban areas frequently prove especially interesting (Adams 1987; Gober 1990).

The filtering concept stresses the economic changes that occur in neighborhoods over time. Popularly known as *trickle-down*, describing the handing down of used houses from rich to poor, the process sometimes works in reverse, as in the case of gentrification. Overall, the filtering process is very complex and has generated substantial political controversy. Readers are directed to Myers (1990b) for a broad historical review. Suffice to say, close monitoring is warranted of changes in housing economic levels for neighborhoods of all ages. The simple calculation shown previously in Table VI, if repeated for all neighborhoods in a city, affords a cost-effective means of measuring local trends in filtering.

☐ 3. Conclusion

The housing stock is the bedrock for the census because it provides the shelter (and address) for the vast majority of the population. The stock expands in response to growing demand, but once built the longevity and immobility of the stock constrains the number of people who can reside there. There are great variations in the type of housing built in different neighborhoods, and these different units attract and hold different types of people.

For these reasons, housing provides the essential substructure for small-area demography. It is important that we grasp the changes in the local housing stock as an important precursor to analyzing changes in the households and population that live in those units.

7 □□□

Marital Status, Family, and Household Relationships

Households are the link between the housing stock and the population. In the previous chapter, we addressed households from the housing side—as occupied housing units. Recall that a household is found whenever a person or group of people occupies an individual housing unit. The present chapter examines local households from the population side—in terms of the relationships among the persons who share a housing unit. Some households are families, consisting of two or more persons related by blood, marriage, or adoption, while others are nonfamily households composed of persons living alone or with unrelated persons.

Changes in marital status have been particularly important for reshaping household relationships in recent decades. During the 1970s and 1980s, in the United States, the percentage of young adults entering marriage fell markedly. In 1990, at age 30, nearly one-third of males and one-fifth of females had never been married. The combination of slower entry into marriage, greater divorce, an aging widowed population, and other factors means that fewer households consist of married couples.[1] As shown in Table I, nonfamily households constitute a substantially rising proportion of all households. In fact, as shown in these data, most of the nonfamily households are made up of persons living alone (24.6 percent of all households).[2]

---

[1]For an excellent, comprehensive assessment of marriage patterns and their relationship to changing families and households see Sweet and Bumpass (1987). A compact assessment of recent trends in singleness, single-parent families, and married couples with children may also be found in a special report by the U.S. Bureau of the Census (1989a).

[2]These data are taken from the U.S. Bureau of the Census (1991a), Table B; and (1991b), Table B.

Table I.

	1970	1980	1990
% never married at age 30			
Females	6.6	11.7	19.0
Males	11.9	23.7	32.2
% of all households headed by married couples	70.5	60.8	56.0
% of all households made up of nonfamilies	18.8	26.3	29.2
% of all households with one person	17.1	22.7	24.6

The linkage among marital status, family composition, and household type is best described through a diagram, as shown in Fig. 1. The most common delimiters of family households involve marital status and the presence of children under the age of 18. By definition, married householders[3] live in family households. Unmarried householders also may live in families, most commonly because their children are living with them. Married households accounted for 56.1 percent of the total in 1990, split roughly half and half between those living with children and those without. The next largest category is households where persons live alone, 24.6 percent of all households. The other household types occur much less frequently.

[] 1. Local Area Data

These national trends in marital status and living arrangements are discovered through the national Current Population Survey and reported annually in *Current Population Reports*, Series P-20, published by the U.S. Bureau of the Census. At the local level, it is only once every 10 years that we can learn how the national trends are reflected in each specific community. The evidence suggests that during the 1970s and 1980s, apartment districts of cities increasingly became the province of young singles. So strong were the nonmarried trends that even suburban, family neighborhoods showed an increase in young singles, an increase in formerly married persons, and a decrease in married couples. The decennial census provides the only means of tracking these trends across the whole array of neighborhoods in our nation.

A rich table of data is available at the census tract level (or higher geographies) for describing local changes in these relationships over the decades. Table II shows part of Table P-1 from the 1980 census (almost identical in content to that for 1970 and 1990). That table summarizes the basic, complete count population information for the population as a whole, and similar tables are repeated for each major racial group in the population.[4] Only the bottom half of P-1 is addressed in this chapter.

[3]The concept and definition of a *householder* was discussed at length in Chapter 3. In brief, the householder is the person (or *one* of the persons) in whose name the housing unit is owned or rented. Among married couples, the householder may be either the husband or wife.

[4]The tables for each separate race group repeat the basic structure shown in Table II. Those tables are listed as P-2 through P-6 in the 1980 census tract books (Series PHC80-2) and as Tables 2 through 7 in the 1990 books (Series CPH-3). Full crosstabulations of these marital and household data by age, sex, and race can be found in STF 2, Tables B5, B7, and B11 in 1980, and Tables PB7 and PB15 in 1990. See Chapter 4 for details of these data sources.

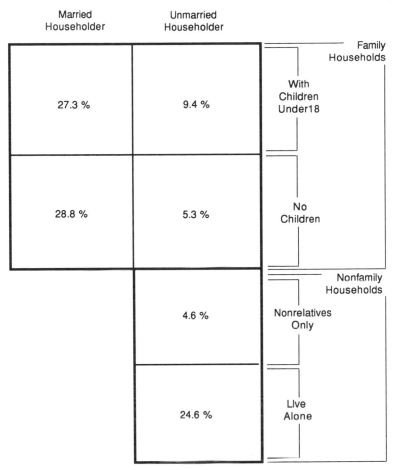

Figure 1. Household types are defined by linkages among family status, marital status, and presence or absence of children. Percentages show the distribution of U.S. households by type in 1990. [From U.S. Bureau of the Census (1991a: Table 16)]

(The upper half of P-1—not shown in Table II—is age data that we will take up in the next chapter.)

Three sections of Table II pertain to the subject of this chapter. The section labeled *Household Type and Relationship* describes each person's relationship to the household reference person (the *head*, or *householder*). The next section describes *Family Type by Presence of Children*, and the last section identifies *Marital Status*, by sex, for persons age 15 and over.

1.1. Crosstabulation by Age at Higher Geographies

More detailed data can be obtained in published reports covering counties or places, as expected from the hierarchy principle of data availability (see Chapter 4). The CP-1 report series provides age breakdowns (10-year wide categories) of married persons, by sex and race, for places of 10,000 or more population, and of householders (sexes combined), by race, for places of 2,500 or more population. These are not

Table II.

Census Tracts

	Los Angeles city, Los Angeles County—Con.				
	Tract 1898	Tract 1899	Tract 1901	Tract 1902	Tract 1903.01
HOUSEHOLD TYPE AND RELATIONSHIP					
Total persons	2 305	7 814	4 250	7 032	5 194
In households	2 243	7 779	4 218	6 848	4 822
Householder	1 221	4 789	2 776	4 304	2 629
Family householder	455	1 568	627	946	897
Nonfamily householder	766	3 221	2 149	3 358	1 732
Living alone	617	2 615	1 775	2 851	1 476
Spouse	337	1 168	400	602	603
Other relatives	405	932	540	1 187	1 204
Nonrelatives	280	890	502	755	386
Inmate of institution	49	33	–	42	175
Other, in group quarters	13	2	32	142	197
Persons per household	1.84	1.62	1.52	1.59	1.83
Persons per family	2.63	2.34	2.50	2.89	3.01
Persons 65 years and over	493	1 836	633	1 163	826
In households	484	1 836	633	1 154	688
Householder	325	1 316	496	993	531
Nonfamily householder	175	780	366	830	388
Living alone	162	725	344	803	373
Spouse	92	380	73	87	64
Other relatives	51	112	45	50	77
Nonrelatives	16	28	19	24	16
Inmate of institution	9	–	–	9	138
Other, in group quarters	–	–	–	–	–
FAMILY TYPE BY PRESENCE OF OWN CHILDREN					
Families	455	1 568	627	946	897
With own children under 18 years	123	303	184	379	364
Number of own children under 18 years	201	409	264	677	571
Married-couple families	337	1 168	400	602	603
With own children under 18 years	91	192	102	224	243
Number of own children under 18 years	162	261	158	435	412
Female householder, no husband present	85	275	160	229	193
With own children under 18 years	28	90	70	129	92
Number of own children under 18 years	34	123	91	204	123
MARITAL STATUS					
Male, 15 years and over	1 082	3 671	2 252	3 746	2 352
Single	518	1 687	1 353	2 225	1 177
Now married, except separated	358	1 270	465	745	723
Separated	27	119	85	168	96
Widowed	40	106	60	137	83
Divorced	139	489	289	471	273
Female, 15 years and over	1 059	3 758	1 769	2 630	2 318
Single	329	1 120	714	933	765
Now married, except separated	363	1 247	432	680	728
Separated	35	91	72	145	78
Widowed	168	622	255	432	410
Divorced	164	678	296	440	337

full crosstabulations of marital status or household status by age, merely breakdowns of one category within each.

The utility of these tabulations is that users can construct percentages of the population in each age group who are married or who are householders. The population counts by age, required for the denominator of the percentage, will be found in different tables in CP-1.

Even more detail is gained in elderly ages for places of 10,000 or more population. Tables in CP-1 identify four age groups: 60 to 64, 65 to 74, 75 to 84, and 85 and older. Information is given on householder status by sex and on household relationship (sexes combined). All of this detail is provided for separate race and Hispanic origin groups.

Discussion in following sections is largely confined to more widely available data found for census tracts and shown in Table II. These data are sufficient for demonstrating the methods of this chapter. If desired, users can expand the categories used in the examples to take advantage of the added detail found at higher geographies.

☐ 2. Marital Status

Given the central importance of marital status in forming families and households, we begin with that subject. There are often substantial differences between the marital statuses of the two sexes. On average, females marry a couple of years younger than males, but they also live longer and so many more females than males are widowed. Given these important differences, the Census Bureau tabulates marital status separately for the two sexes.

Table III presents a useful format for analyzing local changes in marital status. The first two columns report the raw number of persons in each marital status taken from the published data. Note that a *formerly married* category combines persons who are widowed, divorced, or separated. Although this is a useful condensation of the data, for some purposes analysts might wish to distinguish these separate statuses.

The third and fourth columns of the table compute change in each category over time. This measures growth in each category. Alternatively, the final three columns compute the marital composition of the area; i.e., the percentage of male or female residents who fall into each marital category. The last column computes the change

Table III. Change in Marital Status
Hollywood Tract 1899

Males	Number 1970	1980	Change 1970 to 1980 Number	% Change	% of Total 1970	1980	Change in Percentage (a)
Never married (single)	1034	1687	653	63.2	30.2	46.0	15.7
Married, spouse present	1788	1270	−518	−29.0	52.3	34.6	−17.7
Formerly married (b)	598	714	116	19.4	17.5	19.4	2.0
Total over age 14 (c)	3420	3671	251	7.3	100.0	100.0	—
Females							
Never married (single)	908	1120	212	23.3	22.1	29.8	7.7
Married, spouse present	1773	1247	−526	−29.7	43.2	33.2	−10.1
Formerly married (b)	1420	1391	−29	−2.0	34.6	37.0	2.4
Total over age 14 (c)	4101	3758	−343	−8.4	100.0	100.0	—

Notes:
(a) Percentage point difference, or the 1980 percent of total minus the 1970 percent of total.
(b) Includes separated, widowed, and divorced persons.
(c) In 1970, also includes persons age 14.
Source: PHC(1)-117, Table P-1, and PHC 80-2, Table P-1.

in the marital composition of the area by showing the percentage point change in each category's share of the total.

The example chosen to illustrate the analysis is an apartment area in Hollywood, Tract 1899. The number of males who were single (i.e., never married) grew by 63.2 percent, causing the single percentage of males age 15 and older to rise from 30.2 to 46.0 percent—nearly half. Conversely, the percentage of males living with wives fell from 52.3 to 34.6 percent over the decade.

Single females have grown at only about one-third the rate for single males in this neighborhood, and they account for a smaller share of all females in 1980 (only 29.8 percent). A much larger share of females (37.0 percent) are formerly married than males (19.4 percent). Very likely, this difference in marital status reflects differences in the ages of men and women. A sizable block of women in the neighborhood are elderly widows, while the men are skewed toward young singles.

A third of men and women are still married couples, but that number has been dropping (a loss of more than 500 marrieds of each sex). The *number* of husbands and wives does not quite match up, as it should, but is very close. This is due to reporting errors by census respondents. The error is so small as to be inconsequential.

2.1. Scatterplot of Marital Status at Two Points in Time

Not all neighborhoods have such a high percentage of never-married persons as this Hollywood tract. Nor has the percentage increased to such a large degree in most cases. At a minimum we should compare this tract to the county or another regional reference area. It would also be useful to compare other neighborhoods, both in terms of their level of singleness and their degree of change over time.

The relative standing of different places is seen most easily in a graph. Figure 2 demonstrates a graphic technique that might be termed the *before-and-after scatterplot*. The horizontal axis plots status at an earlier point in time, while the ver-

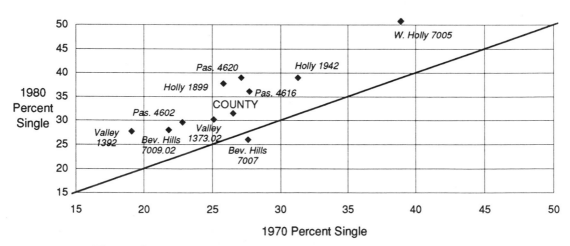

Figure 2. Changes in the percentage single between 1970 and 1980 among selected neighborhood populations. Neighborhoods above the diagonal line in this "before-and-after" scatterplot experienced a rising percentage single.

tical axis shows the status at the time of the most recent census. Any tract with the identical percentage single in both years will fall on the diagonal line portrayed in the graph. Tracts with higher percentages at a later point in time will rise above the diagonal line and those with lower percentages will fall below. This method provides a nice overview of the relative ranking of places in terms of both their status at the latest census and the degree of change. We have labeled each of the places on the chart, but without labels, hundreds or even thousands of places can be plotted.

Los Angeles County is plotted on the graph with 25.1 percent single adults (males and females combined) in 1970 and 30.2 percent single in 1980, so it is located 5.1 percentage points above the diagonal. This constitutes the average degree of change for the county. In contrast, Hollywood Tract 1899 is plotted 12.0 percentage points above the diagonal, showing how much greater the increase in singles was in this area than elsewhere. The highest percentage single in either year is found in West Hollywood Tract 7005, the center of Los Angeles's gay community, and it also increased its percentage of singles substantially over the decade. Another area of large increase is Pasadena Tract 4620, a predominantly black area with growing numbers of Hispanics. The only area in this sample that shows declining percent singles is Beverly Hills Tract 7007, a highly unusual wealthy area. With these high and low exceptions, the other areas reflect a similar degree of change to the county as a whole.

The percentage single can rise for many reasons, only one of which is a growing gay population, as in West Hollywood Tract 7005. Singles also can increase as young persons flock to apartment areas near universities, hospitals, or urban night life (Hollywood Tract 1899); or singles can increase in suburban areas of single-family homes, because many children continue to live at home into their midtwenties (Valley Tract 1373.02),[5] and in zones attractive to immigrants from abroad, as for instance when young male laborers double up in apartments and middle-aged couples exit the neighborhood (Pasadena Tract 4620).

2.2. Age-Expected Calculation of Marital Status

From the foregoing description, it is clear that we expect the proportion single to be highest where very young adults are prevalent, for the never-married marital status declines rapidly with age. For example, in California, the 1980 census showed 97.4 percent of 15- to 19-year-old males were single, falling to 72.4 percent at ages 20 to 24 and 39.5 percent at ages 25 to 29.[6] Among California females, single status declines even more rapidly, for women on average marry two to three years younger than men.[7]

Some neighborhoods may have a lot of singles simply because young college students are living there. Other neighborhoods may have fewer singles overall, but a

---

[5]Nationally, at age 24, as many as 20 percent of males and 12 percent of females are still living in their parental homes (Sweet and Bumpass, 1987: 91–92). The proportion living with parents has been growing over time, in part due to the rising proportion of young adults who have never married.

[6]The published data for 1990 are found in *General Population Characteristics* (CP-1), Table 37, and for 1980, *Detailed Population Characteristics* (PC80-1-D), Table 205.

[7]Detailed analysis for the nation, illustrated by many graphs with rich age detail, may be found in Sweet and Bumpass (1987).

much higher proportion than would be expected among their 30- or 40-year-old population. Neighborhoods of the latter type are overperformers that stand out as bastions of "singledom."

How can we control for the basic age effects to see in which neighborhoods persons are more likely to be single than suggested by their age alone? Two closely related solutions are possible here: indirect and direct standardization.

2.2.1. Indirect Standardization

The best solution with census tract data is to compute the expected percent single based on the neighborhood's age profile, then compare this to the actual number of singles of each sex. The expected marital status is drawn from marital patterns by age and sex that are published for larger reference areas, such as the county or the state.

The procedure is carried out separately for each sex. We simply multiply the proportion single at each age (from the reference area) times the local number in that age group, yielding the expected number of local singles at each age. The expected singles in all age groups are summed, creating an age-expected total of singles, and then compared to the actual total of singles observed for each sex in the locality.[8]

The results for males in our sample neighborhoods are illuminating. The West Hollywood gay neighborhood has single males in residence that far outnumber what would be expected on the basis of age alone. Whereas the age-expected number of single males in 1980 is 1,026, actually 2,387 are in residence, a number 233 percent as large as expected. The other Hollywood neighborhoods also show strong oversupplies of singles. Tract 1942 has a lower percentage single than Tract 7005 (Fig. 2), but its age profile is older. When adjusted for age, Tract 1942 shows 241 percent as many singles as expected. At the other end of the scale, Beverly Hills Tract 7007 has the lowest percent single, but its age distribution is much older. When adjusted for age, its number of singles is 114 percent of what we would expect, slightly above average for California.

2.2.2. Direct Standardization

An alternative to the age-expected calculation described above is direct age standardization. Rather than apply the reference area's marital status percentages to the local area's age distribution, we do the opposite. The effects of age are controlled by applying each local area's marital status distribution in each age group (e.g., the percent single, married, or formerly married at ages 25 to 29, 30 to 34, etc.) to the reference area's age distribution. After summing the expected number of singles in all age groups, the results indicate the number in each marital status that would be found if the same standard age distribution occurred in every location.

The problem with this direct age standardization method, however, is that it imposes data requirements that are less easily met. The necessary age-specific marital status information for local areas is only published for larger areas, as described in

---

[8]The calculations are not shown here, but an example of the method is fully demonstrated in Chapter 13, Table IV.

Section 7.1.1. Even there, the age breakdowns only pertain to married persons, not singles. Accordingly, with these published data, we would have to modify our analysis to focus on married and nonmarried persons (including not only never-married but also widowed and divorced persons). For more complete age-specific marital detail at the level of counties and places, or for census tracts, users would need to use computerized data files.[9] As a result, the indirect method of age-expected calculation is more practical for much local analysis because it requires age-detailed marital data only for the larger reference area.

In general, adjustment for the effects of age composition is important for many different analyses, and the techniques described here can be used widely.[10] In the next chapter, we will address basic skills for analyzing local age structure.

☐ 3. Household Relationship and Family Type

Table IV presents a format for analyzing household and family changes, drawing on two sections of the raw data shown in Table II. The top panel of the table portrays numbers *of persons* in each household relationship, drawing on the raw data section labeled *Household Type and Relationship.* In contrast, the bottom panel portrays numbers *of households* of each family type and also counts numbers of children in each type of family. This is drawn from the raw data section labeled *Family Type by Presence of Own Children.* Considerable analytical power can be gained by understanding how the household relationships in the top half of Table IV relate to the family types in the bottom half.

This exhibit continues with the focus on Hollywood Tract 1899. Rather than emphasize the *story* in the numbers, however, the more important lesson is how the numbers in different categories match up with one another. First, observe that the number of householders in the top line equals the number of households in the bottom line (also equal to the number of occupied housing units). The number of family householders also equals the number of families, and the number of married householders (indicated by the presence of a spouse) equals the number of married couple families. Note that this number is only slightly smaller than the number of married men or women in the tract (shown previously in Table III). The number of households with married couples is smaller because not quite all couples head their own separate households. (A few of the youngest double up with parents or others.)

Single-parent families may be calculated by simply subtracting married couple families with children from all families with children. (Some of these unmarried family children might be living with grandparents or other relatives and not their true parents, but the numbers are very small.) In both 1970 and 1980, in this tract, we see that around 36 percent of families with children are single-parent families. (This is only slightly above average for Los Angeles County.) We can also calculate the percentage *of children* who are living in single-parent families, and that also is

[9]In 1990, crosstabulations of age by marital status of small areas are found in STF 2, Table PB7 and, in 1980, STF 2, Table B5.

[10]For more explanation and examples, see Shyrock and Siegel (1976: 8, 241–243).

Table IV. Change in Household Relationship and Family Status

Hollywood Tract 1899

Household Relationship	Number 1970	Number 1980	Change 1970 to 1980 Number	Change 1970 to 1980 % Change	% of Persons 1970	% of Persons 1980	Change in Percentage (a)
All Householders	4624	4789	165	3.6	57.8	61.3	3.5
Family Householders	2158	1568	−590	−27.3	—	—	—
Spouse of Householder	1720	1168	−552	−32.1	21.5	14.9	−6.5
Other Relative	1182	932	−250	−21.2	14.8	11.9	−2.8
Nonrelative	416	890	474	113.9	5.2	11.4	6.2
Not Living in Households	61	35	−26	−42.6	0.8	0.4	−0.3
Total Persons	8003	7814	−189	−2.4	100.0	100.0	—

Family Status	Number 1970	Number 1980	Change 1970 to 1980 Number	Change 1970 to 1980 % Change	Ratio to Households (b) 1970	Ratio to Households (b) 1980	Change in Ratio (c)
All Families	2158	1568	−590	−27.3	0.467	0.327	−0.139
With Own Children	393	303	−90	−22.9	0.085	0.063	−0.022
Number of Children	576	409	−167	−29.0	0.125	0.085	−0.039
Married Couple Families	1720	1168	−552	−32.1	0.372	0.244	−0.128
With Own Children	251	192	−59	−23.5	0.054	0.040	−0.014
Number of Children	371	261	−110	−29.6	0.080	0.054	−0.026
Total Households (d)	4624	4789	165	3.6	1.000	1.000	—

Notes:

(a) Percentage point difference, or the 1980 percent of total minus the 1970 percent of total.

(b) Number divided by total householders.

(c) The 1980 ratio to households minus the 1970 ratio.

(d) Numbers do not sum to total households; nonfamily households omitted.

Source: PHC(1)-117, Table P-1, and PHC 80-2-226, Table P-1.

around 36 percent each year. Of course, the overall incidence of families with children is very low in this tract. In other neighborhoods, we would expect to find a sizable and growing share of single-parent families.

A major alternative to percentage calculations is introduced in the bottom panel of Table IV. A very useful way to calculate the incidence of different family types and household relationships is to express these *relative to the number of households.* Thus, the number of families in 1970 was equal to a ratio of 0.467 per household, or 46.7 families for every 100 households. By 1980, the ratio fell still further to 0.327 per household. Similarly, in 1970, the number of children in families was 0.125 per household, or 12.5 children for every 100 households in the tract. This is extremely low, but the ratio fell even lower to 8.5 children per 100 households in 1980. This trend is not surprising, given what we have learned about this Hollywood tract. Elderly widows and young singles are not likely to have children living with them.

Returning to the top panel of Table IV, children would be included in the *other relative* category, which includes children under and over age 18, brothers, sisters, and all other relatives, excepting the spouse of the householder. This category decreased over the decade by 21.2 percent in number, but not as fast as did the number of spouses (a decline of 32.1 percent in number). Instead, we see a growth of 114 percent in the number of nonrelatives. By 1980, these roommates or, perhaps, unmarried partners nearly equaled the number of other relatives.

□ **4. Components of Household Size**

The total household size (or persons per household) in Hollywood Tract 1899 was small and declined even further over the decade, from 1.72 to 1.62 persons per household. We can use the information reported previously to help explain why the household size shrank. It will also help to compare the trends in our other case example locations.

4.1. Comparing Overall Change in Household Size

As discussed in Chapter 3, household size is a vitally important concept in local area analysis. Household size is the link between persons in the residential population and the number of occupied housing units. A rising household size will bring a larger population into the same size housing stock, while a falling household size has the opposite effect. What have been the trends in our case example tracts?

Total household size fell in Los Angeles County and in most of the tracts in our sample between 1970 and 1980. A useful overview is gained again through a before-and-after scatterplot. Figure 3 shows that the tract with the largest household size in 1970 (Valley Tract 1373.02) fell by 0.81 persons by 1980. The tracts with the smallest household sizes in 1970, such as West Hollywood Tract 7005 or Hollywood Tract 1899, fell relatively little. In contrast to these widespread declines, two tracts increased their household sizes substantially. Pasadena Tracts 4620 and 4616 are minority-dominated areas and are examined in detail in a later chapter. Tract 4620, in particular, has experienced in-movement of growing numbers of Latinos whose larger household sizes are raising the average.

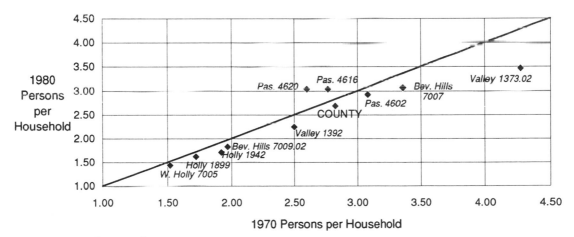

Figure 3. Changes in average household size (persons per household) between 1970 and 1980 in the example neighborhoods. Neighborhoods below the diagonal line in this "before-and-after" scatterplot experienced falling household sizes.

4.2. Decomposing Household Size into Relationships

Using the information presented earlier on household relationships, we can decipher how much of the household size change is accounted for by losses or gains in each category of household relationship. Table V presents a format for carrying out this analysis. The key is that the number of persons in each household relationship is divided by total households (given by total householders), forming a ratio of relationship types per household. The sum of the ratios for all possible household relationships equals total household size.

This simple calculation unlocks the mystery of average household size. In place of speculation about why household size has fallen and how much further it will fall, we now have greater light to shed on the matter. The method of decomposition into relationships explains who it is that is exiting from (and who is arriving in) the area's households. (In the next chapter, we extend this decomposition to the matter of age groups' contributions to household size.)

The minimum household size is 1.0, because every household must have a householder present to exist as an occupied housing unit. If those householders are married, household size must rise to 2.0. Similarly, if they have a child living with them, household size must rise further. Trends in marital status, childbearing, and other family behaviors thus have a direct effect on household size.

Hollywood Tract 1899 exhibits a fairly bare bones household size in Table V. No component other than the householder him- or herself is very large. The ratio of spouses per household fell from 0.37 to 0.24 over the decade, contributing a negative 0.13 to the change in household size. Children fell another 0.04 per household (the difference from 0.03 being due to rounding error), while other relatives fell 0.02 per household. In contrast, nonrelatives increased by 0.10 per household, offsetting much of the spousal decline. Similar decompositions are easily carried out for any neighborhood.

Table V. Household Relationship Components of Household Size
Hollywood Tract 1899

	Number of Persons (a)		Persons per Household (b)		
	1970	1980	1970	1980	Change
All Householders	4624	4789	1.00	1.00	0.00
Spouse of Householder	1720	1168	0.37	0.24	−0.13
Children of Householder	576	409	0.12	0.09	−0.04
Other Relative (excluding children)	606	523	0.13	0.11	−0.02
Nonrelative	416	890	0.09	0.19	0.10
Total (c)	7942	7779	1.72	1.62	−0.09

Note:
(a) Persons in each household relationship are taken from Table IV (children from the family status subtable).
(b) Ratio of persons of each relationship type to total householders.
(c) Total persons per household equals household size. Small discrepancies are due to rounding error.
Source: Table IV and author's calculations.

4.2.1. Cumulative Bar Graph of Household Size

One useful way for comparing the compositions of different places is the stacked bar graph. This graphic device plots the cumulative proportion of cases falling in each category, either summing to 100 percent, or summing to the total number of cases, or summing to some other total measure. In this case, we sum the components of household size, with the total length of the bar reflecting the total household size.

Figure 4 compares the household relationship components of a number of different places, sorted in order of their overall household size but with Los Angeles County at the top of the graph for comparison. In all places, the householder accounts for 1.0 units of household size. It is clear that the tracts with larger sizes achieve this through their high ratios of spouses and children, or other relatives, per household.

Nonrelatives are present in all locations, but they account for a much larger share of total household size in the Hollywood tracts that have the smallest overall household size. There the frequency of nonrelatives approximates that of spouses. Some of these nonrelatives may be roommates; some, persons of the opposite sex sharing living quarters (described by the quaint acronym POSSLQ); and others may be partners of the same sex.

As discussed in Chapter 3, the 1990 census questionnaire for the first time will allow us to decipher which of these nonrelatives are *unmarried partners*. For comparison with the 1980 census, we are still stuck with the previous categories. The wealthy, Beverly Hills Tract 7007 illustrates the pitfall. That tract has a very high incidence of nonrelatives, 0.37 per household, even higher than the Hollywood tracts. Although some might be tempted to conclude that these are the unmarried lovers of

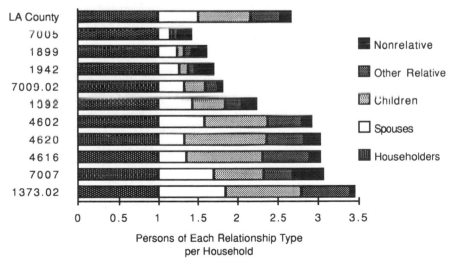

Figure 4. Relationship components summing to total household size in selected neighborhoods. The stacked bar graph indicates the cumulative number of persons per household of each relationship type in 1980.

movie stars, more likely they are simply live-in servants, one more subcategory of nonrelatives.

4.2.2. Comparing Components of Household Size Change

Did the same factors account for the household size changes in all tracts? One means of comparison would be to prepare a before-and-after scatterplot for each separate component of household size. The difficulty comes in comparing the different components. Instead, we will adopt a variation on the stacked bar graph. Figure 5 divides the overall change in household size into categories contributing to size gains—on the right side—and size losses—on the left. The data being plotted were generated from the fifth data column of Table V, repeated for each of our sample places.

The left-hand data, losses to household size, in Fig. 5 are accounted for almost exclusively by losses of spouses and children under age 18. Valley Tract 1373.02 registered the largest decline in household size because it lost more than one whole child per household. In contrast, the Pasadena minority tracts, 4620 and 4616, experienced overall size gains because they alone added children (while still losing spouses).

Overall, the right-hand data, gains to household size, are accounted for by gains in other relatives and nonrelatives. It bears emphasis that the *other relatives* often include children older than 17. This is evident, for example, in Valley Tract 1373.02. As these teenagers and young adults continue to age, most will soon depart their parents' households and the number of *other relatives* will fall. If many parents in such family neighborhoods remain in their homes for a long time, household size will inevitably plummet, as it already has. Only if replacement families enter the neighborhood can household size be sustained. This is one reason why occupancy duration (discussed in the previous chapter) is such an important indicator.

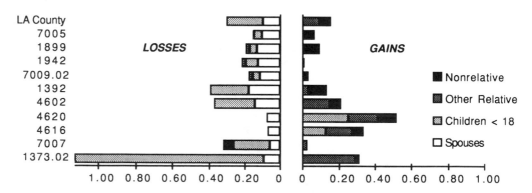

Loss or Gain of Persons of Each Relationship Type per Household

Figure 5. Components of change in average household size in selected neighborhoods between 1970 and 1980. The stacked bar graph displays, on the left, the categories that contributed losses to overall household size and, on the right, the categories that contributed gains.

☐ 5. Conclusion

This chapter has surveyed the uses of local information on marital status, family type, and household relationships. In general, this information provides important insight into who exactly lives in an area's housing units.

Note the linkage between trends in marital status and trends in household relationships. Growth in formerly married persons will be accompanied by declining spousal components of household size. Similarly, growth of other relatives corresponds to increases in the number of single persons. These singles who live in families (often children aged beyond 17) are very different in status from those living in tracts with small households and very few other relatives. In such cases as the Hollywood tracts, it is apparent that the growth of singles is among persons living on their own in apartments.

The household is the prime link between population and housing, and household size is the measure most widely recognized for describing this linkage. Our emphasis here has been to break down average household size into its components. This theme will be continued as part of the age analysis carried out in the next chapter. Age structure is closely related to both marital status and household relationships.

8 □□□

Population Age Structure

Age is the central variable of demography, and the census offers more detail on age than any other variable. In fact, tracking changes in age structure is one of the most important uses for local census data. Every social service and business enterprise is impacted by these changes.

Age is important because so many behaviors are life-cycle related. Marriage and family behavior is strongly correlated with age. In just a few years time, young toddlers enter neighborhood schools, and within a decade they will roam their neighborhoods and shopping malls as teenagers on a mission. In just a few years, today's teenagers become young adults, leaving their parents' home for college or their first apartment, getting a job, entering marriage or another intimate relationship, and possibly buying a home or having children. The probability of each of these behaviors advances sharply from age 20 to 25 to 30.

Aging continues through adulthood, with patterns of labor force participation, mobility, shopping, and home buying all shifting progressively with age. As time passes, older adults find their children leaving home and then face the increasing risk of mortality. The death of one spouse often leaves the survivor living alone, but divorce can produce this status at a younger age.

Age patterns in local areas often reflect a life-cycle model. Particular neighborhoods may attract persons in a particular life-cycle stage, and this will be reflected in a relatively constant age profile for the neighborhood from one census to the next. The population around universities never seems to grow older, and that in retirement areas never gets any younger. Other neighborhoods may tend to hold the *same* people for a long period of time. Persons age in place, with the result that the neighborhood takes on a life cycle that reflects its residents (see chapter 6). Whatever the pattern, census data are well suited to detecting the life-cycle dynamics related to neighborhoods.

Age analysis is a means to understanding many other problems and issues. A great many variables in the census are crosstabulated by age, if not in the printed reports for small areas, then in reports covering larger areas and in the computerized files. With these data, changes in age structure can be used to analyze many other factors, so a sound understanding of local age structure is crucial.

☐ 1. Local Area Data

Age is a complete count variable included on the short-form questionnaire, and the absence of sampling error facilitates fine-grained analysis of changes with this variable. In the preceding chapter, we analyzed the household and family data contained in the lower half of Table P-1. The top half of that rich table holds data on the age distribution by sex for the census tract or locality. As an example, Table I presents

Table I.

Census Tracts	Los Angeles city, Los Angeles County—Con.				
	Tract 1898	Tract 1899	Tract 1901	Tract 1902	Tract 1903.01
AGE					
Total persons	**2 305**	**7 814**	**4 250**	**7 032**	**5 194**
Under 5 years	59	116	86	262	205
5 to 9 years	53	139	83	218	154
10 to 14 years	52	130	60	176	165
15 to 19 years	83	208	196	300	296
20 to 24 years	164	648	608	845	648
25 to 34 years	556	2 096	1 210	1 906	1 282
35 to 44 years	325	992	555	866	620
45 to 54 years	245	807	420	684	552
55 to 64 years	275	842	399	612	446
65 to 74 years	278	1 054	356	656	371
75 years and over	215	782	277	507	455
3 and 4 years	19	27	34	94	68
16 years and over	2 119	7 408	4 002	6 339	4 636
18 years and over	2 090	7 328	3 940	6 239	4 536
21 years and over	2 033	7 144	3 743	5 959	4 296
60 years and over	636	2 257	843	1 455	1 028
62 years and over	579	2 081	760	1 355	955
Median	39.9	40.1	33.5	33.6	33.2
Female	**1 133**	**3 952**	**1 872**	**2 955**	**2 575**
Under 5 years	24	58	38	131	99
5 to 9 years	24	69	40	100	77
10 to 14 years	26	67	25	94	81
15 to 19 years	44	120	100	140	138
20 to 24 years	71	320	246	306	292
25 to 34 years	238	933	467	675	557
35 to 44 years	139	404	181	295	243
45 to 54 years	114	354	188	261	267
55 to 64 years	165	495	203	287	236
65 to 74 years	157	646	202	360	236
75 years and over	131	486	182	306	349
3 and 4 years	10	11	13	47	33
16 years and over	1 044	3 744	1 760	2 612	2 298
18 years and over	1 033	3 693	1 727	2 566	2 249
21 years and over	1 001	3 598	1 634	2 441	2 140
60 years and over	370	1 385	494	816	701
62 years and over	329	1 292	449	763	660
Median	45.1	45.2	35.7	36.4	36.6

the data for Hollywood Tract 1899. Again these data are similar in content for the three most recent censuses (1970, 1980, and 1990), although there are three significant exceptions.

First, the 1970 table contains a small amount of information at the top regarding racial composition. In 1980, this information was greatly expanded and moved to a separate table (to be discussed in Chapter 10). The 1990 table follows the practice set in 1980.

A second exception is that the 1980 and 1990 tables do not directly report age data for males. Instead, the table lists data by age for the total population and then for females. Males are obtained by subtraction. Finally, the exact age categories differ a little between the tables of the different censuses. These differences necessitate care in handling but they are not difficult to reconcile.

Analysts with access to computerized summary tape files, as described in Chapter 4, will find even richer data about age. In addition to the sex and race dimensions offerred in the published tables, age is further reported in single-year categories,[1] or listed separately for persons living in households (instead of group quarters or homeless).[2] Other summary file tables crosstabulate age (by sex and race) with additional variables of interest, such as marital status,[3] household relationship,[4] or labor force status.[5] However, the computerized files have the drawback that they are much less accessible to most users than the published data, so we do not assume their availability for this chapter. It is also important to grasp the fundamentals of age analysis with the simpler published data before delving into the more complex computerized tabulations.

1.1. Extra Detail for Elderly and Young Ages

Although our emphasis in this chapter is on the full age span, from 0 to 75 or older, elderly ages often can be analyzed more fully. The onset of retirement is of such social and economic importance that age data are reported with more detail from 55 years onward. The 10-year age group 55 to 64 is effectively divided into 55 to 59 and 60 to 64 portions by listing a total for persons age 60 and older (see Table I). For reasons of social security eligibility, a total for persons age 62 and older is also listed. With greater numbers of elderly persons living to older ages, there is greater need to distinguish these *old-old* persons from the *young-old*. For this reason, the 1990 census has increased its detail in elderly ages, with the top age category in published sources extended to 85 years and older.

Other age groups also receive special listing in the published data for various legal and programmatic reasons. The total listed for persons age three and four may help

---

[1] The detailed age categories are found in STF 2, Table B8 (for 1980) and Table PB5 (for 1990).

[2] The household population age distribution is found in STF 2, Table B10 (for 1980) and Table PB6 (for 1990). Age of persons who are in group quarters or homeless can be found by subtracting the household numbers from the total population.

[3] The crosstabulation of age and marital status is found in STF 2, Table B5 (for 1980) and Table PB7 (for 1990).

[4] The crosstabulation of age and household relationship is found in STF 2, Tables B7 and B11 (for 1980) and Table PB15 (for 1990).

[5] The crosstabulation of age and labor force status is found in STF 4, Tables PB51 (for 1980) and Table PB50 (for 1990).

with school enrollment planning. A total for persons 16 and older is listed for reasons of labor force eligibility, drivers license eligibility, and exemption from education requirements. A total for those 18 and older defines those eligible for voting or military service, while the total age 21 and older defines those of drinking age in most states.

The elderly categories are much more likely to be used in crosstabulation with other variables. For example, the published data on household relationships discussed in Chapter 7, Table II, contains a special table for persons 65 and older. Even the detailed summary tape files afford extra coverage to elderly categories in crosstabulations. As an example, the 1990 census has expanded the crosstabulation of age by labor force status to include five categories after age 59, in place of the former two. At the same time, only two categories are identifed below age 25. This distribution of age detail suggests greater importance placed on exit from the labor force than entry into it. On the whole, more detailed analysis is possible with these elderly data than for other portions of the age span.

☐ 2. The Population Pyramid

The classic means of representing population age structure is the population pyramid. Each age-sex group is expressed as a percentage of the total population. The percentages falling within each age-sex group are graphed as horizontal bars, with the oldest group at the top of the graph, and the youngest at the bottom. The graph is called a pyramid because in many nations—those with rapidly growing populations—the number of young persons at the bottom of the graph far outweighs the dwindling number of older persons at the pinnacle. In slower growing populations, the graph is more rectangular, with approximately equal numbers of persons at all ages save the oldest.

Population pyramids often taken wildly varying shapes in urban neighborhoods (Coulson 1970). The attractions of a neighborhood may be life-cycle specific, drawing a particular demographic subgroup for a short period of their lives. In most cases, persons do not live out their whole lives in one neighborhood, unlike the case of national populations. Hence, the localized population age profile need not reflect that of the complete population.

Figure 1 portrays a pyramid for Hollywood Tract 1899 (from 1980 data) that is truly inverted! Very small percentages of children are observed at the bottom; instead, the graph is "top-heavy" with older persons. Particularly on the female side, there is very little drop off in the population at older ages. This older population is supplemented by a large concentration of persons in their twenties and early thirties. This bulge is pronounced among women, but more so among men. This neighborhood of apartments (see discussion in Chapter 2) is attractive to both elderly persons and young singles. However, as the oldest population dies off, we might expect more young persons to take their place over time.

For contrast, Fig. 2 portrays a pyramid for San Fernando Valley Tract 1373.02 (1980 data) that has the classic dumbbell shape expected in suburban areas dominated by families with children (Long and Glick 1976). There is a distinct shortage of young adults in their twenties (perhaps some gone to Hollywood, 30 minutes down the freeway), and there also is a shortage of elderly residents. As described in Chapter 2, this neighborhood was newly constructed with large single-family homes prior to the 1970 census. Therefore, the area isn't old enough to have seen its residents *age in place*, nor is its housing especially attractive to elderly in-movers (or to young adults).

A minor curiosity in this tract is that the percentage of women is greater than men in the 30 to 44 age range, whereas the percentage of men is greater than women in the 50 to 64 age range. This reflects the frequent age difference between spouses in married couples, with the husbands' age distribution shifted older (higher in the pyramid) than the wives' age distribution. Other than this slight difference, the male and female halves of the pyramid are virtually identical.

2.1. Method of Constructing a Population Pyramid

The most unusual feature of the pyramid graph is that the percentages for the two sexes at each age are portrayed as left and right running bars extending in opposite directions from the center. Although this effect is not hard to achieve by drawing on graph paper by hand, few general-purpose computerized graphing programs can automatically produce a pyramid graph. The following guidelines may prove helpful in the majority of cases.

First, calculate the age-sex group percentages, dividing the number of each age-sex group by the total population. Then multiply the male figures by minus 1, to transform them to negative numbers. Next, we enter these data into a graphing program as two data series, one for males and one for females. Each series has one percentage value corresponding to each age category.

The basic graph choice is the horizontal bar graph. If possible, select an option for no spacing between the bars. Also select an option for 100 percent overlap between the two data series. (Do not "stack" the two data series.) This will superimpose the two data values for each age, but the male and female values will run in opposite directions because they are opposite signs. One final option to choose is to move the category labels away from this central axis by choosing another position such as "labels low" (i.e., to the left on the negative, or lower valued side of the chart). The chart may be dressed up further by selecting desired shading for the bars, adding labels, and manually erasing the minus signs from the male side of the value axis.

☐ 3. Alternative Formats for Age Data

The percentage breakdown of the population, by sex, portrayed in the pyramid graph is an interesting way of *looking* at age structure. However, there are alternative formats of analysis that prove more useful in practice. These are compared in this section.

We will combine the sexes and analyze the age distribution of the total population in order to simplify the demonstration. With the two sexes combined, twice as much work can be accomplished in the same amount of space. In general, combining the

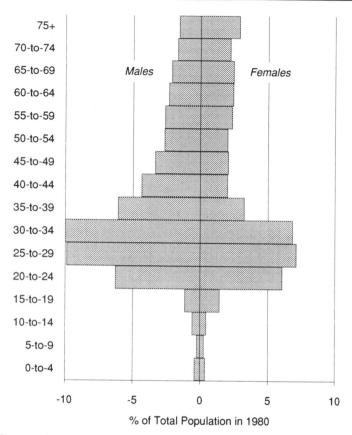

Figure 1. Age–sex pyramid for Hollywood Tract 1899 in 1980.

two sexes saves "a degree of freedom" that permits comparison in the same exhibit of changes over time or differences between places. Given that the numbers of males and females at each age are roughly similar, there is often little need to analyze males and females separately unless the focus is on a sex-segregated activity such as fertility or military recruitment. Of course, if separate analysis of males and females is desired, analysts can carry out parallel analyses for each sex following the methods described below.

Table II presents a format for preparing age data for analysis, demonstrated with data for San Fernando Valley Tract 1373.02, our second pyramid example. This is also the suburban family neighborhood identified in the preceding chapter as experiencing a large drop in household size between censuses. The exhibit analyzes data for this location in four sections.

3.1. Percent of Total Population

The first pair of columns shows the absolute number of persons in each age group, as reported in two censuses These are the raw data processed into alternative formats

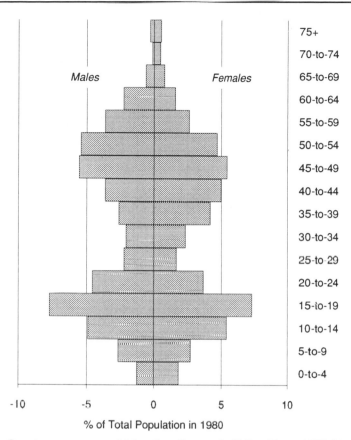

Figure 2. Age-sex pyramid for San Fernando Valley Tract 1373.02 in 1980.

in the remaining sections of the exhibit. The second pair of columns calculates percentage distributions to represent the population age composition, or age structure, for each year. The raw numbers are divided by the total population for that year.

Results of this age analysis are more understandable in a graph, as displayed in Fig. 3. Here we see a distinct aging of the population between 1970 and 1980. The peak adult ages shifted from 35–44 to 45–54, and the number of children plummeted as the children formerly aged 10 to 14 grew up and left home. The life-cycle pattern is revealed by the net loss in certain ages created by fewer in-movers than out-movers.

3.2. Age-Specific Persons per Household Ratios

The third pair of columns in Table II demonstrates an alternative method with important advantages. Instead of dividing age groups by a denominator of total population, we divide them by the total number of households. The result is a ratio of persons per household who are in each age group, or age-specific persons per household (APH).

Table II. Alternative Formats for Age Distribution

San Fernando Valley Tract 1373.02

	Number		% of Total		Ratio of Persons per HH		Change 1970 to 1980	
	1970	1980	1970	1980	1970	1980	in % points	in HH ratio
0 to 4	458	147	7.8	3.1	0.331	0.108	−4.6	−0.223
5 to 9	852	254	14.4	5.4	0.616	0.187	−9.0	−0.430
10 to 14	1003	487	17.0	10.3	0.726	0.358	−6.6	−0.367
15 to 19	643	711	10.9	15.1	0.465	0.523	4.2	0.058
20 to 24	148	387	2.5	8.2	0.107	0.285	5.7	0.178
25 to 29	197	184	3.3	3.9	0.143	0.135	0.6	−0.007
30 to 34	404	208	6.8	4.4	0.292	0.153	−2.4	−0.139
35 to 39	635	319	10.7	6.8	0.459	0.235	−4.0	−0.224
40 to 44	617	407	10.4	8.6	0.447	0.299	−1.8	−0.147
45 to 49	436	519	7.4	11.0	0.315	0.382	3.7	0.067
50 to 54	290	476	4.9	10.1	0.210	0.350	5.2	0.140
55 to 59	108	294	1.8	6.3	0.078	0.217	4.4	0.139
60 to 64	46	182	0.8	3.9	0.033	0.134	3.1	0.100
65 to 69	24	66	0.4	1.4	0.018	0.049	1.0	0.031
70 to 74	20	29	0.3	0.6	0.014	0.021	0.3	0.007
75+	28	40	0.5	0.8	0.020	0.029	0.4	0.009
Total Persons	5909	4710	100.0	100.0	4.276	3.466	0	−0.810
Households					1382	1359		

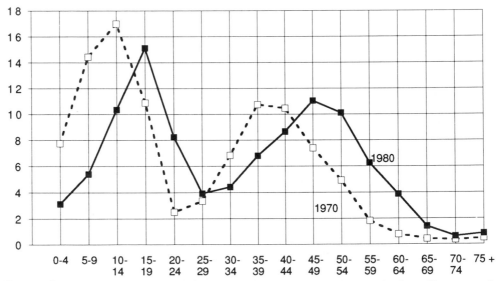

Figure 3. Percentage age distribution of population at two points in time. The example combines the two sexes and uses data for San Fernando Valley Tract 1373.02 in 1980 and 1970.

Figure 4 graphs the resulting age profile for comparison with results from the preceding method. Observe how closely the profiles in the two graphs resemble one another: Both are drawn from the same raw data. The key difference is that the entire 1980 profile seems to be shifted lower relative to the 1970 profile in the APH method (Fig. 4) than in the percentage method (Fig. 3). For example, in the percentage graph, the number of adults age 45 to 49 in 1980 seems higher than those age 35 to 39 ten years earlier. However, in the persons per household graph, the number appears substantially lower. What can explain this difference?

The numerators (the raw age group data) are the same in the two exhibits, but the denominators are different (total population in the percentage graph and total households in the other graph). Inspection of the totals at the bottom of Table II reveals that total population fell by over 1,000 between the censuses. In contrast, the number of households declined by only 23. Thus, the household denominator remained virtually constant between the 1970 and 1980 profiles, while the population total declined markedly. This shrinkage in population means that the 1980 percentages were proportionally larger than if the denominator remained constant.

The discrepancy illustrates one advantage of the APH method. Age groups' ratios to households are independent of one another: a decline in a tract's teenagers, for example, will have no impact on the ratios of babies per household or elderly per household. In contrast, in the percentage method the percentage can go up or down by virtue of declines or increases in *other* age groups. Thus, the trends in an age group's percentage of total population may mean nothing about the age group's own behavior. The number of households, or occupied housing units, provides a more

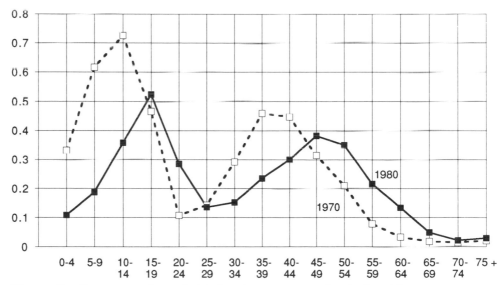

Figure 4. Persons per household of each age at two points in time. The case example is the same one used for Fig. 3.

stable base because rises or falls in the ratio depend on the number of units providing shelter, not the number of persons in other age groups.

A second advantage of the APH method is that it is arithmetically more useful in the modeling of local population changes. Population percentages cannot be combined with other variables to extract greater meaning. In contrast, the APH method has two important arithmetic properties. First, persons per household is the same as household size, and the persons of each age per household, or age-specific persons per household, is a measure that breaks down household size into its age components. This is analogous to the previous chapter where we broke down household size into relationship components. The ratios for all age groups sum to a total that is equal to total persons per household, or average household size.[6] Thus we see in the last column of Table II, or graphically in Fig. 4, that the large household size decline in this tract is accounted for mainly by selected age groups. The modest gains in certain age groups are outweighed by losses in the age brackets of young children (0 to 14) and, to less degree, parents (30 to 44).

In addition to decomposing household size into age segments, the APH method has the advantage of representing demographic multipliers.[7] If 100 households are added to a neighborhood, these can be multiplied by the age-specific ratios to yield estimates of new population. (Other data in the census can be treated in similar fashion, including school enrollment per household or labor force participants per household.) This method becomes even more powerful when we are able to compute APH for separate types of housing units that households live in and by race.[8]

3.3. Cohort Retention Ratios

The preceding section has presented alternative methods for comparing age distributions at two points in time. With these methods the tendency is to inspect the pattern and then infer an underlying behavior. Better insight about the movements of age groups over time can be gained from a cohort perspective, something entirely different from age groups.

Age has a special quality unlike all other variables. Age is closely linked to time, because it measures the elapsed time from the person's birth year. With each passing decade we expect people to grow older by a known amount of time. The predictability of this change is unlike any other variable. In fact, only race and sex, which are not expected to change at all, are more predictable.

3.3.1. Cohorts Versus Age Groups

The terminology and meaning of age groups and cohorts is often confused. These two are not the same thing, although the terminology *age cohort* is frequently used to

[6]This summation requires that all of the persons in the age data reside in households. The summation will overstate actual household size to the degree that group quarters residents are included. Methods to address this potential problem are presented later in the chapter.

[7]This concept was pioneered by urban planning researchers at Rutgers University (Burchell and Listokin 1978) and is developed further in Myers and Doyle (1990).

[8]Systematic variation in household age profiles is demonstrated in Myers and Doyle (1990). This more detailed information may be obtained from tabulation of the Public Use Micro Sample tapes, as discussed in Chapter 14.

mean age group. Age groups are *fixed in age,* like age 21 or age 20 to 24. In contrast, cohorts are *fixed in membership*, defined by their year of birth, like 1970 or 1969 to 1974.[9]

The frequent confusion stems from the fact that at a given point in time, an age group and a cohort both can describe the same persons (hence the term *age cohort*). However, over time cohorts pass out of one age group and into the next as they grow older. As a result, different cohorts, and different people, are in the age group each year. Analysis of changes over time requires a clear distinction between the notion of changes in an age group and the longitudinal experience of a cohort. The term *age cohort* is often confusing, and should be avoided, because it is unclear whether it is one or the other. The unwary should be advised that the term *cohort* is not just a technical sounding synonym for *age group*.

Table III addresses the same data from Table II in two different formulations— age group and cohort. The raw data are repeated in the first two columns, and then alternative age group and cohort changes over time are measured. A good illustration of the differences in result is seen in the 20 to 24 age group. The third column of the table shows a very large 161 percent increase in this age group. However, the next column reports a 61 percent *decrease* in the size of the cohort that now occupies that age group. In other words, compared to when the cohort was 10 years younger, 10 years earlier, it is now almost two-thirds smaller. The cohort decreased in size as its members ceased being teenagers and left home. But the age group grew much larger because the previous cohort that occupied this bracket 10 years earlier was so much smaller.

In some cases analysts may wish to measure change in the size of an age group. Care should be taken, however, not to conclude that more people are moving into the neighborhood in that age group. Those extra persons could just be teenagers or middle-aged parents grown older (i.e., *aging in place*). For analysis of behavior over time, the cohort formulation is generally preferable.

3.3.2. The Retention Concept

The last column of Table III expresses the data for cohorts in a more readable form. Instead of calculating percentage change, we calculate percentage survival, or retention: the net percentage of the original cohort that remains in the location. A *cohort retention ratio* (CRR) is calculated simply as the number of persons in the future age group divided by the number in the cohort's earlier age group, multiplied by 100.

---

[9]In technical usage, a cohort is any group of people or objects that enters a system in the same year or same time period. Birth is the most common means for defining cohorts, but we can extend the concept to reflect entry into marriage (the marriage cohort of 1990), graduation from college (the class of 1990), or even production models of automobiles (e.g., Honda Accord, 1990 model year). In all these cases research may be undertaken to trace the subsequent success of the cohorts over time. We might track divorce rates of the 1990 marriage cohort to see if those unions hold up as well as earlier cohorts, comparing say the marriage cohort of 1980. Or we might track the earnings histories of the college class of 1990 to see whether their salaries increase at the same rate as experienced by the class of 1980. Finally, much research goes into the repair histories of different model cars. Volvo and Mercedes stake their reputation on how well (or how long) their older cohorts have performed. "Cohort" is an extremely valuable concept for tracing changes over time for any group.

Table III. Comparison of Age Group and Cohort Changes
San Fernando Valley Tract 1373.02

| | Number | | % Change 1970 to 1980 | | Cohort Retention Ratio |
	1970	1980	Age Group	Cohort	1970-80
0 to 4	458	147	−67.9	—	—
5 to 9	852	254	−70.2	—	—
10 to 14	1003	487	−51.4	6.3	106
15 to 19	643	711	10.6	−16.5	83
20 to 24	148	387	161.5	−61.4	39
25 to 29	197	184	−6.7	−71.4	29
30 to 34	404	208	−48.5	40.5	141
35 to 39	635	319	−49.7	61.7	162
40 to 44	617	407	−34.1	0.8	101
45 to 49	436	519	19.2	−18.2	82
50 to 54	290	476	63.8	−23.0	77
55 to 59	108	294	172.7	−32.4	68
60 to 64	46	182	294.6	−37.5	63
65 to 69	24	66	172.3	−38.7	61
70 to 74	20	29	46.3	−37.4	63
75+	28	40	42.9	64.5	165
Total Persons	5909	4710	—	—	—

With this method a cohort that holds its size over time has a retention of 100 percent. One that grows may increase above 100, and one that shrinks will fall below that. Thus, our cohort decline of -61 percent in the 20 to 24 age bracket is equivalent to retention of 39 percent.

Some further observations on cohort methods are in order. First, it is important that age groups are properly selected to represent a cohort's changes over time.[10] Second, remember that cohorts may measure persons drawn from the same birth year but these are not necessarily the same persons each year. We cannot actually trace individual persons over time with the census, but we can measure the *net* retention of cohorts. Certainly, more persons will enter and leave a neighborhood than measured by the net change. Nevertheless, the pattern of net cohort retention measures the relative attractiveness of the neighborhood to different age groups.

The cohort retention concept differs from the concept of *censal survival ratios*, despite similarities in computation, because *survival* stresses mortality. In contrast, retention of cohorts in small areas includes the effects of mortality but is much more a function of net migration. Thus, the retention rate for a cohort combines all demo-

---

[10]For such a cohort analysis, the two age groups must be spaced apart by the same number of years in age as the time between the two observations of the cohort. Thus, when comparing census data 10 years apart, we must compare age groups also 10 years apart. Observe also that the age groups at the beginning and end of the observation period must be the same size. For example, we can compare two 10-year age groups or two 5-year age groups. Finally, the age groups must perfectly enclose the cohort to be measured. A 5-year-wide age group cannot represent a 10-year-wide cohort. Nor can a 10-year-wide age group represent a 5-year-wide cohort; it actually would hold two different cohorts of that width.

graphic forces except fertility (which is zero for existing cohorts) and may provide a short-cut means for building population forecasts in small areas.[11]

Figure 5 portrays cohort retention as a set of *cohort trajectories* that represents cohorts' passage from one age group to another that is 10 years older and 10 years later in time. Note how steep the decline of teenagers entering adulthood is in this neighborhood. (This is the 39 percent retention noted at age 20 to 24 in Table III.) Conversely, there is an in-movement of persons between age 25 to 29 and 35 to 39 (162 percent retention), but after age 40 there is renewed departure. These losses are accounted for by some parents leaving the neighborhood after their children have been launched from the home and to a smaller degree by divorce or mortality. Discussion in Chapter 7 emphasized how the household size fell in this neighborhood due to losses in certain family relationship categories. The age changes of this chapter (Figs. 4 and 5) may be usefully compared with relationship changes shown in Chapter 7, Fig. 5. Also compare the cohort trajectories of racial change in Chapter 10 (Figs. 3–6).

3.3.3. Extended Analysis with Retention Ratios

Retention ratios have many additional analytical possibilities because they summarize better than other age methods the underlying behavior shaping age structure in a location. When comparing different locations, retention ratios can be usefully

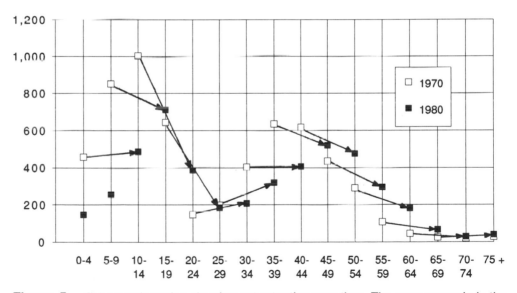

Figure 5. Cohort trajectories showing net retention over time. The case example is the same one used for Figs. 3 and 4.

[11]An early argument for this methodology is that of Hamilton and Perry (1962). For future projections, the assumption is that one cohort's retention ratio observed in X age group will apply to the next cohort that enters X age group. This assumption is violated if special conditions have inflated or deflated the retention ratios in one time period, such as when a sudden housing development adds 1,000 new units on the last vacant land in a community.

graphed and mapped for comparison, or statistically analyzed. A number of possibilities are identified, although not demonstrated here. When comparing only two or three locations, the entire retention profiles (the ratios for a series of cohorts) can be overlayed on one graph by plotting the retention ratios from data shown in Table III. Alternatively, a study might focus on retention of selected cohorts over the past decade, such as young children becoming teenagers, or young adults becoming early middle-aged. In that case, the single retention ratio pertaining to the desired cohort could be plotted for a great many places. A bar graph might be rank sorted from highest to lowest retention. A map of the retention rates would show areas of greater or less attraction to that age group, highlighting any spatial patterns.

The level of retention could also be related to other variables, such as housing prices, size of housing, or racial change. Relationships could be investigated through comparison of two maps or through a scatterplot design that places retention on the vertical axis and the other variable on the horizontal axis. Comparison of a number of such scatterplots may lead to significant insights as to which factors correlate most strongly with the cohort's retention. The scatterplot also has the advantage of showing which tracts retain cohorts more strongly than indicated by the average relationship with the other variable. (They lie above the trend line).[12] Multiple regression analysis could also be carried out with the identified variables to learn which factors are stronger overall. Residuals from the regression analysis would reveal which areas have higher (or lower) than expected retention for certain age groups, and these residuals might be mapped to see if there is a spatial pattern to this overattraction.

☐ 4. Adjusting Age Group Data for Analysis

Published data are not always available for smaller areas in a form required for the analyses described above. Two special tasks of data preparation may be necessary to prepare local census data for the age analyses described above. One involves handling different size age groups; the other involves adjustments of nonresidential population. We discuss each in turn.

4.1. Rearranging the Size of Age Categories

Inspection of the census data published for census tracts reveals that the age groups are reported in different sizes. Some ages are reported in single years, others in 5-year groups, and others in 10-year groups. (The use of mixed 5- and 10-year groups also occurs in computerized files where age is crosstabulated with other variables.) The methods we have presented all require age data distributed in age groups of an even size, preferably five years. The necessary 5-year age groupings from the 1990 census are published in the *General Population Characteristics* (Series CP-1) report for each state, covering individual counties and places down to a size of 2,500

---

[12]The trend line can be estimated through bivariate regression. Expected values can be estimated with this equation. Any tract whose actual retention exceeds that expected has a value lying above the trend line (a positive residual).

persons.[13] Those *places* are municipalities or other *census designated places*, not officially incorporated. Not included are census tracts or block numbering areas, which can only be found in the *Census Tract/BNA* reports (Series CPH-3). (For more explanation of these terms of census geography and the content of different report series, see Chapter 4).

For small areas like census tracts, the published data require reworking (or more detailed data can be accessed from the computerized summary tape files). Reworking the published format of the data is a relatively minor problem, but its solution is not transparent. Solving this simple problem has posed a barrier that, in the past, has prevented most census users from undertaking more useful age analyses in small areas.[14]

The secret to handling published age data is knowing how to convert the 10-year-wide categories, such as 35 to 44, into 5-year-wide categories, such as 35 to 39 and 40 to 44. Four different methods are suggested here, ranked in order of their ease of use. The more difficult methods are presented only because they yield generally more satisfactory results.

The easiest means of splitting 10-year age groups in half is to simply divide them into two equal parts. This is satisfactory as a stop-gap measure only. Any graphs will clearly show a staircase profile formed by the pairs of equal-sized age groups.

A second approach is much better and not much harder. We can interpolate the two halves of the ten-year age group by borrowing information from the bordering groups. In plain English, the first half is estimated as equaling half the 10-year group, plus 6.25 percent of the preceding 10-year group and minus 6.25 percent of the following 10-year group. The second half is estimated in complementary fashion: half the 10-year group, *minus* 6.25 percent of the preceding 10-year group and *plus* 6.25 percent of the following 10-year group. The two interpolated halves will sum to the original ten-year group. This is known as the Karup-King formula for dividing groups into halves.[15] It is generally judged to be quite accurate.

This interpolation method is demonstrated in Table IV. Given three age groups of equal size, G_1, G_2, and G_3, the middle group can be divided into two halves by applying the Karup-King multipliers to the three groups. In the example, utilizing 1980 data from San Fernando Valley Tract 1373.02, note how the preceding age group (25 to 34) is smaller than the 35 to 44 group to be subdivided, while the trailing age group (45 to 54) is larger. Assuming a continuous distribution between the three groups, it stands to reason that the first half of the middle group will be a little less than half of the total, while the second half will be on the high side. The 6.25 percent adjustment achieves that effect. This number is not arbitrarily selected but derives from the coefficients in a third-difference formula used for osculatory interpolation (Shyrock and Siegel 1976: 542). When the formula is expressed as a set of multipliers, it becomes simple to use.

---

[13]The source of five-year age data (by sex and race) for 1990 is state report CP-1, Tables 54 and 55 (for counties), Tables 61 and 62 (for places of 10,000 or more population), and Tables 68 and 69 (for places of 2,500 or more population).

[14]Access to summary tape files avoids this problem because age is reported there in many more categories. See note 1.

[15]See the discussion in Shyrock and Siegel (1976: Appendix C).

Table IV. Karup-King Method Used to Divide the 35 to 44 Age Group

	Age Group	Raw Data	A—First Half (35–39)		B—Second Half (40–44)	
			Multiplier	Part	Multiplier	Part
G₁	25–34	392	× +0.0625 =	24.5	× −0.0625 =	−24.5
G₂	35–44	726	× +0.5 =	363.0	× +0.5 =	363.0
G₃	45–54	995	× −0.0625 =	−62.2	× +0.0625 =	62.2
	Result	726	=	325.3	+	400.7

A third alternative improves on the Karup-King approach because it makes fuller use of the age detail published in the census. Karup-King does an especially poor job with the 25 to 34 age group. That group is preceded in the data by two *5-year* age groups, not the 10-year group called for in the procedure. If we simply add these two together we are destroying information on the slope of the distribution between the 15 to 19 and 20 to 24 age groups. Given the sharp changes in distribution that can occur from ages 15–19 to 25–29, it is important that we use as much information as possible about the shape of the curve. A method described in Shyrock and Siegel (1976: 543) is designed to make use of that information, combining the two five-year age groups with subsequent 10-year age groups.

Figure 6 illustrates the differences among these alternative procedures with data from San Fernando Valley Tract 1373.02 in 1980. After the last published 5-year age group at 20 to 24, the three methods diverge substantially. The stair-step pattern of the simple *divide in half* method is clearly visible. That method ignores the trend between age groups, such as the rising numbers in the middle ages and the declining numbers at older ages. In contrast, the Karup-King and Shyrock-Siegel methods fit a very similar upward and downward sloping curve through these older age group numbers.

As expected, the greatest difference among the three methods lies in the 25 to 29 and 30 to 34 age groups. The population profile in this neighborhood is among the most difficult to model because of the very sharp transition from a large teenage population down to a small group in their twenties and back to a large middle-aged population. By adding the 15 to 19 and 20 to 24 age groups together, the Karup-King method misses the downward trend started between 15 to 19 and 20 to 24. Thus, it estimates the bottom of the curve to be at age 30 to 34. In contrast, the Shyrock-Siegel method does incorporate the detailed information at 15 to 19 and 20 to 24, recognizing the downward trend more quickly, and so it estimates the bottom of the curve to be in the 25 to 29 age group. After the transitional 25 to 34 age range, the two methods track very closely together as they cross the stair-step estimates of the naive *divide in half* method.

Given the superiority of the Shyrock-Siegel method in the 25 to 34 age range, it is worth examining more closely. That method resembles an expanded Karup-King method because it uses a set of five multipliers instead of three. These are applied to the 10-year group being divided, the preceding two 5-year groups, and the trailing two 10-year groups. The coefficients are given in Table V where G₁ and G₂ refer to

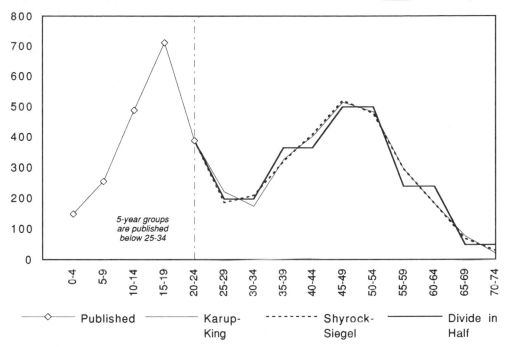

Figure 6. Comparison of alternative methods for subdividing 10-year age groups into two 5-year age groups. The example uses 1980 data for San Fernando Valley Tract 1373.02.

the two preceding 5-year groups, G_3 refers to the 10-year group being divided, and G_4 and G_5 refer to the trailing two 10-year age groups.

In equation form, the first half of G_3 is computed as

$$G_3A = -.0677G_1 +.2180G_2 +.4888G_3 -.0737G_4 +.0097G_5$$

and the second half,

$$G_3B = +.0677G_1 -.2180G_2 +.5112G_3 +.0737G_4 -.0097G_5$$

The Shyrock-Siegel method is most important to apply to the 25 to 34 age group, whereas the Karup-King method may suffice for older ages. To carry out the Shyrock-Siegel interpolation, analysts can arrange the data in a fashion similar to that in Table IV, using five age groups and five multipliers in place of the three shown there.

Table V. Shyrock-Siegel Coefficients

Resulting Interpolated Subgroup	Coefficients to be Applied to:				
	G_1	G_2	G_3	G_4	G_5
A—1st half of G_3	−0.0677	+0.2180	+0.4888	−0.0737	+0.0097
B—2nd half of G_3	+0.0677	−0.2180	+0.5112	+0.0737	−0.0097

The ability to rearrange categories of data is a general skill of considerable value to researchers. It may prove helpful to see exactly how the Shyrock-Siegel method may be applied to a series of age groups using a computer spreadsheet. Table VI illustrates the calculations, showing both the presentation of the spreadsheet and the underlying formulas. The coefficients could have been entered directly into the formulas. Instead, the panels of coefficients are listed separately and each coefficient is given a variable name that can be referenced as a multiplier in the formulas.

The fourth and final method for assembling age data in 5-year groups for tracts is to access it directly from more detailed tabulations stored on the summary tape files. This is the most accurate approach, but it generally is not practical unless one is working with many tracts at a time and has access to a mainframe computer. (An alternative is to order the data as a special tabulation produced by a state data center or another source—see Appendix C.) For most users the Shyrock-Siegel method is more expedient and sufficiently accurate.

4.2. Adjusting for Nonresidential Population

A second adjustment to standard published data may be required for use of the APH method. Despite its virtues, the APH method poses an extra data requirement because it is based on persons per household, by age. Persons not living in housing units (those who are inmates of institutions, in group quarters, or homeless) are excluded from this method. However, the published age tabulations for smaller geographies include both persons in households and the nonresidential population. The smallest geography for which separate age distributions of nonresidential population are published in 1990 is the state level.[16] Without recourse to computerized summary tape files, the local age numbers may require adjustment for nonresidential population prior to using the APH method.

Fortunately, the size of the nonresidential population is negligible in most small areas, although in a few areas it can be quite substantial. Overall, the residential population amounts to no more than 2 or 3 percent in most regions. Obviously, neighborhoods containing colleges, prisons, military bases, and nursing homes will bear unusually high proportions of nonresidential population.

There are four methods for handling the nonresidential population. First, in most neighborhoods, the nonresidential population can be safely ignored with little consequence because the numbers are so small, less than 1 percent. The number of persons who do not live in households is found in the data table that reports household relationships (see Chapter 7).

A second approach is to adjust the published age distributions for local areas by assuming that the larger area's age distribution of nonresidential population applies in each specific tract. For example, in 1980, in the Los Angeles–Long Beach PMSA, the nonresidential population was highly concentrated in two age brackets: 33.4 percent in age 15 to 24 and 35.4 percent in age 65 or older. The remainder is largely spread between ages 24 through 65. These percentages could be applied to the local

---

[16]In 1990, *General Population Characteristics*, CP-1, Table 29.

Table VI. Spreadsheet Layout for Interpolation

Panels of Interpolation Coefficients

A—First Half

Name	G1.A	G2.A	G3.A	G4.A	G5.A
Multiplier	-0.0677	0.2180	0.4888	-0.0737	0.0097

B—Second Half

Name	G1.B	G2.B	G3.B	G4.B	G5.B
Multiplier	0.0677	-0.2180	0.5112	0.0737	-0.0097

Age Group Data

Row/Column:	Group Half B	Raw Data C	Interpolated Results D	Interpolation Formulas in Column D
18	0–4		147	147 =C18
19	5–9		254	254 =C19
20	10–14		487	487 =C20
21	15–19		711	711 =C21
22	20–24		387	387 =C22
23	25–29	A	392	183.99 =D21*G1.A+D22*G2.A+C23*G3.A+C25*G4.A+C27*G5.A
24	30–34	B		208.01 =D21*G1.B+D22*G2.B+C23*G3.B+C25*G4.B+C27*G5.B
25	35–39	A	726	319.05 =D23*G1.A+D24*G2.A+C25*G3.A+C27*G4.A+C29*G5.A
26	40–44	B		406.95 =D23*G1.B+D24*G2.B+C25*G3.B+C27*G4.B+C29*G5.B
27	45–49	A	995	519.31 =D25*G1.A+D26*G2.A+C27*G3.A+C29*G4.A+C31*G5.A
28	50–54	B		475.69 =D25*G1.B+D26*G2.B+C27*G3.B+C29*G4.B+C31*G5.B
29	55–59	A	470	294.47 =D27*G1.A+D28*G2.A+C29*G3.A+C31*G4.A+C33*0.67*G5.A
30	60–64	B		181.53 =D27*G1.B+D28*G2.B+C29*G3.B+C31*G4.B+C33*0.67*G5.B
31	65–69	A	95	64.23 =D29*G1.A+D30*G2.A+C31*G3.A+(0.67*C33)*G4.A+(0.33*C33)*G5.A
32	70–74	B		30.77 =D29*G1.B+D30*G2.B+C31*G3.B+(0.67*C33)*G4.B+(0.33*C33)*G5.B
33	75+		40	40 =C33
34	Total		4710	4710 =SUM(D18:D33)

Notes:

1. Raw data for the 10-year age groups is placed in the first half of the bracket (A), leaving a blank space in the second half (B).
2. Data for the PRECEDING age groups are drawn from column D, while data for the PRESENT and FOLLOWING age groups are drawn from column C.
3. Division of the 65-74 age group requires assumptions about the open-ended category--75+--that follows. We assume here that 2/3 of the open-ended age group is 75-84 and 1/3 is older.

Table VII. Method for Estimating Household Population by Age in Small Areas

	Detailed Data for Larger Area				More Limited Data for the Small Area			
	Total Nonresidential Persons	Percent of All Ages Total	Percent of <65 Total	Percent of 65+ Total	All Persons	Nonresidential Under Age 65	Nonresidential Persons Age 65 and Older	Household Population Estimate
	1	2	3	4	5	6	7	8
Nonresidential Totals: Estimation Down Columns:						148	228	
						[formula A]	[formula B]	[formula C]
0-4	518	0.37	0.57	—	116	0.8	—	115
5-9	706	0.50	0.77	—	139	1.1	—	138
10-14	2,176	1.54	2.39	—	130	3.5	—	126
15-19	21,001	14.89	23.05	—	208	34.1	—	174
20-24	26,143	18.54	28.70	—	648	42.5	—	606
25-29	9,462	6.71	10.39	—	1086	15.4	—	1071
30-34	6,424	4.55	7.05	—	1010	10.4	—	999
35-39	4,798	3.40	5.27	—	580	7.8	—	572
40-44	3,669	2.60	4.03	—	412	6.0	—	406
45-49	3,599	2.55	3.95	—	393	5.8	—	387
50-54	3,876	2.75	4.25	—	414	6.3	—	408
55-59	4,565	3.24	5.01	—	403	7.4	—	395
60-64	4,164	2.95	4.57	—	439	6.8	—	433
65-69	5,065	3.59	—	10.14	548	—	23.1	524
70-74	6,400	4.54	—	12.82	506	—	29.2	477
75+	38,466	27.27	—	77.04	782	—	175.6	606
Total—all ages	141,032	100	100	100	7814	148	228	7438
Total—under 65	91,101							
Total—65+	49,931							

Formulas for Estimation Applied to Each Age Group:

A = Total Column 6 × Column 3/100

B = Total Column 7 × Column 4/100

C = Column 5 − Column 6 − Column 7

nonresidential population and then subtracted from their respective age groups in the total local population.

The accuracy of this estimation can be enhanced substantially by using more fully the information in the published tract data table on household relationships. That table itemizes *two* totals for nonresidential population, one for the total population and one for persons age 65 and older.[17] Thus, we can isolate the sizable group of elderly nonhousehold residents from the bulk of the population, and then estimate separate age distributions within the elderly and nonelderly age segments. Table VII illustrates this method of adjustment. The assumption is plainly that the local age profile for nonresidential population matches the larger area's. In column 8 the estimated nonresidential numbers are then subtracted from the total age distribution of the local area (column 5).[18]

A third approach is clearly preferable when the nonresidential population is of substantial size, say more than 5 percent of the local population. Local research can be performed to see what kind of institutions are contributing this population. County-wide age distributions may deliver a poor estimate if the nonresidential population is entirely associated with a university or entirely housed in a nursing home facility.

The final option is to access detailed tabulations stored in computerized summary tape files. These provide direct numbers of persons living in households by 5-year age category.[19] The best possible solution is probably a combination of this with the former, local research method.

Again, there may be little need to adjust for nonresidential population in most neighborhoods. Census data analysts should exercise judgment about which method is most cost-effective for their needs.

☐ 5. Conclusion

Analysis of age distributions is a vital part of most census analyses. Age data are available in rich detail for small areas and they can be readily analyzed for changes over time.

The reader will note that the term "median age" has not appeared in this chapter. It scarcely makes sense to reduce such rich variation to such a crude summary. Specific age brackets have very different meaning. Compare age 15 to 19 and age 20 to 24, or 55 to 59 and 65 to 69. Median age reduces all this variation, finding the midpoint between parents and children, or between elderly and young adults. This median does not reveal whether the age is the most common in the tract or merely *in*

---

[17]In 1990, CPH-3, Table 1, and, in 1980, PHC-2, Table P-1.

[18]The simple estimation method illustrated in Table VII adjusts the local area's totals in the under 65 and 65 and older age groups to the larger area's age distribution. With this method it is possible that more nonresidential population could be estimated in an age group than is present in the total population. For more accurate estimation, the local age distribution of all persons should be incorporated in the estimation. A more complex, two-dimensional scaling procedure useful for this purpose is introduced and demonstrated in Chapter 13.

[19]See note 2, this chapter.

between important concentrations. (In a later chapter, we approve the use of medians with income data because these are more smoothly distributed and because behavior does not shift as dramatically from one category to the next.)

This chapter has explored a number of methods, some with very different structure. We have contrasted two distributional strategies, the percent of total and the ratio per household. We have also contrasted analysis with age groups versus that with cohorts. These are only basic procedures but they may lead inventive analysts into some rich investigations.

9 □□□

Migration and Residential Mobility

The movement of population between geographic areas can be one of the most potent forces in local demographic change. The basic equation of population change includes migration as a key component:

$$P_t - P_o = B - D + I - O$$

where P_t is the population at the end of the period, P_o is the population at the beginning, B represents births, and D represents deaths. Migration is represented by I, for in-migration, and O for out-migration. The term $I—O$ defines net migration.

The migration factor is increasingly dominant as we move down the geographic hierarchy from larger to smaller places. At the neighborhood level, changes due to population movement far outweigh changes due to births and deaths. However, there are important differences in the causes of population movement that prevail at different geographic levels, and there are very different sources of data available at those levels.

□ 1. Broad Theory of Movement

The subject of migration is one of the most complex in demography because it involves transitions between locations, some of which are more permanent than others, and some of which are longer distance than others. This spatial and temporal complexity, combined with migration's importance for population change, has

189

given rise to an entire subfield known as population geography. In this section we offer only a brief orientation to a rich body of theory.

Population movement is normally conceived of as a relatively permanent change in residential location. Persons looking for work have long traveled to distant locales only to travel back after the work period is ended. More recently, in advanced western nations, many households have begun to maintain second homes, which are visited cyclically, whether on a seasonal basis or on weekends. It is not always clear which home is the main one. All during his years as vice president and president of the United States, George Bush maintained a legal residence (for voting and tax purposes) in a Houston hotel, but slept most nights in Washington, D.C. At the same time, President Bush seemed most permanently attached to a family vacation home frequently visited in Kennebunkport, Maine.

In addition to questions of cyclicality or permanence, there is the matter of distance. *Migration* entails a relatively long distance move; whereas *residential mobility* is a short distance change of residence. The dividing point between the two types of moves is not always clear-cut. In concept, migration is a change in place of residence that is so long distance that the individual's daily and weekly activity system is completely displaced.[1] For example, relocation from central Los Angeles County to neighboring Orange County would be a clear-cut case of migration, conceptually, if the residence change was accompanied by a new job in the new county and if the migrant's shopping and entertainment actitivites were also shifted near the new residence. However, in many cases relocation to a suburban county within the same metropolitan area is merely a residential relocation unaccompanied by job change or other shifts.

1.1. Conventions of Definition

Despite the complexity of movement behavior, we must settle on some simple conventions of definitions that can be easily and consistently applied to movement behavior for record-keeping purposes. These are briefly summarized here.

In terms of *distance*, migration is usually defined as a move that *crosses a county line*. Moves of longer distance, between states and nations, also involve crossing county lines and are also classified as migration. The county-line definition is useful because it also matches the spatial unit used in the record-keeping systems for births and deaths, the other components in the basic demographic equation. Moves *within the same county* are classified as local moves or residential mobility. An alternative distinction between migration and local mobility defines the local area as the *metropolitan area* instead of the county.

In terms of *permanence*, migration and residential mobility are defined in terms of *usual place of residence*. In practice, that definition depends on record-keeping systems from which the data are derived. For example, drivers license data detect migration when persons live long enough in a new location to apply for a license or change their address. The census detects migration through a two-part question:

---

[1]Among the leading proponents of this flexible view of migration as a spatial and temporal process is Roseman (1971, 1991).

"Did this person live in this house five years ago (April 1, 1985)?"

[and if no]

"Where did this person live five years ago (April 1, 1985)"

Blanks are then provided for writing in the state or nation, the county, and the city or town.

In the census, permanence is measured as a five-year interval. Observe, however, that a person could have moved several times during the interval, only the last of which would be recorded. In fact, a mover who left home for four years, returning prior to census day, would be recorded as a nonmover.

One final convention in practice is that migration is most often recorded by place of destination, i.e., as in-migration. Every migrant moves from an origin to a destination, and in theory we can measure both in- and out-migration. However, the departed movers cannot be interviewed at their place of origin. Because it is much more difficult to capture out-migrants and assign them back to their place of origin, that is less often done. An alternative convention is to infer net migration through statistical means, but no person is actually a *net migrant* and the denominator for a net migration rate is artifically constructed (Rogers 1990).

1.2. Correlates of Migration and Mobility

Age is the key variable explaining mobility status: Moves of all types are much more likely among young adults (or children of young adults). However, motivation for moves of the two different types is generally different. Migration is often dictated by the search for new employment, whereas residential mobility is directed by desires for a change of housing or neighborhood living environment. In the latter case, life-cycle changes in household membership often play a major role. Readers may wish to review the discussion of residential mobility and neighborhood change viewed in a housing context (Chapter 6).

The reader who wishes a thorough understanding of migration and mobility has available a wealth of theory and research. A useful overview is provided by Clark (1986), while comprehensive analysis of evidence for the United States is offered in Long (1988). An authoritative summary of general methodologies is Shyrock and Siegel (1976: Chapters 20 and 21). Pittenger (1976: chapter 8) offers explanation of migration analysis within a forecasting framework appropriate at the state and county level, whereas the best work focusing on calculation of net migration for areas as small as census tracts may be that of White (1984).

2. Local Data for Analysis

The mobility status of individuals and households can be tabulated in several different ways, depending on the level of geography to be emphasized and whether we wish to measure the movement of persons or the turnover of dwelling units. The variety of data sources is so rich that it may prove helpful to overview the options.

2.1. Local Tabulations from the Census

The census provides at least five different tabulations of mobility status, as summarized in Table I. Migration researchers are most familiar with data derived from the question asking *where the person lived five years ago* (cited above). Two tabulations in Table I are derived from this source. However, additional data pertaining to migration and mobility is contained in the census, some of which has special advantages. Questions are also asked about place of birth, affording opportunity to measure *lifetime migration* (lower left panel), or about the *year foreign-born persons immigrated* to the United States (lower middle panel). Finally, a housing question asks in what *year the householder moved into the unit* he or she presently occupies (upper middle panel).

Table I. Alternative Census Tabulations of Mobility Status, 1990

Basic Migration (State & County Level Residence 5 Years Ago*)	Occupancy Duration (Year Householder Moved Into Unit)	Metropolitan & Residential Mobility (MSA and House Level Residence 5 Years Ago*)

Persons in same house as in 1985
Different house in
  United States:
    Same county
    Different county:
      Same state
      Different state:
        Northeast
        North Central
        South
        West
Abroad
 	
STF 3: Table P43
	1989 to March 1990
	1985 to 1988
	1980 to 1984
	1970 to 1979
	1960 to 1969
	1959 or earlier
	
	tabulated separately for owners and renters
	
	STF 3: Table H29
		Persons living in an MSA in 1990:
		  Same house in 1985
		  Different house in U.S. in 1985:
		    Same MSA:
		      Central city of this MSA
		      Remainder of this MSA
		    Different MSA in 1985
		      Central city of MSA
		      Remainder of MSA
		    Not living in a MSA in 1985
		  Abroad in 1985
		
		Not living in an MSA in 1990:
		  Same house in 1985
		  Different house in U.S. in 1985:
		    Living in a MSA in 1985:
		      Central city of MSA
		      Remainder of MSA
		    Not living in a MSA in 1985
		  Abroad in 1985

Lifetime Migration (Nativity and Place of Birth)	**Year of Immigration** (Foreign-Born Persons Only)	

Native born persons:
  Born in state of residence
  Born in different state:
    Northeast
    North Central
    South
    West
  Born abroad, at sea, etc.
Foreign born
 	
STF 3: Table P42
	1987 to 1990
	1985 to 1986
	1982 to 1984
	1980 to 1981
	1975 to 1979
	1970 to 1974
	1965 to 1969
	1960 to 1964
	1950 to 1959
	Before 1950
	
	STF 3: Table P36
		STF 3: Table P44

Notes: MSA includes PMSA as well
*Residence 5 years ago only pertains to persons age 5 years and over

2.1.1. A Variety of Temporal Detail

The two tabulations based on residence in 1985 measure movements over a *fixed five-year* time period. There is no question in the census enabling us to measure directly how much movement has occurred over other time intervals, such as a decade (since the last census), or annually. Instead, greater variety of temporal measurement is afforded by the other three tabulations.

Lifetime migration has a variable time interval, depending on the age of person, but this is still a useful concept. It is always interesting to learn what percentage of a local population is native to the state, from elsewhere in the United States, or from abroad. The next alternative, year of immigration, provides richer temporal detail for foreign-born persons than is available for native-born migrants within the United States. For example, we cannot tell from the census how long native-born U.S. citizens have lived in California, but we can estimate this for foreign-born based on year of immigration. We might find that recent immigrants are concentrated in some communities, whereas longer time immigrants may live elsewhere.

Finally, year the householder moved into the unit provides richer temporal detail for residential mobility than is possible with the standard five-year measure. Crosstabulation is also provided by tenure (owner or renter status). As many as 40 percent of homeowners may have resided for over 20 years in their homes. Meanwhile renters can be cycling in and out of the same neighborhood on a more rapid basis. One drawback is that the householder's occupancy duration may differ from that of other household members. Roommates may come and go, or children leave home, but the householder data won't capture that mobility.

2.1.2. Published Versions of the Data

The sources listed in Table I are for tabulations listed in STF 3. The published census tract books provide versions of all the tabulations except year of immigration. First, in the population set of tables there is an abridged version of the metropolitan and residential mobility tabulation (CPH-3, Table 17). This is published in reports covering MSAs. One advantage of this tabulation is that potentially minor moves between counties within the same metropolitan area are not the prime criterion of migration. Unfortunately, however, the state and regional detail of the state- and county-level tabulation is not published. This information would be most interesting for states receiving lots of migrants from throughout the country. Instead, that tabulation is provided as a substitute for metropolitan mobility in published volumes covering geographic areas that are nonmetropolitan.

A second, highly abridged tabulation covers nativity and place of birth (CPH-3, Table 17). The census tract books report all categories, except those identifying in which state and region the person was born. The third tabulation available in the published books is in the housing data section. The tabulation of year householder moved into unit is reported exactly as shown in Table I, with separate detail for owners and renters (except that the two longest duration categories of residence are combined).

More detailed versions of the tabulations may be found in the CP-2 reports. Of particular value, those reports include data that are neglected by CPH-3: the year of immigration and region of migration.

2.2. Data on Gross Migration

All of the foregoing data pertain to tabulations of in-movers at the destination. A much richer perspective on migration is gained by tabulations covering out-movers as well. Migration and mobility place special emphasis on spatial relationships: movers must leave one geographic area and enter another. At higher levels of geography (with fewer spatial areas of accounting) it becomes feasible to record these movements in more detail.

2.2.1. Public Use Microdata

At a higher level of geography—large counties or states—migration data can be accessed in microdata files. (These were briefly introduced in Chapter 4 and will be discussed at length in Chapters 12 through 14.) *Public use microdata area (PUMA)* codes (termed *county group* codes in the 1980 census) are recorded for place of residence, place of work, and place of origin, if a mover since 1985. Thus, with a collection of microdata tapes from across the nation, it is possible to compare characteristics of in-movers and out-movers from an area. All of the extremely rich microdata capabilities can be applied in this migration analysis.

2.2.2. Special Migration Reports/Data Files

Custom tabulation of microdata in that fashion could prove quite demanding. An alternative is to make use of summary tabulations of gross migration flows prepared by the Census Bureau. Each decennial census yields special subject reports addressing migration at the regional, state, and county levels. One book published from the 1980 census records the in-migration and out-migration, by age group, for each county in the nation (U.S. Bureau of the Census 1984). Preparation of those tables required tabulating not only the new arrivals in the county, but also tabulating the out-migrants who lived in the county five years earlier but are now scattered all across the nation. The difference between the gross numbers of in- and out-migrants yields a direct estimate of net migration for every age group. Some of these data are analyzed later in the chapter.

Still more detailed tabulations of gross flows of migrants are available in a multivolume computer tape file. These tabulations break down the migrants to and from counties by such characteristics as education level, employment status, marital status, and race. This permits very detailed analysis of the composition of migration streams. When compared with similar data to be provided from the 1990 census, analysts can study how the nature of migration to a county has changed over time. These detailed tabulations are not available for subcounty areas.

2.2.3. Additional Noncensus Tabulations

Additional, noncensus sources of migration data are briefly noted, as certain data drawn from administrative records can be used to measure migration patterns for intercensal years. These data also allow measurement of gross migration in and out of an area. The Internal Revenue Service provides tabulations of annual migration from state to state in a 51×51 matrix (including the District of Columbia). By special arrangement, data can be tabulated for individual counties if desired. The processing of state or county estimates is time-consuming and the time lag before public release is about two years. The IRS estimates are prepared by comparing tax returns from one year with the next, tabulating the number of returns filed in different locations, and weighting these by the number of dependents. The IRS data can be used to track such trends as the reversal of migration from Michigan to Texas between the early and late 1980s.

Individual states maintain other administrative data files, often derived from drivers license address changes. These state-produced data involve much smaller computations and are produced with less time lag than the IRS data. The State of California Department of Finance utilizes such data in refining the migration components used in preparing population estimates and projections for counties. The department's Demographic Research Unit can provide users with matrices of annual county-to-county migration within California as well as two-way flows between California and other states.

☐ 3. Migration and Mobility in the Sample Neighborhoods

Two tables in the published census tract books provide data relating to the migration and mobility of population. A third addresses mobility of households. We analyze each of these tables in the following sections, assessing their respective results and comparing the different neighborhoods we have been examining.

3.1. Mobility Status of the Population

The most widely available and useful table relating to migration and mobility of the population reports the residence five years before the census (i.e., in 1985 or 1975) of persons aged five years and over. This follows the table outline given in the upper right panel of Table I (omitting and combining some of the categories). The table first reports how many persons have remained in the same house over the past five years (stayers) and how many are movers. The total mobility for the five-year period includes both local mobility and migration. Then the table details the origins of movers in a hierarchical fashion, moving from local mobility (same MSA, whether the central city or suburban portion), to internal U.S. migration (movers from outside the MSA), and then to international migration from abroad.

We can analyze these data in two stages. First, we compute a five-year mobility rate as the percentage of the census population that is living in a different house than five years earlier. Comparing this mobility rate across neighborhoods will show how much variation there is in population turnover. In a second stage of analysis we then compute the percentage distribution of origins for the movers to see what share of the movers are local, from other U.S. origins, or from abroad.

The results of these two calculations are combined in Table II. The table's header is designed to represent the hierarchical structure of the different categories. This design resembles versions employed in various reports from the Bureau of the Census. The first data column reports a mobility rate, describing movers into the area as a percent of all persons aged 5 and over. The second column then establishes the subset that are movers as the new denominator, emphasizing this with the placement of the 100 percent totals in that column. From there the table distributes the share of movers coming from different origins.

Table II. Mobility Rate and Origins of Movers Since 1975 (1980 Census)

			Percentage Distribution of Origins of Movers					
			Lived Elsewhere in United States in 1975					
					Outside Current MSA			
	Mobility Rate	All Movers	Total	Same MSA	Total	Diff. MSA	Other U.S.	Abroad
Total Population								
Los Angeles County	51.9	100	87.0	67.1	19.9	17.7	2.2	13.0
San Fernando Valley								
Tract 1373.02	34.5	100	82.9	53.0	29.9	27.3	2.5	17.1
Tract 1392	63.2	100	89.7	66.2	23.5	22.0	1.5	10.3
Hollywood								
Tract 1899	65.9	100	78.4	45.5	32.8	31.2	1.6	21.6
Tract 1942	48.8	100	94.5	60.5	34.0	30.5	3.5	5.5
Tract 7005	71.9	100	93.8	50.7	43.1	40.3	2.8	6.2
Beverly Hills								
Tract 7007	34.3	100	85.3	72.6	12.7	12.7	0.0	14.7
Tract 7009.02	60.5	100	87.9	60.0	27.9	27.0	0.9	12.1
Pasadena								
Tract 4602	48.9	100	93.1	73.2	19.9	18.7	1.2	6.9
Tract 4616	47.2	100	92.8	75.1	17.7	16.3	1.4	7.2
Tract 4620	61.8	100	84.1	59.2	24.9	21.4	3.5	15.9
Black Population								
Los Angeles County	47.5	100	98.0	76.5	21.5	19.1	2.4	2.0
Tract 4602	44.0	100	92.8	84.5	8.3	8.3	0.0	7.2
Tract 4616	43.1	100	100.0	78.0	22.0	20.9	1.1	0.0
Tract 4620	53.5	100	97.7	63.3	34.3	28.5	5.9	2.3
Hispanic Population								
Los Angeles County	57.6	100	77.1	68.9	8.2	7.4	0.8	22.9
Tract 4616	70.8	100	71.3	71.3	0.0	0.0	0.0	28.7
Tract 4620	82.3	100	60.2	50.7	9.5	9.5	0.0	39.8

Source: PHC 80-2-226, Table P-9.

The rows in Table II report the results for each of our example neighborhoods. At the bottom we have added detail for the black population and Hispanic population in selected neighborhoods, previewing cases to be developed in the following chapter on changing racial composition.

The mobility rate varies substantially across neighborhoods, ranging from a high of 71.9 percent in West Hollywood Tract 7005 down to 34.3 percent in the neighboring Beverly Hills Tract 7007. Inspection of housing data from previous chapters shows that the variation in mobility corresponds in large part to the percentage of the housing stock that is owner-occupied, although this is not the sole factor. Racial change also may contribute to population mobility, but the reverse causation is at least as strong: mobility is the vehicle for racial change. In fact, mobility rates in the Pasadena tracts with racial minorities lie both above and below the average for the county. Blacks show substantially lower mobility than the Hispanic population, but differences between the races in their rate of owner-occupancy are again important.

3.2. Origins of the Movers

From where do the movers to different neighborhoods come? The remainder of Table II focuses on movers' origins. Local origins are measured under the column labeled *Same MSA*. Beverly Hills Tract 7007 has a high share of movers, 72.6 percent, coming from the local area, but the Pasadena tracts, and black population in general, have even a higher share. Blacks moving in Pasadena Tract 4602, a middle-class owner-occupied neighborhood changing from white to black, are 84.5 percent from the local area.

In contrast, other neighborhoods draw many migrants from elsewhere in the United States, as measured in the column labeled *Outside Current MSA, Total*. The Hollywood tracts are especially high, with West Hollywood Tract 7005 leading with 43.1 percent of its movers coming from elsewhere in the United States.

Still other neighborhoods draw migrants from abroad. Table II shows a high share of all movers are from abroad in Hollywood Tract 1899 and especially in the Hispanic portions of the Pasadena tracts. Tract 4620 is undergoing rapid change from black to Hispanic occupancy and is analyzed in detail in a following chapter. An extremely high proportion (82.3 percent) of all Hispanic persons has moved in within the previous five years, and fully 39.8 percent of the movers are from abroad (Mexico and Central America). This computes to nearly one-third (32.8 percent) of the entire Hispanic population (age 5 and older) living outside the United States five years earlier.

3.3. Lifetime Migration

Further insight into movers' origins can be gained through the published census tract table reporting *Nativity and Place of Birth*. This follows the table outline given in the lower left panel of Table I (but omits the regions of persons born in different states). Unfortunately, details of state of birth are not published in the census tract books for blacks and Hispanics.

Table III reports results pertaining to the lifetime migration of current residents. With these data we can ask how high a proportion of the neighborhood residents are

Table III. Lifetime Migration of Population in 1980

Percentage Distribution	Total	Native-Born Citizens				Foreign Born
		Total	Born In:			
			Same State	Diff. State	Abroad/at sea	
Total Population						
Los Angeles County	100	77.7	40.6	36.3	0.9	22.3
San Fernando Valley						
Tract 1373.02	100	91.3	44.0	47.3	0.0	8.7
Tract 1392	100	82.3	32.6	48.4	1.2	17.7
Hollywood						
Tract 1899	100	64.3	14.0	49.6	0.7	35.7
Tract 1942	100	85.1	26.3	58.2	0.6	14.9
Tract 7005	100	79.0	19.2	58.6	1.2	21.0
Beverly Hills						
Tract 7007	100	75.9	33.6	41.4	0.9	24.1
Tract 7009.02	100	68.9	22.6	45.8	0.5	31.1
Pasadena						
Tract 4602	100	89.7	41.0	47.4	1.3	10.3
Tract 4616	100	89.8	45.5	44.0	0.3	10.2
Tract 4620	100	79.0	41.4	37.1	0.5	21.0
Black Population						
Los Angeles County	100	97.2	0.0	0.0	0.0	2.8
Tract 4602	100	89.7	0.0	0.0	0.0	10.3
Tract 4616	100	97.7	0.0	0.0	0.0	2.3
Tract 4620	100	97.8	0.0	0.0	0.0	2.2
Hispanic Population						
Los Angeles County	100	54.4	0.0	0.0	0.0	45.6
Tract 4616	100	47.5	0.0	0.0	0.0	52.5
Tract 4620	100	41.2	0.0	0.0	0.0	58.8

Source: PHC 80-2-226, Table P-9.

native Californians (the state of residence), how many are natives of another state, or how many are foreign born. Native Californians are a minority in this high migration state, amounting to only 40.6 percent of the county's population. In our sample, the percentage is highest in the upscale San Fernando Valley Tract 1373.02 and in the Pasadena tracts.

Native Californians are fewest in the Hollywood tracts. Instead, the proportion born in a different state is highest there. This finding is consistent with the pattern of migration observed above. Hollywood attracts movers from far away.

The foreign born population is greatest in the Hispanic population, amounting to over half in the Pasadena tracts. Note that this calculation divides the number of foreign born by persons of all ages, including young children who may have been born after their parents' arrival in the United States. Hence, the percent foreign born is likely to be even higher among the population of older ages. In fact, the high percent foreign born in Hollywood Tract 1899 might reflect the distant origins of the sizable elderly population living there.

In general, the problem with lifetime migration analysis is that it measures the cumulative effects of migration occurring in different decades, often long ago if the residents are old enough. It is not a good measure of recent migration trends. Place of residence five years before the census is better for that purpose.

3.4. Occupancy Duration of Householders

There are two perspectives on local mobility. One asks what percentage of persons have moved to a different housing unit from elsewhere in the local area. The other asks what percentage of local housing units have been recently occupied. The latter, housing unit perspective has advantages and disadvantages. The major disadvantage is that we do not know where the movers came from. We also do not know the number of *persons* living in newly occupied units. There is also the potential contamination that the year the householder moved into the unit may differ from the year other household members moved in.

However, the advantages of the housing unit perspective on mobility are considerable. As noted previously, the question about occupancy duration is not restricted to a fixed five-year interval; instead, we gain insight into how long the householder has resided in the unit. This permits direct measurement of continued occupancy from one census to the next (one decade) or between censuses separated by two decades.

A second major advantage is that the housing unit perspective permits cross-tabulation by housing unit characteristics, including the potent factor of housing tenure. Renters are far more likely to move than owners, and so our local understanding is greatly improved if we can calculate mobility rates separately for owners and renters. The housing unit perspective has the further advantage that it draws attention to another major factor underlying mobility into dwellings: new construction. More will be said about this later.

The basic data are reported in census tract books following the format shown in the upper middle panel of Table I. Year moved in is given separately for owner- and renter-occupied households. First, we can lump all households together to construct an overall five-year mobility rate that might be compared to that from the population data. Next, we can break out the separate durations for owners and renters, comparing the different neighborhoods in our sample.

The second column of Table IV depicts the overall five-year mobility rate of households in each neighborhood. The highest mobility is among the Hispanics in the Pasadena tracts, but the highest overall mobility for a tract is the West Hollywood Tract 7005 at 78.3 percent. The lowest is Beverly Hills Tract 7007 at 32.6 percent, followed by San Fernando Valley Tract 1373.02 at 38.8 percent. In general, when compared to Table II, we see that the distribution of housing mobility rates tracks around five percentage points above the distribution of population mobility rates. This may be because the households with fewer members tend to be more mobile, both because they are younger and renters. In two neighborhoods the population mobility is higher than the housing mobility, but these are unusual cases: the extremely wealthy Beverly Hills Tract 7007 and Pasadena Tract 4602 with its rapid racial change.

Table IV. Occupancy Duration of Householders (Year Moved In), by Tenure, in 1980.

Percentage Distribution	All Households		Owner-Occupied Units			Long-Term			Renter-Occupied Units			Long-Term
	Total	1975-80	Total	1975-80	1970-74	Total	1960-69	<1960	Total	1975-80	1970-74	<1970
Total Population												
Los Angeles County	100	57.4	100	37.3	18.3	44.4	23.4	21.0	100	76.4	13.5	10.1
San Fernando Valley												
Tract 1373.02	100	38.8	100	36.7	18.7	44.6	43.6	0.9	100	90.4	9.6	0.0
Tract 1392	100	71.3	100	50.0	13.5	36.5	14.8	21.7	100	87.5	8.6	3.9
Hollywood												
Tract 1899	100	69.1	100	57.1	8.7	34.2	8.4	25.8	100	70.0	16.5	13.4
Tract 1942	100	54.8	100	42.2	21.1	36.7	17.3	19.4	100	68.5	12.3	19.2
Tract 7005	100	78.3	100	58.1	12.4	29.4	17.7	11.8	100	83.4	11.0	5.6
Beverly Hills												
Tract 7007	100	32.6	100	32.0	18.1	49.9	26.2	23.7	100	41.0	0.0	59.0
Tract 7009.02	100	61.2	100	34.3	23.0	42.7	24.8	17.9	100	67.7	14.2	18.1
Pasadena												
Tract 4602	100	45.9	100	37.2	26.3	36.5	20.9	15.6	100	80.7	11.8	7.5
Tract 4616	100	50.1	100	28.9	11.3	59.8	18.4	41.3	100	66.7	24.1	9.1
Tract 4620	100	66.1	100	30.5	17.1	52.4	23.3	29.0	100	75.5	15.1	9.4
Black Population												
Los Angeles County	100	57.1	100	30.7	23.2	46.1	27.4	18.7	100	74.2	15.3	10.6
Tract 4602	100	NA	100	NA	NA	NA	NA	NA	100	NA	NA	NA
Tract 4616	100	48.4	100	24.0	12.4	63.6	18.7	44.8	100	64.7	24.8	10.5
Tract 4620	100	62.5	100	22.0	24.3	53.7	18.9	34.7	100	71.9	18.2	9.9
Hispanic Population												
Los Angeles County	100	66.8	100	44.5	19.7	35.8	21.3	14.5	100	79.7	13.1	7.2
Tract 4616	100	80.5	100	63.6	0.0	36.4	0.0	36.4	100	93.2	6.8	0.0
Tract 4620	100	79.9	100	61.3	0.0	38.7	20.0	18.7	100	84.3	5.3	10.4

Source: PHC 80-2-226, Table H-7.

Subsequent columns in Table IV report the five-year mobility rate of owners and renters (householders who have moved in during the five years preceding the census). How closely do these rates correspond to one another? Figure 1 graphs the mobility rates of owners and renters in each tract, superimposing on this the population mobility rate reported in Table II.

Mobility rates of renters are much higher than those for owners, but the variation between the tracts in owner and renter mobility has very little correspondence. The levels of owner or renter mobility also bear little relation to the overall population mobility. Overall mobility depends greatly on the mix between renters and owners, with tracts dominated by renters to the right of the graph. We also note that the mobility rate of owners seems elevated in the rightmost tracts. These are tracts with higher rates of construction prior to the census, particularly of condominiums and town houses, whose new owners moved in recently. For lack of a much larger, randomly selected sample, we will not pursue further explanation of the variation in mobility rates.

A final observation from Table IV: The data also report the number of long-term occupants, some for longer than two decades. This is especially prominent among owners. Compare the findings for the two San Fernando Valley tracts. Tract 1373.02 has a higher overall proportion of long-term occupants than Tract 1392 (44.6 versus 36.5 percent), but it has a much smaller fraction of its homes occupied for two decades or more (0.9 versus 21.7 percent). The explanation is that very few (6.1 percent) of the homes in Tract 1373.02 were built more than 20 years ago. In contrast, fully 44.3 percent of the housing units in Tract 1392 are of that vintage, making them eligible to hold occupants for two or more decades. We might expect the proportion of very long-term occupants to rise in Tract 1373.02 when its housing is 10 years older in the next census.

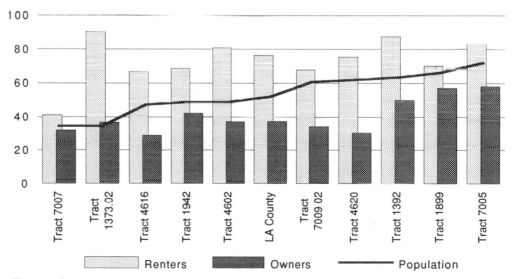

Figure 1. Comparison of the overall population mobility rate with owner- and renter-occupied mobility rates in selected areas for 1980.

3.5. Pitfalls of Construction and Dwelling Age

In general, new construction and the subsequent aging of housing units creates a fundamentally important context for mobility analysis. The housing unit perspective highlights factors that are overlooked by the population perspective on mobility. Areas with high rates of housing construction, whether for rent or for sale, will accrue larger numbers of movers. Yet new construction may yield only a temporary concentration of movers. A reduction in construction, for example, as vacant land is "built out," will lead to reduced in-movement, thus making hazardous any forecast based on the *prior* mobility rate. In principle, the local mobility rate should be adjusted for the temporary effects of new construction.[2] Conversely, the settling of large proportions of long-term stayers depends on the aging of the housing stock and sufficiently long exposures in the market. Analysts who overlook these factors of construction and age of housing may be deceived in their conclusions as a result.

☐ 4. Measuring Gross and Net Migration

All of the preceding analysis has been based on observations of in-movers. The out-movers have departed the local area and cannot be counted there. Net migration is logically the difference between in- and out-movements. However, a direct count of out-movers is available only for counties or higher geographic levels. Small-area analyses require indirect estimation of out-movers and of net migration.

4.1. County-Level Analysis

Gross and net migration can be computed directly for counties by tabulating all of the out-movers scattered across the country and comparing these to the number of locally observed in-movers. The fascinating conclusion to be drawn from data on gross flows in and out of counties is how large the volume is relative to the net change. Population turnover recorded five years apart is several times larger than the net change over the interval. But this is an understatement of the true turnover because many persons may have moved in and out annually during the five-year interval. The census question merely asks where persons lived five years earlier, not how many different places they may have lived during that interval.

The gross flows between 1975 and 1980, by age, are reported for every U.S. county in a special report based on the 1980 census.[3] Figure 2 shows the pattern of migration by age group for Los Angeles County. The most striking aspect of the graph is how large the out-migration is in every age group. In-migration is especially

---

[2]The census does not report a tabulation jointly of year dwellings were constructed and year householders moved in. That must be inferred from the numbers of new units produced in a period and the number of movers in a period. Data on the year built of owner- and renter-occupied housing is available in STF 4, Table HB3 (in 1980, STF 3, Table 109). This can be matched with the readily available data on year moved in for owners and renters. If the newly built units are subtracted from both the numerator and denominator, we can construct a local mobility rate adjusted for the effects of new construction.

[3]Source: U.S. Bureau of the Census, PC80-S1-17. A similar 1990 volume will also be released as a special report.

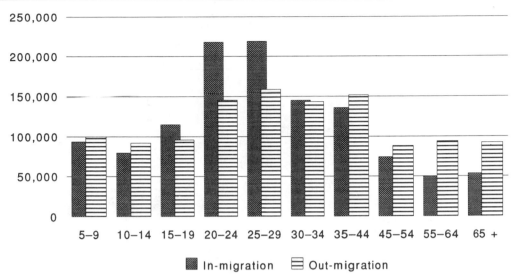

Figure 2. Migration in and out of Los Angeles County, 1975–1980. [*Source:* U.S. Bureau of the Census (1984).]

strong in ages 20 to 29 (age measured at the end of the migration period). The large surplus of in-movers over out-movers translates to a net in-migration of about 140,000 persons in this age range. Conversely, after age 35 out-movers exceed in-movers by a substantial proportion. Children aged 5 to 14 also register net out-migration from Los Angeles County. This pattern suggests that families with children, retirees, and pre-retirees are leaving Los Angeles County, on balance, for greener pastures. The only substantial growth is in the age range where young persons are seeking to establish their employment careers.

A second graph depicts the gross migration pattern for the bordering county of San Bernardino (Fig. 3). This is a rapidly growing area of housing tracts and, to a lesser degree, burgeoning employment centers. Migration into this county is also strongest in the 20 to 29 age range—the most migration prone years—but we also note the relatively strong in-migration at ages 30 to 44 (and the accompanying child age groups). This county has the cheapest housing in the Los Angeles area and is very attractive to families with children. In fact, there is a large net in-migration in every age bracket. Nevertheless, the volume of out-migration is still striking.

Analysis of gross migration flows emphasizes how large the volume is of population turnover. Businesses oriented toward growth have much larger markets than indicated by the net changes alone. Even with a constant population level, persons are buying and selling houses, renting apartments, buying furniture, using employment services, and all the rest. However, it may be that the in-movers at age 20 are not substitutable with the out-movers at age 50, creating a potential imbalance. For example, young persons are more often renters and unable to buy the out-movers' houses. A shift in the age pattern of migration could have important economic consequences over and above the effects of total population change.

Graphs similar to those shown here deserve to be prepared for every local area. Analysis by Long and Glick (1976) has shown that central city and suburban coun-

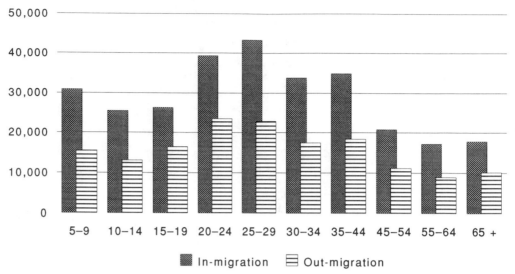

Figure 3. Migration in and out of San Bernardino County, 1975–1980. [*Source:* U.S. Bureau of the Census (1984).]

ties are often mirror images of one another.[4] Central city locations attract persons in their twenties, whether for jobs or the bright lights of entertainment, but lose persons after age thirty who seek locations they consider more conducive to family life. In contrast, suburban areas have tended to lose young adults and gain older adults with their children. This pattern may be breaking down as suburban areas gain more apartment housing and stronger employment concentrations. Comparison with the migration data to be reported from the 1990 census should prove very interesting in most local areas.

4.2. Small-Area Analysis

Tabulation of gross out-movers for census tracts is impractical. There are simply too many origins and destinations across the nation. Moreover, most out-movers do not know what census tract they were living in five years ago. Counties are better recognized.

We must infer out-movers and *net migration* in these small areas from the data available. Two local facts are known and one national fact. Locally, we know how many persons have moved into the census tract. And we know how much the total population has grown between censuses in each age group or cohort, specific for sex and race. Nationally, we know what fraction of each age-sex-race group has survived (not died) between one census and the next.

In barest outline, we can calculate net migration by applying national survival rates to the initial local population. This estimates how many persons in each age-sex-race group should be counted at the next census (assuming that local mortality rates for subgroups match the nation's). If we subtract this expected survived local

---

[4]See also the migration pattern typology developed by Pittenger (1976: 187-193).

number from the actual census count, we can estimate how many persons on net moved in or out of the area. Working backward, it is then a simple calculation to estimate gross out-migration (out = in − net). White (1984) provides an excellent, concise explanation of this research strategy.

Following this method we might construct migration profiles resembling those shown for counties in the previous section. However, White (1984) cautions that sampling error will cause substantial fluctuation if the analysis is carried out for each separate age-sex-race group. He advocates combining categories as needed, preferring first to collapse the sex distinction because their migration patterns are likely to be most similar. Next preferred is to combine bordering age groups. The concerns expressed above about the pitfalls of fluctuations in new construction also bear some emphasis.

The reader may recognize that the cohort retention ratio methodology of the preceding chapter provides an alternative perspective on net migration in census tracts. That method does not attempt to separate out mortality; rather it is combined as a relatively minor portion of the overall retention ratio expressing a cohort's (not age group) size in one census to its size in the preceding census when it was 10 years younger. For many purposes, the pattern of retention ratios for all cohorts serves the same function of describing neighborhood attraction as would knowledge of net migration by age group. Yet the retention ratios are much easier to calculate and can be linked to changes in the housing base more readily (see Chapter 8).

☐ 5. Conclusion

Migration and mobility are powerful forces shaping local areas. They operate most frequently through the younger members of the population. This chapter has revealed how impressive the volume is of gross movement. Specific portions of the metropolitan area are found to attract movers with different origins. We have also seen that in the smallest geographic areas mobility also depends on the nature of the housing stock—how many dwellings are newly constructed and what fraction of all units are renter-occupied. Still further lessons related to mobility are revealed in the racial change analysis that follows.

10 □□□

Racial Composition and Change

One of the most important characteristics of local populations is race and Hispanic origin. To many people this is the most visible and significant demographic factor present in cities. Race and Hispanic origin define major subgroups in the population, often differing markedly from neighborhood to neighborhood. The consequences of change in this composition may have great social, political, and legal significance. Fortunately, analysis of this subject benefits from extremely rich data, which in the 1990 census, surpass even the detail regarding age.

In this chapter, we often use the terms race or racial change to also include variables that define Hispanic origin. As described below, Hispanic origin is distinct from race, but the two topics are closely related in practice. Unless otherwise stated, for convenience we will refer to the entire set of race and Hispanic origin variables under the rubric of race.

The interpretation of race data is different from that of age, marital status, or household relationship data. Unlike those other demographic factors, racial characteristics do not lend themselves to behavioral interpretation. Being black or white does not imply particular behavior in the way that being a certain age or marital status does. Thus, we cannot say that a different racial group will act differently. Racial differences take on greater meaning when they are measured in terms of their associated characteristics: in terms of age, marital and household status, and the like.

Much greater insight into racial change can be gained by linking these changes to other demographic factors and to housing characteristics. In essence, census data allow us to carry out parallel demographic analyses for each separate racial subgroup, including Hispanics. These parallel demographic changes unfold within the context of a fixed geographic space and within a competitive housing market. Thus, we can trace how each group captures its share of housing and space, and then analyze the differences in composition between the groups.

☐ 1. Local Data for Analysis of Race and Hispanic Origin

Race and Hispanic origin are complete-count variables recorded by both the short- and long-form questionnaires. Free of any random sampling error, these data are as reliable in small areas as age, sex, and other complete-count variables. Data crosstabulations by race are extensive, more so than with any other variable, and this emphasis has been increasing each decade. In 1990, a maximum of 52 categories of race and Hispanic origin are reported in census files (STF 2, Table PA2), including 23 more Asian and Pacific Islander subcategories (such as Sri Lankan or Northern Mariana Islander) than shown in 1980. However, this extensive list is not included in most other census reports, nor is it ever crosstabulated with other variables (such as age or sex).

Table I lists the two most often reported sets of categories in 1980 or 1990 census products. The short list summarizes a set of five major groupings. The one change from 1980 is that the white category excludes Hispanics in order to eliminate the overlap that produced double counting in 1980 tables (see following discussion). The short list is used to repeat whole tables by race in the census tract publications of 1980 and 1990 (if 400 or more persons of that race are present). It also is used to crosstabulate a few other basic characteristics.

The long list is used more extensively in 1990 than in 1980, and is even used more often than the short list. As shown in Table I, both specific categories and subtotals for larger groupings are itemized, for a total of 33 entries, over six times larger than the short list. Four particular subgroups of Asians were added in 1990 to this longer list of racial groups: Cambodian, Hmong, Laotian, and Thai. The number of persons in each category of the long list is tabulated in published census tract books as Table P-7 in 1980 and as Table 8 in 1990. More significant, the long list is also used to crosstabulate most tables published in 1990 in the *General Population Characteristics* and *General Housing Characteristics* reports. These extensive crosstabulations are printed for areas as small as places with 2,500 or more population (but not census tracts or block numbering areas). Even more extensive crosstabulations are recorded in 1990 in the summary tape files (STF 2 and STF 4, *record Bs*), where many tables are reproduced multiple times, once for each racial subgroup in the long list. The 1990 census reports introduce an extraordinary amount of racial detail, often at the expense of age or sex detail that is crowded out. (See the discussion in Chapter 4.)

The repetition of tables by each separate racial category multiplies the volume of data until the data files become quite lengthy. The advantage for racial analysis is that this repetition permits detailed analysis, such as that demonstrated in preceding chapters for the total population, for every local racial group that has sufficient numbers to warrant recording in a table.[1] By studying change over time in each

---

[1] The threshold for reporting details on a racial group in a specified area was 400 cases both in 1980 and in 1990. For more discussion on the suppression of detail in small subgroups, see Chapter 4.

Table I. Race and Hispanic Origin Groups Specified in 1990 (with Changes from 1980)

Short List		Long List	
	All persons		*All persons*
1	Black	1	White
2	American Indian, Eskimo, or Aleut	2	Black
3	Asian or Pacific Islander	3	*American Indian, Eskimo, or Aleut*
4	Hispanic origin (of any race)	4	American Indian
5	White, not of Hispanic origin*	5	Eskimo
		6	Aleut
Notes:		7	*Asian or Pacific Islander*
	Italicized categories are totals or	8	*Asian****
	subtotals broken down into more	9	Chinese
	detail in categories that follow.	10	Filipino
		11	Japanese
	*In 1980, a "white" category was	12	Asian Indian
	also reported, but that did not	13	Korean
	exclude Hispanics.	14	Vietnamese
		15	Cambodian**
	**New category reported in 1990	16	Hmong**
		17	Laotian**
Source:		18	Thai**
	"Short list" contains categories	19	*Pacific Islander****
	used in brief tabulations by race,	20	Hawaiian
	and for replication of tables in	21	Samoan
	census tract books.	22	Guamanian
	"Long list" contains categories	23	Other race
	used in STF-2 (record B)		
	tabulations, and in extensive	24	*Hispanic origin (of any race)*
	crosstabulations for 1990	25	Mexican
	published reports CP-1, CP-2,	26	Puerto Rican
	CH-1, and CH-2.	27	Cuban
		28	Other Hispanic
	Not shown here is the full list of		
	52 categories reported in 1990,	29	White, not of Hispanic origin
	but never cross-tabulated, in	30	Black, not of Hispanic origin
	STF-2, table PA2.	31	American Indian, Eskimo, or Aleut
			not of Hispanic origin
		32	Asian or Pacific Islander
			not of Hispanic origin
		33	Other race, not of Hispanic origin

group's size and characteristics, we gain a better understanding of the changing racial makeup of local areas.

☐ 2. Problems in Defining Race and Hispanic Origin

The definition of race and Hispanic origin is less clear-cut and less stable over time than other variables we have studied. Determination of race in the census is through *self-identification*, as shaped by the category choices presented to the respondents. Both the questionnaire phrasing and data reporting have been growing better, but this creates some difficulty with continuity over time, and other problems remain.

2.1. Changing Terminology of Racial Categories

The terminology used in this book conforms to that used in the 1990 census reports. Those terms of racial and ethnic identity were selected in planning sessions prior to 1990 because they reflected the most commonly recognized terms at the time. Races are represented in the census in two ways—through category choices listed on the questionnaire and through categories listed in reports of census results.

The Census Bureau does not view its role as validating group identifications; rather, its goal is to generate as complete a count as possible—by making participation in the census attractive to minority groups who have been historically undercounted—and to use terminology describing racial categories that will elicit a well-understood response on the questionnaire. Also, the practical requirements for identification of groups in reports are such that only brief descriptors can be used, not strings of alternative identifiers. Coining of the summary term *Hispanic* reflects that need. Groups of widely diverse origins are covered under that label (see following discussion), none of whom may embrace it in particular.

Conventions have evolved so rapidly that different reporting terminology has been used in each census (Table II). Several different preferred terms may coexist in different regions, among different generations, or among different political constituencies.[2] Inevitably, not all current usages work their way into the census. For example, *Afro-American* was a term used by many blacks around 1970, but it did not come into vogue soon enough to gain widespread acceptance prior to the 1970 census, nor did it increase and sustain its acceptance leading into the 1980 census. Thus, that term never appeared on the questionnaire or in the reports. It is too soon to learn whether or not the term *African-American*, introduced into popular usage just prior to the 1990 census, will build sufficient recognition to warrant use in the year 2000 census. At the same time, older terms may continue in popular use. As late as 1990, older blacks, such as the great Supreme Court Justice Thurgood Marshall, an authority on race matters, still used the term *Negro* for describing their race. For this reason, the 1990 questionnaire retained Negro as an alternative in a category choice labeled *Black or Negro*, although census reports only identified the category as black.

Table II. Comparison of Census Reports and Alternative Terminology

Post-1990	African-American	Latino or Latin
1990 Census	Black	Hispanic
1980 Census	Black	Spanish/Hispanic
1970 Census	Negro	Spanish

[2]On the internal differentiation of self-identification among blacks, see Petersen (1987: 208–212). On internal differentiation among Hispanics, see Petersen (1987: 223–229) and Bean and Tienda (1987: Chapter 1).

2.2. Problems of Identifying Race

Beyond terminology alone, additional problems accompany racial identification. First, racial identification is often confused by its seeming overlap with national origin (e.g., persons from China or African-Americans), with language (e.g., persons who are Spanish-speaking), or with religion (e.g., persons who are Jewish or Muslim). The census disallows all three of these criteria for establishing race. Persons of Chinese race can come from any country; not all persons from China are Chinese. Also language and religion cannot be used to specify a race. For more on definitional issues, and the attendant politics, see Petersen (1987).

A second problem is that census respondents may not be able to match their racial identity to one of the standard categories provided in the census. This may be due to confusion by the respondent regarding the meaning of *race*, or increasingly because the respondent has parents from two different races and doesn't know which single category to choose.[3] The 1990 census questionnaire listed 14 categories, followed by a slot for *other race* where respondents could write in a different racial identity (see the questionnaire reproduced in Appendix A).

A further problem posed by multiracial families is that the household and housing data are tabulated according to the race of the householder. If other members of the household are not of the same race, this creates some ambiguity in the data. For example, as we will see in one of our examples, the average household size of black households may be different from the average persons per household of black persons. Some black householders have persons living with them of a different race, augmenting the household size of *black* households.

Thus, racial identity is blurred and ambiguous at two levels: Persons have two parents that may not be of the same race. Then households may have multiple members that are not all of the same race. Conventionally, we have assumed that these two problems are negligible, but they are growing in magnitude, particularly in the western United States and in the largest cities. Nevertheless, for analytical purposes, if we are to make use of tabulated data, we must assume that the potential error created by racial ambiguity is inconsequential.

2.3. Overlap of Racial and Hispanic Identification

A final problem in measuring racial identity has broad practical consequences. Above we emphasized that language and nationality cannot define a race. Yet the category termed *Hispanic* is treated similar to a race in popular usage and in government programs for the disadvantaged. However, in the 1980 and 1990 censuses Hispanic is not a racial category available to respondents; instead Hispanic is the subject of a separate question asking "Is this person of Spanish/Hispanic origin [1980 only: or descent]?" The response choices are:

---

[3]See Lieberson and Waters (1988: Chapter 6) on the growing incidence of intermarriage.

No;

Yes, Mexican, Mexican-Am., Chicano;

Yes, Puerto Rican;

Yes, Cuban; or

Yes, other Spanish/Hispanic.

We see here how language and nationality jointly define membership in a multifaceted category labeled *Hispanic*. Hispanic is a broad umbrella definition that may apply to none of the groups in particular but has the merit of favoring none of the major constituent groups (Bean and Tienda, 1987: Chapter 1).

A procedural problem is that Hispanic persons are double-counted in the data tabulations, appearing once in the racial tabulations and again in the tabulation of persons of Hispanic origin. The 1980 census for the first time provided a table to sort out this overlap in census tracts,[4] making available at the tract level data as detailed as that published at the state level. The data for Pasadena Tract 4620 (PHC-2, Table P-7) is reproduced here as Table III. The top panel lists the racial composition of the tract for all categories listed in the 1980 questionnaire. The bottom panel then lists persons of Spanish origin by category, breaking these out by race. The 1990 cen-

Table III.

Census Tracts	Pasadena city, Los Angeles County—Con.					
	Tract 4615	Tract 4616	Tract 4617	Tract 4619	Tract 4620	Tract 4621
RACE						
Total persons	8 775	4 477	1 592	5 268	5 983	4 954
White	3 459	552	985	2 310	1 016	2 193
Black	4 168	3 392	526	1 162	3 779	1 766
American Indian, Eskimo, and Aleut	45	19	10	30	17	40
American Indian	43	19	10	30	17	40
Eskimo	–	–	–	–	–	–
Aleut	2	–	–	–	–	–
Asian and Pacific Islander[1]	218	220	38	156	106	150
Japanese	56	80	27	25	29	25
Chinese	9	35	9	52	16	12
Filipino	114	70	1	29	23	74
Korean	8	4	1	15	–	6
Asian Indian	13	22	–	19	1	11
Vietnamese	10	5	–	8	31	20
Hawaiian	7	4	–	4	6	–
Guamanian	1	–	–	3	–	2
Samoan	–	–	–	1	–	–
Other	885	294	33	1 610	1 065	805
SPANISH ORIGIN[2]						
Persons of Spanish origin	2 030	585	64	2 667	1 761	1 811
Mexican	1 533	513	41	2 285	1 521	1 438
Puerto Rican	43	7	4	23	9	29
Cuban	21	5	4	9	4	24
Other Spanish	433	60	15	350	227	320
Persons of Spanish origin	2 030	585	64	2 667	1 761	1 811
White	1 090	257	34	1 030	648	975
Black	73	51	6	33	84	66
Other races	867	277	24	1 604	1 029	770

[4]The table is P-7 in the published census tract books, Series PHC-80-2.

sus tract data (CPH-3, Table 8) follows the same format, adding the starred categories shown previously in Table I and also itemizing the numbers of persons of each race that are not of Hispanic origin.[5] (Those numbers can be obtained from the 1980 table also, of course, through simple subtraction.)

In the example, Pasadena Tract 4620 has 5,983 persons of whom 1,016 are white, 3,779 are black, and 1,065 are other. The remaining categories only have small numbers reported, the largest of which is 31 Vietnamese. In the lower panel we find 1,761 persons are labeled as of Spanish origin, the vast majority of whom are Mexican. (This is typical from Texas to the west coast, even in the major midwestern cities. In Florida or parts of New Jersey the vast majority would be Cuban; in New York City, Puerto Rican.)

The key data describing the racial overlap with the Spanish origin or Hispanic category consist of the three rows at the bottom of Table III. Those show that 648 Hispanics are of white race, 84 are black, and 1,029 are of other race. These latter persons comprise the bulk of the *other* category recorded in the top, race section of the table. This illustrates the confusion with census categories of race. Apparently much of the Mexican-American population of this tract views their racial identity as different from the categories provided: not white and not Indian, but *Mexican*, which is not allowable as a racial category. Overall, even with the detailed data added in 1980 and expanded in 1990, accounting for nonoverlapped categories requires judgment, as detailed below.

2.3.1. Defining Nonoverlapped Racial and Hispanic Categories

Often census users need to define a set of demographic groups that includes Hispanic population along with racial minorities. This is required for affirmative action programs or other government and legal decisions. As published, the census data double count Hispanics among the racial subgroups. There is a clear need to develop a nonoverlapped set of racial and Hispanic categories. The easiest solution is to follow categories produced by the Census Bureau, but this is not the only option or necessarily the best.

Table IV presents a method for disentangling the overlap between racial and Spanish-origin (or Hispanic) data. The top panel shows the raw data drawn from Table III in columns A and C. Column E calculates the non-Hispanic portion of each race by subtracting the Hispanic numbers from the respective race categories. This subtraction makes a large difference in some categories. We see in column D that 86.6 percent of other races in this tract are Hispanic, as are 63.8 percent of white persons. Only 2.2 percent of blacks are Hispanic.

The bottom panel of Table IV rearranges the data into a set of nonoverlapped categories, such that the sum of all races and Hispanic group members equals the total number of persons in the tract. Two alternative schemes are shown here. The first makes the Hispanic definition dominant in the categorization, following the practice of the Census Bureau. The Hispanic total is preserved as reported in the census

[5]Series CPH-3, Table 8.

Table IV. Overlap of Race and Hispanic Origin
Pasadena Tract 4620: 1980

	A	B	C	D	E
				(=C/A×100)	(=A−C)
	Total	% of total	Hispanic	% Hispanic	Non-Hispanic
Total Persons	5983	100	1761	29.43	4222
White	1016	16.98	648	63.78	368
Black	3779	63.16	84	2.22	3695
Other	1188	19.86	1029	86.62	159
Total Races	5983				
Total Hispanic	1761				

	Alternative Definitions of Racial/Ethnic Composition				
	Hispanic Dominant (1)		Black Dominant (2)		
	Number	% of total	Number	% of total	
Total Persons	5983	100	5983	100	
Non-Hispanic white	368	6.15	368	6.15	
Black	3695	61.76	3779	63.16	
Other races	159	2.66	159	2.66	
Hispanic	1761	29.43	1677	28.03	
Total Races and Hispanic	5983	100	5983	100	

Notes: (1) Hispanic number is retained as given in raw data; Hispanic numbers are subtracted from all race categories.
(2) Black number is retained as given in raw data; black numbers are subtracted from Hispanic; Hispanic numbers are subtracted from all race categories except black.
Source: PHC80-2-226, Table P-7

(1,761 persons), and the other races are reduced in size by subtracting the Hispanic numbers (column E in the top panel). The numbers in all categories sum to the total for the tract—5,983. This definition implicitly states that Hispanic status is more important than racial status, because those numbers are maximized while the racial subtotals are minimized through subtraction.

An alternative definition focuses on the overlap problem in a different manner. The heart of the problem is that Hispanics are double counted in two particular categories—white and *other* races. We wish to separate out these numbers, finding the number of non-Hispanic whites (sometimes termed *Anglos*) and finding the number of Hispanics who resorted to *other* as their racial identity. But we do not necessarily wish to subordinate the black race to the Hispanic definition. For historical and legal reasons, blacks or African-Americans have a disadvantaged minority status that is at least as significant as that of Hispanics. Therefore, the alternative, black-dominant definition retains the full number of black residents reported in the data, removing the black number from the Hispanic category where they overlap (84 persons in this tract). The Hispanic numbers are then subtracted from all other racial categories *except black*. (In a similar fashion, an Indian-dominant definition could be fashioned that gives Native Americans quantitative emphasis over Hispanics.)

The Hispanic-dominant definition is most convenient to use, but alternative definitions are easily constructed, either using methods shown in Table IV or using 1990 data published in Table 8 of report series CPH-3. Analysts should make a conscious and informed choice about which method of separation—black-dominant or Hispanic-dominant—is more appropriate for their purposes. In the present case, the practical differences between the two methods are relatively minor because there is so little overlap between black and Hispanic in this tract. Greater overlap will be found among Puerto Rican or Cuban populations in the eastern United States. The black-dominant definition slightly increases the black percentage of the population and slightly reduces the Hispanic percentage. Non-Hispanic whites and other races remain the same under either method.

2.3.2. Comparability Over Time of Racial and Hispanic Tabulations

Attention to definitions of data tabulations prior to the 1990 census is essential if an analyst wishes to measure change over time. The 1980 and 1990 censuses handle race and Hispanic definitions in almost identical fashion, although more extensive tabulations by race and Hispanic origin are available for 1990. Of special importance, the identification of a white, non-Hispanic category in the short list facilitates analysis with 1990 data.

Unfortunately, changes from the 1970 census are much more of a problem. Greater change probably has occurred over time in the definition of Hispanic than any other variable in the census. The appendix to the 1980 census tract books warn that: "The 1980 data on Spanish origin are not directly comparable with those of 1970 because of several factors. . . ." (U.S. Bureau of the Census 1983: PHC80-2, B-15). Some procedural changes in 1980 yielded higher counts of Hispanics through better coverage. Other changes in the questionnaire and coding procedures yielded a slightly different definition of Hispanic, and the overlap with racial categories was much different.

In 1970, separate tables of data for Hispanics were produced for persons of Spanish mother tongue, combined with other persons designated as Hispanic. In five southwestern states (Arizona, California, Colorado, New Mexico, and Texas) these other persons were those with Spanish surnames. In New York, New Jersey, and Pennsylvania these additional persons were identified as those of Puerto Rican birth and parentage. Thus, inconsistent definitions were used across the country.

Race data were also affected by the differential handling of the Hispanic origin population in 1970 and 1980. A much larger portion of the Hispanic population identified themselves as *other races* in 1980. Back in 1970, these persons were more likely to be coded as white population. Nationally, only 1 percent of Spanish origin population were classified as *other* race in 1970, but this grew to 38 percent in 1980. Thus, the numbers of white persons were inflated in 1970 and the *other* races were deflated. The numbers of Negro (1970) and black (1980) population remained relatively unaffected by the handling of Spanish origin population. At this writing it appears that the 1990 data closely resemble the reporting patterns of 1980.

The improvements in the 1980 questionnaire were a big step forward. Also the publication of Table P-7 (in 1980) and Table 8 (in 1990) for every tract is essential for sorting out the overlap of race and Hispanic origin. The consistency between the 1980 and 1990 censuses is to be welcomed. For purposes of illustrating methods in this book, we will use the 1970 data as if there is no significant difference with the 1980 definitions. However, analysts should exercise caution if they wish to compare changes involving Hispanics or the white population between 1970 and 1980.

☐ 3. Racial Concentration and Change in the Sample Neighborhoods

For work with censuses prior to 1990, we must define a method for separating the overlap of racial and Hispanic categories in the crosstabulations. Knowing from 1980 data that the overlap is relatively small in our case example neighborhoods, we can address three groups: all blacks, all Hispanics, and a residual category of all other persons.[6] This *other* category is predominantly made up of white population, although in some other locations Asians or American Indians could predominate in the *other* category.[7]

Racial patterns in our sample neighborhoods are summarized in Table V. This table reports the percent black and percent Hispanic of the population in both 1970 and 1980. In addition, the table gives the percent of households headed by black or Hispanic persons. In Los Angeles County the percent of population that is black grew from 10.8 percent to 12.6 percent, while the percent Hispanic grew from 18.3 percent to 27.6 percent.

Different results are found for the racial composition of households. In the county, the black population percentage is very similar to the percent of households that are headed by black persons. However, the Hispanic share of households (19.8 percent) is substantially lower than the Hispanic share of population. This is because Hispanic household sizes are greater than the rest of the population. The consequence is that a small shift in the number of housing units occupied by Hispanics is magnified into a larger shift of population. As we will see, in some neighborhoods black population may also be "magnified" relative to the number of occupied housing units.

The numbers of black and Hispanic residents are low and show little change in all neighborhoods except the three Pasadena tracts. We will focus on those three tracts for the remainder of the chapter. The percent black has fallen in Tracts 4616 and 4620, while the percent Hispanic has grown. Tract 4620 is especially noteworthy, because the Hispanic percentage has nearly tripled from 10.9 percent to 29.4 percent. Note, however, that the percent of units occupied by Hispanics is much lower, only

---

[6]This "other" category is formed by subtracting blacks and Hispanics from the total population. To the extent that blacks and Hispanics overlap in identity, they are double counted, leaving an underestimate of the size of the "other" category.

[7]A new category of white, non-Hispanics is presented in 1990 crosstabulations. Any table reported for blacks or Hispanics is now also reported for the new category instead of for the entire white population as in the past.

Table V. Summary of Racial and Hispanic Composition

		Population		Households	
		1970	1980	1970	1980
Los Angeles County	% Black	10.8	12.6	9.9	12.2
	% Hispanic	18.3	27.6	—	19.8
1373.02	% Black	0.2	0.8	0.1	0.9
	% Hispanic	7.6	2.5	—	1.5
1392	% Black	0.1	1.2	0.2	1.3
	% Hispanic	7.0	11.1	—	7.8
1899	% Black	0.9	3.8	0.8	3.3
	% Hispanic	6.6	6.1	—	4.4
1942	% Black	2.1	2.4	1.7	2.2
	% Hispanic	5.8	3.1	—	2.3
7005	% Black	2.2	5.5	2.2	5.6
	% Hispanic	7.1	6.0	—	4.6
7007	% Black	2.9	2.4	0.3	1.1
	% Hispanic	9.9	7.4	—	1.5
7009.02	% Black	0.3	1.6	0.2	1.3
	% Hispanic	1.7	3.3	—	2.6
4616	% Black	83.0	75.8	83.5	79.3
	% Hispanic	8.7	13.1	—	8.8
4620	% Black	74.2	63.2	73.7	71.3
	% Hispanic	10.9	29.4	—	20.2
4602	% Black	29.0	57.4	23.8	50.2
	% Hispanic	7.6	6.6	—	4.9

Source: PHC(1)-117, Table P-1, and PHC 80-2-226, Tables P-7 and H-1.

20.2 percent. In fact, the percent of units occupied by blacks has hardly changed over the decade. Blacks have held their share of the housing stock but Hispanics are outgrowing them in population.

In the more suburban, middle-income Tract 4602, Hispanics have not grown in number, but the black population has doubled its share of the total from 29.0 percent to 57.4 percent. Note that blacks' share of occupied units remains at a lower level than their share of population. As we will see, this indicates that black household sizes are greater in this tract than the *other*, largely white, non-Hispanic population. It is also likely that the change in black population in certain age groups exceeds the pace of household change by an even greater margin.

3.1. Spatial Patterns in the Pasadena Area

Racial change is one phenomenon where spatial patterns are extremely significant. Racial concentrations, particularly of black population, tend to expand contiguously around existing pockets, not spreading evenly or "leap frogging" randomly around a region. Therefore, the existing pattern of concentration has powerful influence on the pattern of change over the next decade. Racial patterns can be mapped at one of two levels: census tracts and blocks. In this section, we will illustrate this mode of analysis with census tract data for blacks in the Pasadena area.

A series of maps can summarize a lot of data very succinctly. Repetition of the same map layout, but with different data, is a technique of *small multiples* that Tufte (1990) recommends for envisioning and communicating information. Figure 1 presents four maps of the Pasadena area, including South Pasadena and the wealthy community of San Marino bordering on the south.

Map A depicts a spatial pattern of homeownership in 1980 that resembles a doughnut pattern. There is very low homeownership (and a consequent high degree of renting) at the center of Pasadena, surrounded by a ring of high homeownership neighborhoods. Map B shows a somewhat similar pattern in terms of median household income in 1980. A large area of relatively low income occupies the center of Pasadena and stretches to the north and west. Note the concentration of very high income neighborhoods on the west side of Pasadena and in San Marino to the south. Of particular interest, the high homeownership neighborhoods to the north and east have relatively modest income levels.

Against this housing and economic backdrop, maps C and D then display the pattern of black occupancy in 1970 and 1980. The black population lies largely in the northwestern quadrant, centered just north of the center of Pasadena and just east of the Rose Bowl. The north-south line limiting the black population in 1970 is Lake Avenue.

By 1980 the black population has begun to spill across Lake Avenue and has increased its concentration, particularly in the northernmost neighborhoods. In those neighborhoods, blacks are replacing the white population. Conversely, in the southernmost part of the black area, that nearest to central Pasadena, the percentage black began to fall slightly by 1980. In those neighborhoods, blacks are beginning to be replaced by Hispanics. Data from the 1990 census will provide yet another snapshot of this process of racial change, and we would expect the in-movement of Hispanics to accelerate.

This broad spatial overview is useful for providing context for more detailed analysis of specific census tracts. Tract 4602 is an area of middle-class homeowners changing from white to black located at the top of the map, just west of Lake Avenue. We can contrast this area with Tract 4620, a poorer area of early black concentration that is beginning to incorporate Hispanics. That tract is visible as the notch-like, eastern protrusion of high black concentration. In contrast to middle-class Tract 4602, this area is less than 40 percent homeowners.

3.2. Housing Occupancy Trends

To learn more about the course of racial change we can study what types of housing are being acquired, or vacated, by each group. Following that we may ask what types of people, within each racial subgroup, are occupying those units.

In principle, an entire housing analysis may be carried out for each racial subgroup, following procedures outlined in Chapter 6. However, the analysis of multiple groups places emphasis on a scaled-down investigation of housing factors. A further consideration is that the smaller size of racial subgroups within a tract may not sup-

Figure 1. Pasadena-area census tract patterns of homeownership, income, and percentage black population. The location of the Rose Bowl is indicated by a small circle visible in two of the maps.

Map A

1980 % Owners
- 0% to 20%
- 20% to 40%
- 40% to 60%
- 50% to 80%
- 80% to 100%

Map B

1980 Median Household Income
- $8702 to $16877
- $16877 to $25053
- $25053 to $33228
- $33228 to $41404
- $41404 to $49579

Map C

1980 % Black
- 0% to 5%
- 5% to 20%
- 20% to 40%
- 40% to 60%
- 60% to 100%

Map D

1970 % Black
- 0% to 5%
- 5% to 20%
- 20% to 40%
- 40% to 60%
- 60% to 100%

port as detailed an analysis. Data also are not easily accessible in as full a form for the racial subgroups as for the whole tract. Our data to be analyzed are derived directly from published census tract books for 1970 and 1980, but even within those limitations a remarkable amount of detail and insight can be generated.

3.2.1. Working with Data Limitations

Prior to 1990 there was no table of housing data reported for the white, non-Hispanic population. Instead, we must subtract the tables for blacks and Hispanics from the total tract table, yielding an *other* (i.e., non-black and non-Hispanic) table that is a residual. However, this method yields incomplete housing data for the *other* racial subgroup. If applied to the entire housing stock, and not just occupied units, this method will place vacant units in the other category as well. In view of the small size of the *other* racial subgroup in some tracts, this contamination by vacant units is unacceptably large.

In addition, the data items are drawn from different tables, some reporting data collected from the complete count and some from the sample questionnaire. The totals from the two different sources are not in perfect agreement. As a practical procedure, data items from the sample questionnaire (or the complete count) should be percentaged against totals drawn from the same source. When items from the two sources must be combined, the complete-count total is preferred over that reported from the sample data. Both sets of totals are recorded in Table VI.

3.2.2. Occupancy Trends in Tract 4620

Table VI presents a summary of housing changes for racial subgroups in Pasadena Tract 4620. The whole census tract has experienced an increase of 209 (complete count) households over the decade. All of this increase has been among renters: owners declined by 3 households. Among blacks, 18 out of the 107 household increase were owners. Among Hispanics, 27 out of the 223 household increase were owners. The *other* racial subgroup (composed largely of non-Hispanic whites) experienced a loss of 48 owners and a loss of 100 renters.

The rapid growth of Hispanics in this neighborhood is of greatest interest. From the housing data in Table VI, we learn that Hispanics are living in small units—median size of 3.2 rooms. In contrast, black households live in units with a median of 4.0 rooms. This size differential is quite significant in view of Hispanics' substantially larger household sizes (discussed in the next section).

Data on persons per room reflect a substantial increase in crowding. In 1970, only 10.2 percent of households in the whole tract were living with more than 1.0 person per room. By 1980, this had increased to 22.1 percent. An additional 253 households were overcrowded in 1980. Of this number, an additional 53 black households were overcrowded, and an additional 221 Hispanic households were overcrowded, while the "other" population group recorded a decrease of 21 overcrowded households. Clearly, most of the increase in overcrowding is contributed by Hispanic households. Their likelihood of overcrowding[8] increased from 7.1 percent in 1970 to

---

[8]"Likelihood" (also known as "incidence") is normally defined as households in a status divided by all households at risk for that status. In 1970, 10 overcrowded Hispanic households are divided by 141 total Hispanic households for an incidence of overcrowding of 7.1 percent.

Table VI. Changes in Housing by Race: Example of Black to Hispanic
Pasadena Area Tract 4620

	Whole Census Tract			Black Occupied			Hispanic Occupied			Other Occupied		
	1970	1980	Change	1970	1980	Change	1970	1980	Change	1970	1980	Change
100% Count Totals												
Year-round units	1932	2066	134									
Total occupied units	1731	1940	209	1276	1383	107	141	391	250	314	166	−148
Owners	406	403	−3	241	259	18	54	81	27	111	63	−48
Renters	1325	1537	212	1035	1124	89	87	310	223	203	103	−100
Sample Count Totals												
Year-round units	1932	2066	134									
Total occupied units	1738	1940	202	1282	1383	101	141	393	252	315	164	−151
Structure Type												
1 (det + att)	963	789	−174	618	514	−104	63	173	110	NA	NA	NA
2 to 4	245	345	100	164	212	48	29	85	56	NA	NA	NA
5+ units	724	932	208	500	657	157	49	135	86	NA	NA	NA
Mobile home	NA	0	NA	NA	0	NA	NA	0	NA	NA	NA	NA
Total	1932	2066	134	1282	1383	101	141	393	252	NA	NA	NA
Year Structure Built												
Built last 10 years	513	262	−251	426	230	−196	16	12	−4	NA	NA	NA
Vintage:												
1970s	NA	262	262	NA	230	230	NA	12	12	NA	NA	NA
1960s	513	402	−111	426	297	−129	16	60	44	NA	NA	NA
1950s	342	362	20	244	235	−9	22	77	55	NA	NA	NA
pre-1950	1077	1040	−37	612	621	9	103	244	141	NA	NA	NA
Total	1932	2066	134	1282	1383	101	141	393	252	NA	NA	NA
Size of Housing												
Median rooms	3.9	3.8	−0.1	3.9	4	0.1	NA	3.2	NA	NA	NA	NA
Persons per Room												
1.00 or less	1555	1511	−44	1140	1194	54	131	160	29	284	157	−127
1.01 to 1.49	123	222	99	96	123	27	10	93	83	17	6	−11
1.50 or more	53	207	154	40	66	26	0	138	138	13	3	−10
% overcrowded	10.2	22.1	11.9	10.7	13.7	3.0	7.1	59.1	52.0	9.6	5.4	−4.1
Housing Costs												
Median value	$16,700	$57,100		$16,800	$56,600		$15,200	$57,500		NA	NA	NA
% of county average	69	65	−3	69	65	−4	63	66	3	NA	NA	NA
Median contract rent	$91	$189		$93	$186		$87	$190		NA	NA	NA
% of county average	83	77	−5	85	76	−8	79	78	−1	NA	NA	NA

Note: NA indicates data not obtainable from published census tract reports.
Source: PHC(1)-117, Tables H-1 through H-5, and FHC 80-2-226, Tables H-1, H-3, H-6, H-7, H-11, and H-13.

59.0 percent in 1980. Viewed another way, fully 88.4 percent of the total additional Hispanic households were living in overcrowded conditions.

One surprise is that Hispanics are more likely to live in single-family structures than blacks or the whole tract population. While blacks experienced a decreased occupancy of 104 single-family units, Hispanics *increased* their single-family occupancy by 110 units. This high degree of single-family living is unexpected in light of three facts already observed about Hispanics: (1) a low degree of homeownership, (2) a very small average number of rooms per unit, and (3) a very high degree of crowding. The surprising level of single-family occupancy might be explained if Hispanics are occupying small detached (or attached) structures that were formerly garages, sheds, or ground-floor portions of other structures. This kind of creative housing conversion has become very common in parts of the Los Angeles region.

Data on housing vintage provides further insight into structure characteristics. Very few Hispanic households claim to be in new housing, and a remarkably high proportion identify their housing as pre-1950 vintage. Fully 56.0 percent of all the Hispanic growth is in the oldest vintage; whereas, the whole tract showed an actual decline in the oldest vintage and black households showed a growth of only nine units (9 percent of black household growth).

The final evidence to be drawn from Table VI concerns housing costs. In the tract as a whole, median values of owner-occupied homes and median contract rents are both very low relative to the county average. Both have also been declining slightly in relative terms. There is little appreciable difference between the racial subgroups. On average, all pay approximately the same freight to live in the neighborhood (although the Hispanics seem to have less housing to show for it).

In sum, the housing stock in Pasadena Tract 4620 is shifting slowly to Hispanic occupancy. Those households are entering the older housing units, many of which are very small, even if they are single-family structures. With their large household sizes, the Hispanic households experience much higher levels of crowding per room, contributing to a doubling of the neighborhood's overall crowding rate over the decade.

3.2.4. Occupancy Trends in Tract 4602

A very different pattern of occupancy change is observed in Tract 4602, a neighborhood attracting increased numbers of black households and very few Hispanics. Here the *other* population is white, non-Hispanic. The basic data on housing changes in this area are presented in Table VII.

The neighborhood is predominantly single-family housing, and the great majority of residents are homeowners. Housing cost data at the bottom of the table show that the area's home values were almost equal to the county average in 1970, but these fell relatively by 8 percentage points over the decade. Median rents fell by an even greater amount. Some might conclude that racial change is lowering property values relative to the sharp upward trend in the rest of the county. However, we note that in both 1970 and 1980 the black households paid *higher* housing costs than the other population residing there. The lagging prices are more likely due to the fact that this part of Los Angeles county is relatively remote from the booming areas on the west side and closer to the coast. A further inhibitor to housing prices could be the widespread publicity given to unhealthy air conditions from smog in the Pasadena area.

Table VII. Changes in Housing by Race. Example of White to Black

Pasadena Area Tract 4602

	White Census Tract			Black Occupied			Other Occupied (a)		
	1970	1980	Change	1970	1980	Change	1970	1980	Change
100% Count Totals									
Year-round units	2031	2060	29						
Total occupied units	1970	2002	32	469	1005	536	1501	997	−504
Owners	1528	1603	75	411	862	451	1117	741	−376
Renters	442	399	−43	58	143	85	384	256	−128
Sample Count Totals									
Year-round units	2031	2060	29						
Total occupied units	1970	2002	32	459	1005	546	1511	997	−514
Structure Type									
1 (det + att)	1890	1840	−50	459	925	466	1431	915	−516
2 to 4	58	67	9	0	16	16	58	51	−7
5+ units	83	153	70	0	64	64	83	89	6
Mobile home	NA	0	NA	NA	0	NA	NA	0	NA
Total	2031	2060	29	459	1005	546	1572	1055	−517
Year Structure Built									
Built last 10 years	325	31	−294	143	18	−125	182	13	−169
Vintage:									
1970s	NA	31	31	NA	18	18	NA	13	13
1960s	325	252	−73	143	191	48	182	61	−121
1950s	700	606	−94	169	379	210	531	227	−304
pre-1950	1006	1171	165	147	417	270	859	754	−105
Total	2031	2060	29	459	1005	546	1572	1055	−517
Size of Housing									
Median rooms	5.5	5.7	0.2	5.9	5.9	0	NA	NA	NA
Persons per Room									
1.00 or less	1898	1916	18	436	951	515	1462	965	−497
1.01 to 1.49	51	61	10	22	42	20	29	19	−10
1.50 or more	21	25	4.0	11	12	1.0	10	13	3.0
% overcrowded	3.7	4.3	0.6	7.0	5.4	−1.7	2.6	3.2	0.6
Housing Costs									
Median value	$23,300	$77,100		$26,700	$80,500		NA	NA	NA
% of county average	96	88	−8	110	92	−18			
Median contract rent	$123	$240		$145	$275		NA	NA	NA
% of county average	112	98	−13	132	113	−19			

Note: NA indicates data not obtainable from published census tract reports.
(a) "Other" category is largely white, non-Hispanic; because it is defined as a residual, it also includes vacant units in some data categories.
Source: PHC(1)-117, Tables H-1–H-5, and PHC 80-2-225, Tables H-1, H-3, H-6, H-7, H-11, and H-13.

The gain of 536 black households is almost completely matched by the loss of 504 other, largely white non-Hispanic households. Of the black growth, 84 percent is by homeowners; whereas, of the losses of other households, 75 percent is by homeowners. On net, the transition to a black population is increasing owners by 75 and decreasing renters by 43 households.

Black households occupy much larger units than found in the other Pasadena tract we examined. In fact, the black-occupied median unit size of 5.9 rooms even exceeds the average for the tract as a whole. Similarly, in both 1970 and 1980 blacks are even more likely than the other population to live in single-family units. Not surprisingly, the crowding level within households is much less than in the other Pasadena tract.

The greatest difference between the black and other population is found with regard to the ages of occupied housing. Overall, we note that about half of the stock was built before 1950. This number has grown somewhat between the censuses, either because of conversions of older structures into additional units or because respondents misjudge how new their units really are. Sampling error is also a factor because age of housing is a sample count variable.

Between the censuses, the percentage of black households residing in pre-1950 housing grew from 32.0 to 41.5. However, the other households remaining in the area resided in older housing in even greater concentration: 54.6 percent in 1970 and 71.5 percent in 1980. (All of these differences are so large that they are statistically significant given this sample size.) This is a curious pattern. Unlike the example of the previous neighborhood, where the growing population of Hispanics were entering the oldest units, in this case the growing population of blacks are entering the newest units. The other population (non-Hispanic whites) is relinquishing the newer units and holding on to the older units. This transition by vintage is graphed in Fig. 2 using the data reported in Table VII. The figure shows that blacks had captured over 40 percent of the newest housing by 1970, extending this to just over three-quarters by 1980. In contrast, in the oldest housing, blacks occupied less than a 15 percent share in 1970, extending this to just over 35 percent by 1980.

Explanations for this intriguing pattern may be as follows. First, blacks moving out toward this suburban neighborhood may desire to leave behind older housing

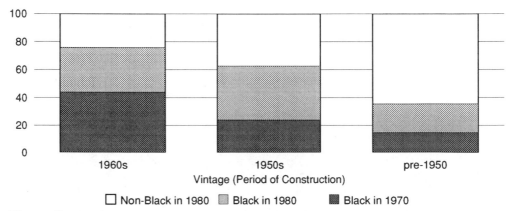

Figure 2. Black percentage occupancy of the housing stock, by vintage, as derived from the censuses of 1970 and 1980.

and therefore seek newer structures. By paying larger amounts for housing they should be able to afford their pick of the neighborhood's stock. Conversely, whites who are giving up the newer housing have many more choices about where to live in similar aged structures. Communities to the east offer newer housing in a similar environmental setting and much of it is priced competitively. Instead, whites may be remaining in the older housing because it has greater aesthetic appeal to them and it is a more limited resource. There are few alternative sources of older housing that are as inexpensive and have such an attractive setting by the mountains.

3.3. Household Components of Racial Change

Racial and ethnic change in a local area largely occurs because one group or another captures a larger share of the area's housing stock. The group may be growing in such numbers that additional housing is required, with those needs met by expansion from one neighborhood or another.

However, we saw in Table V that a group's share of the total population may be magnified beyond its share of occupied housing. Part of the racial change is due to changing housing occupancy and part is due to the different size of households that move into those units. As we will see later, the household size component of racial change can be further broken down by age or household relationship.

3.3.1. Components of Change Analysis

The problem of breaking down change into its component parts is often addressed through a *components of change analysis*. This general procedure has wide applications beyond the immediate application used to demonstrate it here. In this procedure we carry out a set of hypothetical calculations to ask how much change would have occurred if all factors held constant except the one being tested. In this manner we "decompose" the overall change into its component parts.

Once each of the factors has been tested, we sum the hypothetical amounts of change and compare the total to the actual change. A residual amount of change then remains to be explained. This residual occurs because the addition of separate components often omits additional change contributed through the interaction of components. We then distribute the residual amount across the separate components in relation to the relative magnitude of each component.

In formal terms we may express the components of change as follows:

$$\text{Change} = [(A_{70} \times B_{80}) - (A_{70} \times B_{70})] \quad [\text{change in B}]$$
$$+ [(B_{70} \times A_{80}) - (B_{70} \times A_{70})] \quad [\text{change in A}]$$
$$+ [(A_{80} - A_{70}) \times (B_{80} - B_{70})] \quad [\text{interaction of A \& B}]$$

where:

Change is the overall change in number of persons, or another aggregate such as change in the overall number of deaths in an area,

A and B are two linked factors, such as household size and number of households, or number of persons by age and mortality rates by age,

The 70 and 80 represent two time periods of observation.

Components of change analysis is well suited to the question of how much racial change is contributed by household size versus the number of occupied housing units. In the formulas, factor A can represent household size and factor B the number of occupied units.

In the first component of change, a group's total population change due to number of households is calculated by multiplying its 1980 household number times its 1970 household size, subtracting from this the product of the 1970 household number times the 1970 household size. Second, the change due to household size is calculated in complementary fashion, holding constant the number of households and calculating change due to variation in household size alone. Finally, the residual or interaction component is calculated as the difference between the 1980 and 1970 household size times the difference between the 1980 and 1970 numbers of occupied units. Alternatively, the interaction component is calculated as the residual after subtracting the two other components from the actual change.

In Table VIII, these calculations are worked through for racial change in Pasadena Tract 4620. The equation specified above would explain changes in the household population, but to extend this to the total population, including those in group quarters, requires only minor extension. The first portion of the table gives the raw input data used in the calculations. The middle panel shows the calculation of the two components, and the final panel provides a summary of the components of change.

Tracing the calculations for Hispanic population change, we begin with a total population change of 1263, an increase in households of 255, and an increase in average household size of 0.768. Also, group quarters population grew by 29 persons. The first component, change in number of households, holds the growth in household size to 0 (by using the 1970 value twice) and accounts for a change of 933.75 persons (rounded to 934). The second component, change in household size, holds the growth in households to 0 and yields a change of 104.43 persons.

The last panel of the exhibit rearranges the results of the calculations. The summary shows actual and hypothetical changes. Note that the sum of the two factors falls short of the total change in household population by 195.82 persons (rounded to 196). This shortfall is the residual that results from the interaction of the two changes. Following the last portion of the equation given above, if we multiply the change in the number of households (255) times the change in the average household size (0.768), we get 195.82 persons.

Assigning portions of the residual to each of the other main components increases each proportionately. Household size changes now account for 124 added Hispanic persons in the tract. Finally, we compute what proportion of the total population change in the tract is due to the different factors. Change in group quarters population amounts to 2.3 percent of the total change; change in household numbers (occupied units) accounts for 88.0 percent; and, change in household size accounts for 9.8 percent growth in population. The larger share of housing stock captured by Hispanics is of greatest importance, but growing household size is also significant.

This estimate may underestimate the contribution of household size. The residual change could be assigned differently. Recall that this is interaction between changes in household numbers and household size, defined as the product of the change in household size and the change in the number of households. If we wished to stress

Table VIII. Components of Racial and Hispanic Change

Pasadena Tract 4620

Race Change = (1) Change in Number of Housing Units Occupied by Each Group
+ (2) Change in the Average Household Size of Each Group
+ (3) Change in Group Quarters Population of Each Group

INPUT DATA

Number of Persons

	1970	1980	Change
Black	3403	3779	376
Hispanic	498	1761	1263
Other	658	443	-215

Number of HHs

	1970	1980	Change
Black	1268	1383	115
Hispanic	136	391	255
Other	327	166	-161

Average HH Size

	1970	1980	Change
Black	2.684	2.717	0.033
Hispanic	3.662	4.430	0.768
Other	1.835	2.440	0.605

Population in Group Quarters

	1970	1980	Change
Black	0	22	22
Hispanic	0	29	29
Other	58	38	-20

COMPONENTS OF CHANGE

Number of Households Factor
Population Expected if Group's Household Size Remained Constant:

	1970(a)	1980(b)	Change
Black	3403.00	3711.63	308.63
Hispanic	498.00	1431.75	933.75
Other	600.00	304.59	-295.41

a-1970 HH Size × 1970 Number of HHs
b-1970 HH Size × 1980 Number of HHs

Household Size Factor
Population Expected if Group's Number of Households Remained Constant

	1970(a)	1980(b)	Change
Black	3403.00	3444.60	41.60
Hispanic	498.00	602.43	104.43
Other	600.00	797.80	197.80

a-1970 HH Number × 1970 HH Size
b-1970 HH Number × 1980 HH Size

SUMMARY OF RESULTS

	Actual Change			Hypothetical Change Due to:		
	Change in Population	Change in Group Quart.	Change in HH Pop.	Change in HH Number	Change in HH Size	Residual
Black	376	22	354	309	42	4
Hispanic	1263	29	1234	934	104	196
Other	-215	-20	-195	-295	198	-97

With Residual Allocated Proportionately Among Hypothetical Components:

	Hypothetical Change Due to:			Actual Change in Group Quart.
	Change in HH Number	Change in HH Size	Residual	
Black	312	42	0	22
Hispanic	1110	124	0	29
Other	-354	159	0	-20

Proportionate Change Due to Each Component:

	Percentage Change Due to:			Total Change in Population
	Change in Group Quart.	Change in HH Number	Change in HH Size	
Black	5.85	82.97	11.18	100
Hispanic	2.30	87.88	9.83	100
Other	-9.30	-164.53	73.83	100

227

the importance of household size, instead of distributing the residual proportionately, we could allocate all of it to the household size category. In that case, the sum of the household size and residual effects (104 + 196 = 200) would account for an even larger percentage of the total population gain among Hispanics (200 / 1263 = 15.8 percent).

Household size can have different effects in different locations. In some cases growing household size may offset declining numbers of occupied units; more often, declining household size will amplify declines in a group's occupied housing. An example is found in Pasadena area Tract 4602, the suburban, middle-income residential area undergoing change from white to black. Calculations similar to those in Table VIII were performed for that location (not shown here). In the case of the other, non-black population, shrinking household size reinforced the population losses due to fewer occupied housing units. An additional 524 persons were lost through this shrinkage, fully 28 percent of the entire decline among the non-black population.

3.3.2. Relationship Components of Household Size

Following procedures presented in Chapter 7, we may investigate why household size has expanded or contracted for each racial subgroup. One handicap in this analysis is that the 1970 published tabulation for racial subgroups is not as detailed as the 1980 or 1990 data. We cannot separate children under age 18 from relatives other than the spouse.

Tables IX and X break down the changes in household size for Tracts 4620 and 4602, respectively. Household sizes grew between 1970 and 1980 for all groups in Tract 4620. Change was slight for blacks: a substantial drop for spouses (−0.16) was offset by an increase in nonrelatives (0.07) and in other relatives (0.13). Among Hispanics, there was a very large increase of 0.77 persons per household. Spouses declined here as well (−0.18), but nonrelatives increased by 0.22 and other relatives increased by 0.73. Even among the *other* population we find large increases in household size. All categories grew by 0.11 or more.

Table IX. Changes in Relationship Components of Household Size: 1970–1980
Tract 4620

	Black Persons per Householder			Hispanic Persons per Householder			Other Persons per Householder		
Household Relationship	1970	1980	Change	1970	1980	Change	1970	1980	Change
All householders	1.00	1.00	0.00	1.00	1.00	0.00	1.00	1.00	0.00
Spouse of householder	0.40	0.24	−0.16	0.79	0.61	−0.18	0.25	0.36	0.11
Other relative (including children <18)	1.21	1.33	0.13	1.76	2.49	0.73	0.46	0.80	0.34
Nonrelative	0.08	0.14	0.07	0.10	0.32	0.22	0.13	0.28	0.15
Total	2.68	2.72	0.03	3.66	4.43	0.77	1.83	2.44	0.60

Source: PHC(1)-117, Tables P-1, P-5, and P-7, and PHC80-2-226, Tables P-1, P-3, and P-6.

Table X. Changes in Relationship Components of Household Size: 1970–1980
Tract 4602

	Black Persons per Householder			Other Persons per Householder		
Household Relationship	1970	1980	Change	1970	1980	Change
All Householders	1.00	1.00	0.00	1.00	1.00	0.00
Spouse of Householder	0.75	0.62	−0.14	0.72	0.56	−0.17
Other Relative (including children <18)	1.85	1.67	−0.18	1.12	0.79	−0.34
Nonrelative	0.02	0.09	0.07	0.06	0.14	0.08
Total	3.63	3.38	−0.25	2.90	2.48	−0.42

Source: PHC(1)-117, Tables P-1, P-5, and P-7, and PHC80-2-226, Tables P-1, P-3, and P-6.

A different picture emerges for Tract 4602. Despite the large size of units in this area, household sizes dropped for both blacks and the other population. Among both groups spouses declined, by 0.14 for blacks and by 0.17 for the other group. Unlike the other tract in Pasadena, other relatives also declined for both groups, by 0.18 for blacks and by 0.34 for the other population. However, like all of the tracts we have examined, among both groups nonrelatives increased, by 0.07 for blacks and by 0.08 for the other group.

This decline in household size resembles what we found for other tracts in Chapter 7. The *other* population group already had much lower household size than the black population in 1970, but this declined at an even faster pace from 1970 to 1980, by 0.42 versus 0.25. The consequence was to magnify the effects of housing transition on the racial population change.

An analysis of the changing age profiles can shed more light on the household size factor in racial change. The age-specific persons per household (APH) method developed in Chapter 8 is a fruitful line of analysis. In the interest of brevity we must forego it here and, instead, view age and race through the alternative, cohort retention method.

3.4. Cohort Trajectories and Racial Age Structure

Figure 3 traces the growth and decline of black cohorts residing in Pasadena Tract 4620. Lying near the heart of Pasadena's black community, this area has grown increasingly congested as the housing stock continues to increase slowly and as black and Hispanic households move in with ever larger household sizes. Cohorts of blacks in their twenties or younger in 1980 received increases from the age group 10 years younger in 1970.[9] However, the number of 30 to 34 year olds was less than three-quarters the number of 20 to 24 year olds in 1970. The graph reflects the downward trajectory of this cohort and all those that are older. This is a curious pic-

___

[9]This increase is more remarkable in that it comes in an age group, and race, for which undercounts are known to be greatest. See Chapter 4 for more about undercounts.

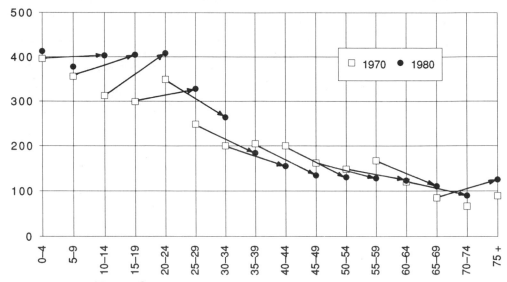

Figure 3. Black cohort trends, for Pasadena Tract 4620.

ture of departing middle-aged blacks combined with growing teenagers and adults in their twenties, plus increasing numbers of children. It is evident that many of the young blacks are residing with children in the area. In fact, analysis following the methods of Chapter 7 shows that 44 percent of households have children present under age 18, and two-thirds of these households are single-parent families (data not shown).

The strong growth of Hispanic population is reflected in sharp upward cohort trajectories for cohorts under age 45 in 1980 (Fig. 4). Above that age, virtually no in movement occurs. Some of the cohorts show increases of 200 or 300 percent over the decade. The cohort aged 20 to 24 in 1980 soared upward from a starting base as the largest age group in 1970, aged 10 to 14. If the 1980 age groups were to serve as the base for similar cohort trajectories from 1980 to 1990, Hispanic population growth would be tremendous.

Turning to the suburban middle-class tract, 4602, Fig. 5 portrays the sharp downward trajectories of the non-black *other* population that is largely white, non-Hispanic. Many of the cohorts fell by more than half. Particularly steep drops were experienced by cohorts starting the decade under age 14, by adults starting out at ages 25 to 34, and by older adults in their fifties and sixties. This is a pattern of wholesale departure from the neighborhood. Only for the cohort entering the 30 to 34 age bracket is there evidence of a more gentle out-migration. As discussed earlier in the chapter, the declines in *other* population were apparently much less in the older, pre-1950 housing, but we have no way of knowing which cohorts originated in those homes.

Black cohorts follow a complementary pattern of in-migration characterized by sharp upward trajectories, but only in selected age brackets. Figure 6 shows the greatest trajectories entering ages 30 to 44 and 10 to 19 in 1980. This pattern reflects the family attractions of the neighborhood for blacks. Growth entering the twenties

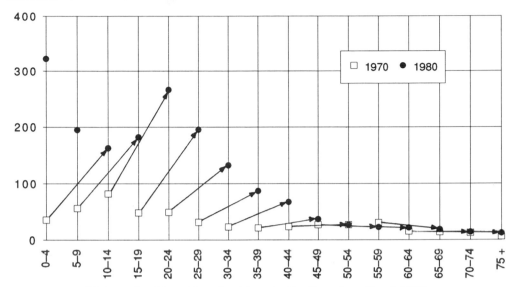

Figure 4. Hispanic cohort trends, for Pasadena Tract 4620.

is relatively flat, as is growth of cohorts entering ages above 44. We might speculate that some of the growth entering the 30 to 44 age bracket is fed by the departure of cohorts of the same age from Tract 4620. However, the census provides no data for inferring the prior neighborhood locations of residents (except whether five years earlier they lived inside the central city or outside the MSA entirely).

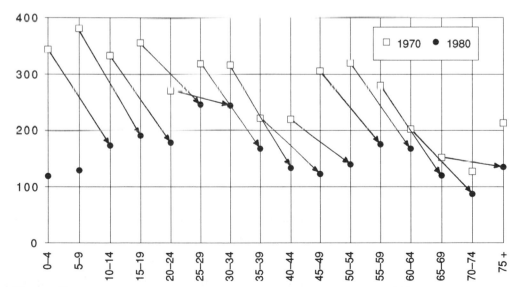

Figure 5. White cohort trends, for the suburban Pasadena Tract 4602. (*Note:* "White" is defined as the residual after subtracting black population from the total. Few persons of other races live in the area.)

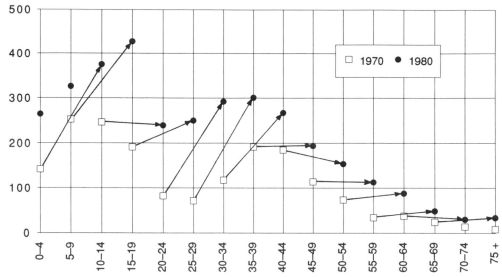

Figure 6.   Black cohort trends, 1970–1980, for the suburban Pasadena Tract 4602.

☐ 4. Conclusion

The analysis of racial composition and change calls into play many of the methods developed in previous chapters. Integrated analysis of census data is required, using both population and housing data for each race. With the profusion of repeated, detailed tabulations from the 1990 census for each of a long list of racial groups, methods demonstrated here may take on even greater utility.

In the case examples of local analysis, we have emphasized how racial change occurs through competition for housing. In one tract Hispanics were squeezing into older housing in a black neighborhood. In the other tract, blacks were moving into newer single-family homes, with the largely white, non-Hispanic other population retaining occupancy of the older housing built before 1950.

The occupancy of housing units by different racial subgroups leads to greater or lesser amounts of population depending on the household sizes of the different groups. In one of our examples, all subgroups were experiencing growing numbers of persons per household, so that population was increasing much faster than indicated by the number of occupied housing units. In the other example, the subgroups were both experiencing shrinking household size, one faster than the other.

The changing age profiles of local racial groups are easily tracked with census data, bringing methods shown in Chapter 8 into prominent use. When combined with a household and family analysis, racial change takes on even richer interpretation.

At the outset we described how definitions of racial categories were problematic, with census tabulations frequently "double-counting" persons who belong to both a racial category and an Hispanic category. Hispanic is not technically a racial subgroup, although it is frequently treated as equivalent to one. Of particular problem, prior to the 1990 census, was definition of a white, non-Hispanic category for detailed analysis. There still remains a question of how best to separate blacks and Hispanics. We have recommended some methods for sorting out the overlap between Hispanic status and the different racial subcategories, but the overlap problem still requires cautious handling.

11 ☐☐☐

Measuring Local Income Levels

One of the most important uses of census data is measurement of the income level of a local population. Several different census data items are often used, including household and family income, per capita income, and poverty status. This chapter discusses the differences between these alternative variables and presents methods of calculation.

Income has one difficult characteristic that is unlike most other census variables: Because it is measured in dollars, the value of income changes over time due to inflation. As a result, a family with $20,000 income measured in one census is much poorer, in real terms, than if that income had been obtained in an earlier census year. Over 10 years time, inflation may cut the value of that income in half. Most of the methods to be described for handling income data are designed to respond to this problem of shifting dollar values over time.

A number of specific questions may be addressed with local census data. The most common question is: How rich (or poor) is an area compared to others around it? This is closely followed by the logical followup: Has the area been growing richer or poorer over time? Some very simple and direct methods can be used to answer these basic questions.

At a deeper level of analysis we can ask about the distribution of income in an area. For example, how many persons in a wealthy place like Beverly Hills would be classified as poor? Although it is clear that the average income level is high, we may wish to know how many persons are in the upper income category and how many are in middle or lower income categories. How poor can you be and still live in Beverly Hills?

Special concern is often directed toward lower income persons that are the subject of government programs. With census data we can measure persons in poverty, as well as persons who fall below a specified percentage of the regional median income. The latter portion of the chapter addresses the question of lower income analysis.

We must emphasize that the census collects information only on current income, not on wealth. Hence, our usage of the terms *rich* and *poor* does not reflect savings, equity acquired in homes, or other financial assets. The discrepancy between wealth and income grows larger at older ages, particularly after retirement, and wealth differences between groups or places may vary even more than income.

Overall, analysis of income requires special skills in handling interval data grouped in categories. This chapter addresses the calculation of medians and other breakpoints in distributions. We also stress how to set up useful comparison standards for measuring the relative income status of local areas. In addition, because income is a sample variable, statistical significance must also be addressed.

☐ 1. Income Recorded by the Census

We begin by defining and contrasting the alternative measures of income that are available in the census. All income measures begin with the raw data collected in the sample, long-form questionnaire (items 32 and 33 in both the 1980 and 1990 questionnaires). From these raw data, four principal alternative measures of income are constructed.

1.1. Defining Income

In the census, income is defined as total money income received by persons in the calendar year preceding the census (e.g., 1989 for 1990). The U.S. Bureau of the Census (1982) defines total income as the sum of amounts reported separately for several different factors:

wages and salaries;
nonfarm self-employment;
farm self-employment;
interest, dividends, and rentals;
Social Security;
unemployment compensation;
welfare or other public assistance;
and all other sources.

These income amounts are *before tax*, representing the gross figures prior to any subtractions for taxes, social security, or any other payroll deductions. Not counted as income are such items as receipts from the sale of property (unless that was the purpose of an ongoing business enterprise), gifts, inheritances, or tax refunds.

1.2. Alternative Measures of Income

The income accruing to individuals is aggregated and processed into a number of alternative summary measures. Each has a particular meaning and different use.

1.2.1. Family Income Versus Household Income

Perhaps the most commonly referenced measure is *family income*, the total money income received by all family members 15 years old and over who live in the same housing unit. It is important to note that families are only a subset of households, excluding persons and groups who are not related by blood, marriage or adoption. That means persons living alone, such as the elderly or students, are not included in family income measurement.

As a result of this family definition, family income is substantially higher than *household income*. For example, the 1980 census reported a median household income for Los Angeles County of $17,551 and a median family income of $21,125. It is erroneous to compare the family income from one area with the household income from another area. It is also erroneous to compare the family income reported in one census year to the household income reported for the same place in the next census year.

Clearly, these comparisons mix apples and oranges; nevertheless, census users frequently fall into the trap of selecting income numbers indiscriminately. In part, this is due to confusion over the word *family*. Many persons use that term when they really mean *household*. Another cause of poor choice is convenience: Users often select the first income number presented in a table, regardless of whether that pertains to families, households, or persons.

Using the definitions presented here, users can be more deliberate in their choices. Most often, household income is the preferred choice, although consistency of choice from census to census, or place to place, is the prime consideration.

1.2.2. Median Versus Mean Income

Median income is used most commonly to summarize average income levels. However, census reports also make *mean* income available, a second measure of average income. In most cases the median figures are preferred, because they are not affected by extreme values. Such extremes could be recorded due to various errors, or they could be true values that simply skew the distribution. In either event, the median is a more reliable measure of *average* income.

The one advantage of mean income, however, is that it can be manipulated arithmetically and, therefore, used for a wide variety of analyses. Mean household income can be multiplied by the number of households to arrive at a measure of total household income in an area. However, for *describing* the income level of an area, median household income is preferable.

1.2.3. Per Capita Income

Per capita income is a handy summary measure often used, but prone to errors of interpretation. This measure is the mean income computed for every man, woman, and child in an area. It is derived by dividing total income by total population. Per capita income shares the same arithmetic advantages as other forms of mean income.

The major disadvantage of per capita income is that it does not reflect relative income levels well. It is not equivalent to the earnings level of workers because per capita income includes children and other nonworkers in the denominator. In particular, per capita income is a poor means of comparing income between places, or over time, when there are major differences in the number of children per household. For example, based on per capita income, some lower middle-class Hispanic areas of California may be classified as poor because they have more children per earner than other locations.

1.2.4. Poverty Status

A greater number of children certainly should have some weight in the determination of poverty. Those calculations are conducted by the Census Bureau and reported as *poverty status*. The incomes of families and unrelated individuals are classified as above or below poverty by comparing their total income to a cutoff or *poverty threshold*. These thresholds are carefully constructed to reflect the needs of persons, based on family size, number of children, and age of the family householder or unrelated person.[1]

Poverty status is determined for all families, and therefore all family members. When combined with the assessment of poverty status for unrelated persons, the result is a tabulation of poverty status for all persons in an area. This status has several levels: below 75 percent of the poverty threshold; from 75 to 99 percent of poverty level; from 100 to 149 percent; from 150 to 174; from 175 to 199; and 200 percent of the poverty level and above.

This measure of income is better than per capita income for measuring the needs of the lower income population. However, for a comparison of overall income levels between places, or over time, the median household income may be preferable.

☐ 2. Adjusting for Inflation Between Censuses

All dollar values increase in size from one census to the next. Because of inflation, over 10 years time, an area of declining income will appear to have growing incomes. For example, Pasadena Tract 4620, predominantly composed of black and Hispanic renters, reported a median family income of $9,836 in 1980, up from $6,451 in 1970. Does this truly represent an increase in the area's income level? Consider that the median family income for Los Angeles County as a whole rose from $10,972 to $21,125 in the same time period.

Two alternative strategies can be followed to adjust income for inflation: one is deflation by the Consumer Price Index (CPI) or a comparable deflator; the other is

---

[1]The concept of poverty and the calculation of poverty thresholds is more subjective, and politically contentious, than many realize (see Jencks 1987). For an explanation of Census Bureau procedures, see U.S. Bureau of the Census (1982).

direct measurement relative to a regional standard. Table I demonstrates both procedures for addressing the case of Pasadena Tract 4620.

The first column of Table I shows the census-reported income figures for the tract and the county. Also shown is the CPI pertaining to each year. Note that the CPI is given for 1969 and 1979 because the incomes reported in the census are for the tax year preceding the census year. The ratio of the 1979 to 1969 CPI gives an inflation factor that can be used to blow up the 1969 incomes to represent 1979 dollars. Some uncertainty exists over what is the proper deflator to use. Should the index be selected for the nation as a whole or for the region in question? Alternatively, should some other measure such as a GNP deflator be used in place of the CPI? The index chosen here is the CPI for "all items" at the national level because this is easily obtainable from the annual *Statistical Abstract of the United States.*

The second column shows the first adjustment alternative, A, which uses the ratio of 1979 to 1969 CPI data to inflate 1970 census data to be comparable with 1980 data. The results for the regional median income show a slight decline in real dollars over the decade. However, the results for Pasadena Tract 4620 show a very substantial decline in 1979-equivalent dollars. In constant dollars, median family income fell from $12,761 to $9,836 in 1980 (a decline of 22.9 percent), far below the regional average of $21,125 (which fell only 2.7 percent).

The last column of the table shows the second alternative, B, which adjusts income relative to the regional standard. The tract's medians are simply divided by the regional medians for each respective year and then multiplied by 100. This procedure standardizes income into an index where 100 equals the regional average. Thus, in 1970, we can say that the tract's median family income equaled 59 percent

Table I. Alternative Strategies to Adjust Income for Inflation

	Census-Reported Median Family Income	A Adjusted by CPI to 1979$	B Relative to Region (% of Regional Median)
Pasadena Tract 4620			
1970 census	$6,451	$12,761	59
1980 census	$9,836	$9,836	47
% change	52.47%	−22.92%	−20.81%
% point change	NA	NA	−12.23
Los Angeles County			
1970 census	$10,972	$21,705	100
1980 census	$21,125	$21,125	100
% change	92.54%	−2.67%	0
% point change	NA	NA	0
National Consumer Price Index (CPI)			
1969	36.7		
1979	72.6		
1979/1969	1.978		

Note: Income data reported in the census are for the preceding tax year. Therefore, the 1980 census is reporting 1979 dollars.

of the regional median. By 1980 the median had fallen to only 47 percent of the regional median. Meanwhile, the region remained constant at the 100 level because it is the reference standard.

Alternative B is often preferred as a strategy because of its simplicity and the conciseness with which it expresses two essential ideas: relative level and change over time. *Only two numbers*, with a total of four to six digits, are required to express the tract's income relative to the county's in two years. In contrast, to express the same two ideas with alternative A, we would need to report more numbers, each with many more digits. (Compare the last sentence of the second preceding paragraph.) This efficiency of communication is of particular value whenever *more than one place* is being compared over time.

In summary, adjustment by the CPI (or another similar measure) is useful if we need to know the exact dollar level or the actual change in real income for some purpose. However, the percent-of-region approach is better whenever comparisons of relative income status are the goal. The methods illustrated in Table I can be carried out equally with median family or household income, as well as with per capita income. In fact, the same methods can be applied to any dollar distributions, including housing values and rents.

2.1. Reporting Trends over Time

Two graphic methods for presenting data on income trends are illustrated here. Both methods make use of the percent-of-region standardization method, but incomes adjusted to constant dollars by CPI could also be used.

Figure 1 is a simple line graph portraying the slope between two census years for three Hollywood tracts. A horizontal line at 100 represents the level of regional income. Hollywood Tract 1942, up in the hills, has slipped somewhat, but remains

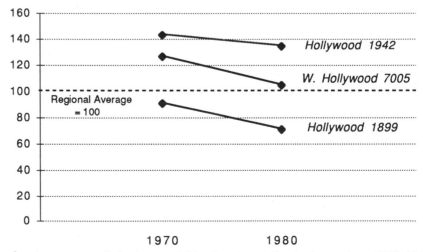

Figure 1. Income trends in three neighborhoods relative to the region, 1970–1980. The income level each year is measured as a neighborhood's median family income divided by the median for the region and multiplied by 100.

more than 30 percent above the regional average. West Hollywood Tract 7005 has slipped more sharply, most likely because a large volume of apartment construction has added households at middle and lower income levels. Hollywood Tract 1899 was already below average in 1970 and fell even further by 1980.

When more than three or four places are to be compared over time, a scatterplot representing before and after status is most efficient. Figure 2 focuses on a diagonal line of equal status. Points on that line have the same relative income level in 1970 and 1980. Five tracts lie below the line, meaning that the 1980 income level lies below that for 1970. The large drops in Hollywood Tracts 1899 and 7005 are clearly depicted, as are the decreases for Pasadena Tract 4620 and Beverly Hills Tract 7009.02 (more will be said about that tract shortly). None of our tracts shows substantial relative increases in income over time, probably because newer tracts located in outlying portions of the county are entering the distribution with higher median incomes.[2]

The scatterplot not only traces changes over time, but also displays the level of income at two different points in time for each place. Tracts plotted in the lower left corner are the poorest and tracts in the upper right corner are the richest. One tract, Beverly Hills 7007, literally ran off the chart; its 1980 relative income level was over 400 (more than four times the regional average).

The scatterplot format has the potential to support a large *cloud* of data points representing a very large number of tracts. Different plots could be constructed of incomes in tracts with different characteristics (e.g., newer versus older, or minority versus other). Several different statistical and graphic computer software programs

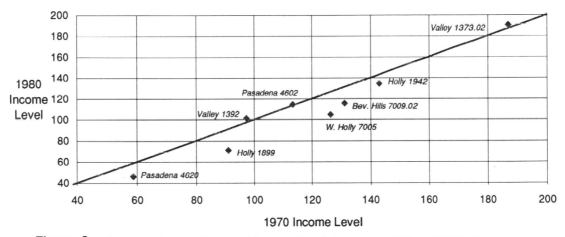

Figure 2. Changes in neighborhood income levels between 1970 and 1980. Neighborhoods above the diagonal line in this "before-and-after" scatterplot experienced rising income relative to the region as a whole. Income levels are measured as in Fig. 1.

[2]In general, neighborhoods experience declining economic status over time because newer neighborhoods are developed with newer, more expensive housing and with residents of higher socieconomic status (Guest 1974). Occasionally, neighborhoods will plunge in economic status; at other times, neighborhoods will gentrify and rise in status counter to the downward trend. The method here serves to spot the places that depart from the average trend.

can even plot three-dimensional scatterplots, allowing the user to rotate the data cloud for different perspectives on the distribution.[3]

☐ 3. Computing Breakpoints in Income Distributions

Computation of breakpoints in an income or other dollar-based distribution is often necessary. Breakpoints have two major uses: one for comparison of levels, and the other for counting the share of subarea cases that falls under a regionally defined level.

In the first usage, we may wish to determine how high is the median income, or how high is the income at another designated point in the distribution. In addition to the median (halfway point), there are quartiles (one-fourth steps), quintiles (one-fifth steps), or decile points (one-tenth steps) in a distribution. (Collectively, these various breakpoints are known as *quantiles*.) Ratios between the tract and the regional distributions can be calculated at each breakpoint just as illustrated above for the median. Still other breakpoints of frequent use are the income levels corresponding to 50, 80, and 120 percent of the median level. These levels are proscribed by the federal government and many states and localities as the definition, respectively, of very low income, lower income, and moderate income households.

The second use of breakpoints builds on the first, applying it in reverse. Once we have determined the relevant breakpoint in a regional income distribution, we may ask what proportion of local households falls under this level. For example, we may first calculate the dollar breakpoint under which 25 percent of county households fall (the bottom quartile). Then applying this breakpoint to local areas, we can ask how large a share of local households falls within the regionally defined bottom quartile. This procedure allows us to compare local distributions to a common regional standard. The second use of breakpoints is demonstrated in a following section.

3.1. Data Available on Income Distributions

Income distributions of household or family income are presented in more or less detail, depending on the source. Published census sources report income grouped into fewer categories, for reasons of limited space, than shown in the summary tape files. Nine income categories were published in 1980 and 1990, both in census tract books and other sources. In contrast, in the summary tape files, 17 categories were listed in 1980, and 25 categories were listed in 1990. Table II contrasts the two different sources and also shows how the categories have shifted between the 1980 and 1990 censuses to accommodate the effects of inflation.

As explained below, the finer-grained tabulation of income permits more accurate manipulation of the data than in a coarser-grained distribution using wider income

---

[3]The earliest such program for microcomputers was MacSpin, appearing in 1986. This three-dimensional scatterplotting function, with the ability to rotate the axes, has since been adopted by most statistical programs that produce graphical displays.

Table II. Alternative Tabulations of Income, 1980 and 1990, by Source

1990 Census		1980 Census	
Summary Tape Files	Census Tract Books	Summary Tape Files	Census Tract Books
Less than $5,000	Less than $5,000	Less than $2,500	Less than $5,000
$5,000 to $9,999	$5,000 to $9,999	$2,500 to $4,999	
$10,000 to $12,499	$10,000 to $14,999	$5,000 to $7,499	$5,000 to $7,499
$12,500 to $14,999		$7,500 to $9,999	$7,500 to $9,999
$15,000 to $17,499	$15,000 to $24,999	$10,000 to $12,499	$10,000 to $14,999
$17,500 to $19,999		$12,500 to $14,999	
$20,000 to $22,499		$15,000 to $17,499	$15,000 to $19,999
$22,500 to $24,999		$17,500 to $19,999	
$25,000 to $27,499	$25,000 to $34,999	$20,000 to $22,499	$20,000 to $24,999
$27,500 to $29,999		$22,500 to $24,999	
$30,000 to $32,499		$25,000 to $27,499	$25,000 to $34,999
$32,500 to $34,999		$27,500 to $29,999	
$35,000 to $37,499	$35,000 to $49,999	$30,000 to $34,999	
$37,500 to $39,999		$35,000 to $39,999	$35,000 to $49,999
$40,000 to $42,499		$40,000 to $49,999	
$42,500 to $44,999		$50,000 to $74,999	$50,000 or more
$45,000 to $47,499		$75,000 or more	
$47,500 to $49,999			
$50,000 to $54,999	$50,000 to $74,999		
$55,000 to $59,999			
$60,000 to $74,999			
$75,000 to $99,999	$75,000 to $99,999		
$100,000 to $124,999	$100,000 or more		
$125,000 to $149,999			
$150,000 or more			

Sources: 1980 STF 3, Table 68 (Household),* Table 73 (Family)*
1980 Census Tracts, PHC-80, Table P-11 (Household)**
1990 STF 3, Table P80 (Household),*** Table P107 (Family)***
1990 Census Tracts, CPH-3, Table 19 (Household)**
*Equal detail by race on STF 4 (PB71 and PB85)
**Equal detail by race in the census tract books
***Less detail by race on STF 3; equal detail on STF 4 (PB58 and PB71)

categories. The importance of this difference is greater in 1990 than in 1980, because the same number of categories (nine) is used to cover an income span that has doubled.[4]

3.2. How to Estimate a Breakpoint

The Census Bureau usually computes medians as a matter of course, but users must estimate their own quartiles, quintiles, or deciles. Breakpoints in the dollar scale are estimated in two basic steps.

[4]The categories shown in Table II for census tract books (Series CPH-3) are the same ones used in books reporting income for larger areas as well (Series CP-2).

In the first part of the procedure, we add the number of cases in successive whole categories until the cumulative number of cases exceeds target shares of cases, such as one-fourth, two-fourths, or three-fourths. For example, if an area has 10,000 households, the third quartile is the income point below which three-quarters or 7,500 households fall. The last added category that causes the cumulative total to exceed the target share (e.g., 7,500 cases) is the category within which the breakpoint lies.

Only a portion of the cases in that category lies beneath the target share, and accordingly only a portion of the dollar range is required to reach the breakpoint. The second step is then to prorate the portion of the dollar range in that category that corresponds to the number of cases lying beneath the target share.

The heart of the procedure for finding a dollar breakpoint is that we want to use the proportion *of cases* lying below the target share in the category to define the proportion *of the dollar range* in the interval that lies below the breakpoint. Given the target cases breakpoint, we arrive at the corresponding dollar range breakpoint. For example, if 16 percent of the cases in a category are needed to reach a cumulative total of three-fourths of all cases, then the third-quartile dollar breakpoint equals 16 percent of the dollar range of the category added on top of the beginning dollar level of the category.

3.2.1. Formula for Breakpoint Calculations

The above ideas can be expressed algebraically, as follows:

$$X = \left(\frac{\text{target} - N_1}{N_2 - N_1}\right) \cdot (C_2 - C_1) + C_1 \, ,$$

where

X = the desired breakpoint in the income distribution,
target = the target number of cases (e.g., median at 50 percent of total cases),
C_1, C_2 = the lower and upper bounds, respectively, of the dollar interval containing the breakpoint, and
N_1, N_2 = the cumulative number of cases with incomes less than or equal to C_1 and C_2, respectively.

The formula uses linear interpolation to prorate portions of categories. This is the method the Census Bureau uses in cases where the width of the dollar interval ($C_2 - C_1$) is no greater than $2,500. For wider intervals, they use a Pareto formula based on logarithms; however, the difference between the two methods is small.

Of greater consequence for accuracy, the Bureau calculates its medians from more detailed income distributions, those given in the summary tape files, than published in printed reports. As shown previously in Table II, fewer categories, with wider intervals, are published in the upper income range. This may lead to slight discrepancies between the Bureau's calculation of medians and users' calculations from the published data. The discrepancy is greater in the higher income brackets (where medians are rarely situated). If required, more precise calculation of breakpoints is possible with the more detailed income distributions reported in summary tape files or on CD-ROM (see sources listed in Table II).

3.2.2. Procedure Using a Calculator or Spreadsheet

The procedure for finding a breakpoint is demonstrated in Table III. The steps also are defined as required for spreadsheet calculation using *lookup table* functions. (For those unfamiliar with lookup tables, the syntax is explained in the notes at the bottom of Table III.)

The example here uses the linear method. The calculations for each step are carried out using separate lookup tables. In step A we search down the column of cumulative cases until we find the category within which the breakpoint resides. Step B operates off a different initial column, that of category numbers, returning variable values for the category identified as containing the breakpoint in step A.

3.3. Results Comparing Beverly Hills and Los Angeles County

To illustrate the results from a quartile breakpoint analysis, let's compare Beverly Hills with the whole of Los Angeles County. The published median household income for Beverly Hills City is 43 percent higher than that for Los Angeles County. Are each of the quartiles for Beverly Hills also higher by the same amount?

Table IV summarizes the findings from computing quartile breakpoints for the region and the city. Beverly Hills's quartile points are substantially higher than the county's, but the gap grows proportionally greater at higher quartiles. In fact, the third quartile for Beverly Hills lies in the open-ended top income category reported in published census books ($50,000 or more). That means that the third quartile point is at least as high as $50,000, a good 67 percent above the county's third quartile. At least a quarter of all Beverly Hills households had 1979 incomes of $50,000 or greater.

With access to more detailed tabulations in summary tape files, where the top category is listed as $75,000 or greater, we can reach a more definitive judgment (see Table II). We find that the third quartile for Beverly Hills is $59,348, approximately double that of the county. We also are able to replicate the published medians exactly by applying the method in Table III to the more detailed data. The reason that our calculation of the county median varies more from the published median than for Beverly Hills is twofold: (1) the width of the dollar category where the median lies is fairly large, from $15,000 to $19,999; and (2) the county median falls near the middle of that category, the area most subject to variation due to method. Note that more fine-grained income categories, such as $1,000 increments, would yield a still more precise estimate of the median, but the Census Bureau is satisfied with $2,500 increments. In fact, many census users may be satisfied to estimate their breakpoints from the larger categories in the published books.

Table III. Calculation of a Dollar Breakpoint in a Distribution

Logical Steps and Calculations Using Spreadsheet Lookup Tables

STEP_A	Result	Cell	Spreadsheet Formula
1. Determine "Target" breakpoints in cases (e.g., for quartiles, .25 x Total, .5 x Total, .75 x Total)	Target = 3728.5	D1	1/4 * 14914
2. Find category where cumulative cases just exceeds the target cases breakpoint	4	D2	VLOOKUP(STEP_A,D1,2) + 1

STEP_B	Result	Cell	Spreadsheet Formula
1. Find number of cases in that category	1774	D3	VLOOKUP(STEP_B,D2,2)
2. Find cumulative total prior to that category	3057	D4	VLOOKUP(STEP_B,(D2-1),5)
3. Determine the number of cases in that category under the breakpoint	671.5	D5	D1 - D4
4. Determine what proportion of cases in the category lies below the breakpoint	0.379	D6	D5 / D3
5. Find the dollar range of the category	$5,000	D7	VLOOKUP(STEP_B,D2,4)
6. Apply proportion of cases to this dollar range	$1,893	D8	D6 * D7
7. Add this value to the bottom limit of the category's dollar range	$11,893	D9	D8 + VLOOKUP(STEP_B,D2,3)

Layout of the Spreadsheet Lookup Tables

STEP_A————[2]————.

	cumulative cases	category number	category cases	category bottom limit	category dollar range	cumulative cases
		STEP_B—[2]—	—[2]—	—[3]—	—[4]—	—[5]—
	1,597	1	1,597	$0	$5,000	1,597
	2,314	2	717	$5,000	$2,500	2,314
	3,057	3	743	$7,500	$2,500	3,057
Target--->>	4,831	4	1,774	$10,000	$5,000	4,831
[3728.5]	6,249	5	1,418	$15,000	$5,000	6,249
	7,448	6	1,199	$20,000	$5,000	7,448
	9,019	7	1,571	$25,000	$10,000	9,019
	10,613	8	1,594	$35,000	$15,000	10,613
	14,914	9	4,301	$50,000	open-ended	14,914

Notes:

1. Two Lookup table ranges are defined:
   STEP_A consists of the numbers under the column headings within that box.
   STEP_B partially overlaps and consists of numbers under its column headings.

2. The syntax for reading a Lookup formula has three parts.
   Example : VLOOKUP(STEP_B,D2,2)
   1. Read down the first column of the range "STEP_B"
   2. until the value reported in "cell D2" is reached.
      Note: Cell D2 is not part of the Lookup table but exists elsewhere on the spreadsheet.
   3. Then read across to the "second column" and find the result.
      Note: Some spreadsheet programs count the first column as 1 and some as 0.
      In Lotus 1-2-3 the second column is found with a "1" not a "2".

Table IV. Summary of Results

	Bottom Quartile	Second Quartile–Median	Third Quartile	Published Median
Los Angeles County	$9,057	$17,767	$29,935	$17,551
Beverly Hills City	$11,893	$25,057	Open-ended	$25,046
Ratio of BH City to Los Angeles County	1.31	1.41	over 1.67	1.43

4. Computing the Local Share Falling Below a Regionally Defined Breakpoint

A second major use exists for breakpoints in addition to the direct comparison of breakpoints between areas. The alternative is to estimate what proportion of each local area's households falls below the regionally defined breakpoint. For example, if the regional bottom quintile is $6,000, what proportion of households in a local area falls into this range? A 20 percent share in the bottom quintile means that the local area distribution matches the region's (since quintiles are one-fifth steps). A larger (smaller) share in the bottom quintile means the local area is poorer (richer).

4.1. What Local Share is "Very Low Income"

Perhaps the clearest illustration of this approach is found in federal definitions of income categories based on regional averages.[5] *Very low income* households have incomes of less than 50 percent of the regional median. *Low-income* households have incomes between 50 and 80 percent of the regional median. (*Lower income* households consist of low and very low incomes combined.) *Moderate income* households have incomes ranging between 80 and 120 percent of the regional median.

The regional breakpoints are clearly defined; the question is: How many *local* households fall below the breakpoints? The calculations are reversed, essentially, from those required to find a median or other breakpoint. In this case, we start with the dollar breakpoint and work toward the number of cases that falls under that point. For example, how many "very low income" households reside in Beverly Hills?

The necessary sequence of calculations is defined in Table V. Here we begin with the 50 percent level of the regional median income, $8,776, and then we seek the number of corresponding households in Beverly Hills. Step A finds that the breakpoint lies in the third income category. Step B then calculates what proportion of the cases in this category lies below the prescribed breakpoint. A total of 2,693 households in Beverly Hills qualify as very low income, or 18.1 percent of the households in the city.[6] Although this number is smaller than the 24.1 percent of all households in the county that fall below 50 percent of the regional median, it seems remarkably high for this local area. With the high-income reputation of Beverly Hills, it is very surprising to find so many very low-income households.

---

[5] Federal definitions make additional adjustments for household size, something not considered here.

[6] Calculation of a 95 percent confidence interval for this figure indicates variability due to sampling error of only +/− 1.4 percentage points.

Table V. Calculation of Local Share Below a Regional Breakpoint

Logical Steps and Calculations Using Spreadsheet Lookup Tables

STEP_A	Result	Cell	Spreadsheet Formula
1. Identify target dollar breakpoint	Target = $8,776	D1	Derived from other analysis
2. Find category that includes dollar breakpoint	3	D2	VLOOKUP(STEP_A,D1,2)
STEP_B			
1. Find the bottom dollar limit of the category	$7,500	D3	VLOOKUP(STEP_B,D2,2)
2. Determine number of dollars under the breakpoint in the category	$1,276	D4	D1 - D3
3. Find the dollar range of the category	$2,500	D5	VLOOKUP(STEP_B,D2,3)
4. Determine what proportion of the dollar range lies below the breakpoint	0.510	D6	D4 / D5
5. Find number of cases in the category	743	D7	VLOOKUP(STEP_B,D2,4)
6. Apply proportion of the dollar range to the number of cases	379.1	D8	D6 * D7
7. Add this number of cases to the cumulative number of cases prior to this category	2693.1	D9	D8 + VLOOKUP(STEP_B,D2,5)
8. Percent of local households under the breakpoint	18.10%	D10	D9 / VLOOKUP(STEP_B,9,5)

Layout of the Spreadsheet Lookup Tables

	STEP_A ——[2]—		STEP_B ——[2]———[3]———[4]———[5]—			
	category bottom limit	category number	category bottom limit	category dollar range	category cases	cumulative cases
	0	1	0	$5,000	1,597	1,597
	$5,000	2	$5,000	$2,500	717	2,314
Target--->>	$7,500	3	$7,500	$2,500	743	3,057
[$8,776]	$10,000	4	$10,000	$5,000	1,774	4,831
	$15,000	5	$15,000	$5,000	1,418	6,249
	$20,000	6	$20,000	$5,000	1,199	7,448
	$25,000	7	$25,000	$10,000	1,571	9,019
	$35,000	8	$35,000	$15,000	1,594	10,613
	$50,000	9	$50,000	open-ended	4,301	14,914

Notes:

1. Two Lookup table ranges are defined:
   STEP_A consists of the numbers under the column headings within that box.
   STEP_B partially overlaps and consists of numbers under its column headings.

2. The syntax for reading a Lookup formula has three parts.
   Example : VLOOKUP(STEP_B,D2,2)
   1. Read down the first column of the range "STEP_B"
   2. until the value reported in "cell D2" is reached.
      Note: Cell D2 is not part of the Lookup table but exists elsewhere on the spreadsheet.
   3. Then read across to the "second column" and find the result.
      Note: Some spreadsheet programs count the first column as 1 and some as 0.
      In Lotus 1-2-3 the second column is found with a "1" not a "2".

248

4.2. The Puzzle of Low-Income Households in Beverly Hills

Three approaches may be taken to investigate this puzzling outcome. First, we should examine the entire income distribution, not just the very low end. Second, we should look at income data for portions of Beverly Hills, not just the city as a whole. Finally, we should also look closely at characteristics of different tracts in addition to income to see if these help us understand the income pattern.

4.2.1. Quintile Distribution of Incomes

A clear picture of the overall income distribution in Beverly Hills is gained by calculating what share of the local households falls into each of the region's five quintile groups. The regional breakpoints are first calculated, as shown previously in Table III. After calculating the four breakpoints that define the five quintiles, we then calculate the local share under each breakpoint as shown in Table V.

The results are graphed for three areas: the whole of Beverly Hills City and two neighborhoods (Fig. 3). Tract 7007 lies up in the hills and is the most exclusive part of Beverly Hills. Tract 7009.02 lies down on the *flats* and is much more densely built up. The results from the quintile analysis are compared to the regional standard of 20 percent share in each quintile.

For the city as a whole, each of the bottom quintiles has at least a 15 percent share of the households in Beverly Hills. This reflects the earlier finding of a sizable number of lower income households. Tract 7009.02 has an even larger proportion of lower income households than the city as a whole and even approaches or exceeds the regional average.[7]

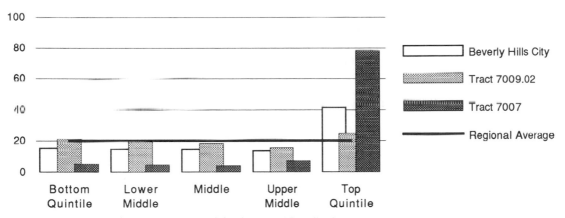

Figure 3. Percentage of households in Beverly Hills and two of its neighborhoods that have incomes falling into each quintile (fifth) of the regional income distribution in 1980.

[7]Significance tests performed on the difference between the regional norm (20 percent) and the share in each quintile reveal that the bottom three quintile shares in Tract 7009.02 do not differ from the region. All other quintile shares for this tract and the other places portrayed in Fig. 3 are significantly different from the region.

Despite the number of lower income households, approximately 40 percent of the households in Beverly Hills City are in the top income quintile, twice the regional average. However, Tract 7009.02 has a top quintile share that only slightly exceeds the regional average. More in keeping with expectations, Tract 7007 has a top-quintile share of approximately 80 percent, four times the regional average. Beverly Hills clearly has at least one neighborhood dominated by the wealthy, but it also has at least one other neighborhood that is very ordinary in its income distribution.

4.2.2. Relation to Other Characteristics

In the interest of brevity, we cannot analyze at length the differences between the various neighborhoods in Beverly Hills. Nevertheless, attention to a couple of key characteristics may help shed light on the reasons for the apparent income differences. Three plausible candidates are the percentage of minorities in each neighborhood, the number of elderly, and the type of housing.

As shown in Table VI below, the percentage minorities (black or Hispanic; Asians excluded) is very low throughout the city. However, the highest proportion of minorities is found in the richest area. Although some of these minority residents may be highly paid professionals or entertainers, a number may also be among the sizable number of live-in servants in this wealthy neighborhood. (Recall the discussion from Chapter 7 of the high proportion of nonrelatives living in this tract, a clear indication of either servants, live-in lovers, or roommates.)

A high proportion of elderly also might contribute to lower incomes, and that group does make up a sizable portion of the population. However, the difference in elderly proportion between the highest and lowest income neighborhoods is not great.

Finally, we observe a surprisingly high proportion of multifamily units (apartments and condominiums) in Beverly Hills. Nearly two-thirds of the city's housing units are in these higher density structures, a substantially higher proportion than the 40 percent figure for the region as a whole. The multifamily concentration is even greater in the poorest Beverly Hills tract, but the concentration is exceptionally low in the richest tract.

Apparently, it is the prevalence of multifamily housing units that best explains the surprisingly ordinary income distribution through much of Beverly Hills. It is only in Tract 7007 and one of its neighbors in the hill section that we find the wealthy dominance often associated with Beverly Hills.

Table VI. Summary of Other Characteristics

	Beverly Hills City	Tract 7009.02	Tract 7007
Percent black or Hispanic	5.7%	4.9%	9.7%
Percent elderly	21.3%	24.5%	17.3%
Percent apartments/condos	62.1%	84.5%	4.9%

☐ **5. Assessing Statistical Significance**

Income is a sample variable collected only from the respondents to the long-form questionnaire. As discussed previously in Chapter 4, the sample variables require some attention to possible sampling variability. Chances are that the estimates of income levels for small areas may be higher or lower than the true population value.

Procedures for testing statistical significance of the sample data are as outlined in Chapter 4. One notable difference encountered with income data is the heavy reliance on medians or other measures such as quartiles. A second difference is the use of standardization techniques relating local areas to a regional reference. Each of these is addressed in turn.

5.1. Significance of a Breakpoint

The sampling variability of medians or other breakpoint estimates can be evaluated in the same manner as estimates of subcategories of cases. Simply divide the total number of cases (households in the example of household income) by the number of groupings to be formed by the breakpoint. For example, calculation of the median depends on the total cases divided by 2 (call this $N/2$). The first quartile would be $N/4$ and the second quartile would be $N/4*2$, or the same as $N/2$. The estimated lower and upper bounds of each breakpoint can then be found by the confidence interval calculations shown in Chapter 4.

To find the corresponding variability in the dollar value associated with the cases' breakpoints, the Census Bureau recommends carrying out three repeated calculations. First, calculate the median for the initial cases-breakpoint estimate. Then repeat the calculation with the lower bound for the estimate and again with the upper bound of the estimate. The result of the three trials is an estimate of the median income, accompanied by an upper and lower bound to the median that represent a confidence interval.

5.2. Significance of a Share Under a Breakpoint

The local percentage of households in a regionally defined category has some statistical uncertainty. Not only is sampling error associated with the local percentage, but sampling error also exists around the regional reference number.

The easiest and most direct method for evaluating this combined sampling error is to treat the local share as a difference from the regional standard. For example, in the assessment of income distribution in Beverly Hills, we contrasted the share in each quintile to the region. By definition, the region had a share of 20 percent in each quintile. Therefore, the appropriate significance test is a difference of percentage, following the procedure outlined in Chapter 4. By this test, we find that the bottom three quintiles of Tract 7009.02 are not statistically different from the regional standard of 20 percent (see Fig. 3).

□ 6. Conclusion

Income is certainly one of the most useful measures provided by the census. It provides one of the best means for describing the socioeconomic status of different neighborhoods or cities. Income is also an essential factor in business marketing plans.

This chapter has summarized the several different measures of income that are available in the census. Several pitfalls to avoid have been identified. One is the potential confusion between family and household income. Another is the misleading nature of per capita income that makes places with more children look poorer.

Analysis of changes over time faces the extra difficulty that income is not measured in stable units: Dollars fall in real value due to inflation. Two major alternatives have been presented for dealing with this problem: adjustment by a CPI or other deflator, or standardization relative to the regional income distribution of the same census year.

This chapter has given major attention to the problem of how to manipulate income data, or other interval data, reported in tables. We have seen how to calculate medians, quartiles, or other breakpoints in the distribution. Conversely, we have demonstrated calculation of the local share of households that falls under a specified breakpoint derived from regional data. With these two skills, analysts should be able to address a great many tasks involving income data.

Part 3

Extracting Information from Local Microdata

12 □□□

Structuring Analysis with Microdata

Previous chapters have concentrated on analysis of summary tabulations provided for small areas. We now proceed to consideration of the added information that can be derived from microdata, specifically the Public Use Microdata Samples (PUMS). As first introduced in Chapter 4, microdata afford a highly flexible alternative to the limited and predefined summary tabulations. With access to a sample of the individual records from the census long-form questionnaire, analysts can combine variables in any combination desired. However, for effective use, this flexible freedom to tabulate necessitates a disciplined structure to the analysis. Guidelines for such a structured analysis are the primary focus of this chapter.

The one limitation of microdata is that, for reasons of confidentiality, these individual household records are made available only for geographic areas of 100,000 or more population, areas that are roughly 25 times the size of census tracts. In larger urban areas, microdata are available for counties, cities, and even portions thereof. However, it would be better if we could link the information contained in microdata to our tract-level analysis. A suitable method for this linkage is explained in the next chapter.

Poverty is the census subject selected for demonstrating the methods of this chapter. Questions to be addressed include how we should describe the incidence of poverty and how poverty data structured in tables can be manipulated in different ways. The lessons shown here would apply equally to a great many other topics, ranging from fertility behavior, to educational attainment, and to homeownership.

The chapter is organized as follows. First, we examine the fundamental task of selecting the proper *denominator* for percentages and ratios, and we make a distinction between behavior *rates* and *shares* of a total. The question of denominators would seem to be a simple one, but it is a frequent source of error. Our discussion here builds the basis for more detailed analysis to follow.

The second section takes up the issue of *disaggregation*: how to construct multidimensional tables and how to calculate rates for ever more detailed subgroups of the population. We illustrate this with the basic demographic procedure for handling the composition of data groups and interacting this with behavior rates applied to the subgroups. The third section then proceeds to a discussion of the fundamentals of manipulating *marginals* in multidimensional tables. This skill is the key for bringing detailed information from microdata tabulations down to the tract level and is illustrated in detail in the next chapter.

The final section addresses the issue of constructing tabulations from alternative universes: population, housing, or employment. While this issue is treated at length in Chapter 14, we introduce it here and describe the basic techniques for extracting different data sets from the hierarchical PUMS database.

1. Selection of Percentage Denominators

We begin with the humble percentage, the most widely used, and often abused, tool of quantitative analysis. The semantic ambiguity of the word *percentage* often leads to fuzzy conceptualization and flawed analysis, even among some highly skilled researchers. Everyone knows how to total a column of numbers and compute each number's percentage of the total. The problem comes when a second dimension is added to the table, crosstabulating the numbers by time, space, or some other factor.

Any number in a two-variable table can serve as the numerator for *three different* percentages using three different denominators: the total for the table, the column total, or the row total. Each of these has very different meanings. This is illustrated here with an analysis of poverty among whites and blacks in the U.S. population.

1.1. Black and White Poverty

Consider Table I, a representation of poverty status by race, based on the Census Bureau's Current Population Survey in March 1989. More than nine million black persons were estimated to be below the poverty line. That is the numerator, but what denominator should we divide this by?

Division by the table total would tell us 3.87 percent of all persons were black and in poverty—not a very useful fact. Division by the column total tells us that 29.57 percent of persons below the poverty line were black, and we can compare this share *of total persons in poverty* with that for other races. In fact, the number of non-black persons, termed "white" here, in poverty is much greater—over 22 million and comprising 70.43 percent of total persons in poverty. From these data some might conclude that poverty is much more a white problem than a black problem.

Table I. Distinguishing Rates and Shares—A Black and White Poverty Analysis

Persons by Race		Poverty Status of Persons		TOTAL
		Below Poverty Line	Not in Poverty	
Black	Number	9,426,000	20,421,000	29,847,000
	% of poverty status group	29.57	9.65	
	% of racial group	31.58	68.42	100.00
	% of all persons	3.87	8.39	
White*	Number	22,452,000	191,227,000	213,679,000
	% of poverty status group	70.43	90.35	
	% of racial group	10.51	89.49	100.00
	% of all persons	9.22	78.52	
Total		31,878,000	211,648,000	243,526,000
		100.00	100.00	100.00

Source: Estimated distribution of the U.S. population in March 1989; Current Population Reports, P-60, no. 166, Table 21. Identical data can be assembled from PUMS tapes for each local area with 100,000 or more population.
Note: *"White" population consists of non-black persons; 3.6% of this group are races other than white.

The third possibility is division by the row total, in this case measuring what percent *of each racial group* is in poverty. Using this set of denominators we learn that 31.58 percent of black persons are in poverty, versus 10.51 percent of white persons. In contrast to the earlier conclusion, when viewed from this perspective, the data seem to show that poverty is much more a black problem than a white problem.

Which is the right percentage to use? The answer calls for judgment about one's intended analysis. What is the comparison that is most relevant? Computer programs are of no help in making this judgment. In fact, the leading statistical analysis programs, SAS and SPSS, print out all three percentages in each cell as part of a standard crosstabulation, throwing the problem right into the lap of the analyst. Sometimes analysts choose whatever percentage looks largest or makes their political case look stronger. Other times they choose the column percentage, because many persons are more comfortable with percentages that total to 100 at the bottom of a column. However, that is a very inflexible and unreliable strategy.

Variables on the row and column dimension are easily reversed in a computer-generated table. Note that most programs determine the column variable (that across the top of the table) by whichever variable is specified *last* in the tabulation command. The example in Table I would be generated by the sequence: RACE * POVERTY. Had we specified the variables in the other order—POVERTY * RACE—the race categories would have appeared across the top and poverty down the side. In that case, row and column percentages would be reversed in their meaning from what is shown in Table I.

1.2. Selecting the Appropriate Denominator

Analysts must develop a firmer basis for deciding which percentage to use. That basis rests on a solid grasp of the denominators that define the most desirable comparison. The reader may wish to return to Table I for a closer look—to scrutinize the three different denominators that can form a percentage with the same numerator. Take as our numerator the 9.4 million blacks in poverty and circle that number. Now circle each of the alternative denominators. Label those denominators in any order—A, B, C. What basis of comparison does each imply?

1. Percentages of the table total (243 million) tell us what share *of the total U.S. population* is *both* of a given race *and* in a given poverty status. Blacks in poverty are by far the smallest group in the table. This combined view depends both on the small size of the black racial group and its likelihood of being poor, thus combining two different issues.

2. Percentages of the column total (31 million in poverty) tell us what share *of the total poverty population* is black or white. This view compares the size of the two races within the poverty status. This tells us the relative size of the two groups facing poverty. Whites account for more than twice the amount of poverty population as blacks.

3. Percentages of the row total (29 million blacks) tell us what is *the probability that a black person* faces poverty. This can be contrasted with the likelihood that a white person faces poverty. According to this comparison of "poverty rates," blacks are three times as likely to face poverty.

What accounts for the radically different interpretation of the three different denominators? The first two denominators both combine the size of the group and its likelihood of experiencing poverty. Given that blacks are only 29 million out of a total population of 243 million, their share of any total is naturally very small. However, a percentage expressed relative to the size of the black population adjusts for the overall size differences and gives us a truer measurement of black experience.

1.3. Naming the Percentages

By itself each of these percentages is very clear, but the differences between the alternatives are difficult to make clear. If extreme care is not taken, the analyst may use one percentage when another is intended or preferred. A key problem is that the same word—percentage—can be used in describing all three measurements. Both the analyst and the audience can become fuzzy on the analytical logic very quickly. One solution is to select the denominator by comparing each of the alternatives.

A further solution is a semantic one: Provide each percentage with a different name. It may help to adopt a terminology of *rates* and *shares*, a distinction consistent with much professional usage. For example, poverty rates, unemployment rates, or fertility rates are calculated as the percentage of a group that engages in a behavior, experiences an event, or experiences a status. The key principle in *defining a rate* is that the denominator is the *group at risk* for the event, such as in alternative 3 above, and we may specify denominators in very detailed fashion (such as unemployment rates by age, sex, and race).

By contrast, a poverty share consists of the total subject behavior *accounted for by a group*, as in alternative 2 above. The contributions of all groups sum to 100 percent of the subject behavior. This is a useful formulation for breaking down a total, but it is not clearly linked to the behavior of specific subgroups.

In practice, shares are calculated far more often than deserved. This happens because of overly narrow problem definition. Analysts define a behavior of interest, such as buyers of compact disk (CD) players, and then *look inside only that behavior group,* subdividing those actors into subgroups. The share of all CD buyers that are under age 30, or black or under $20,000 income, is of some interest. However, this calculation omits from the denominator critical information: the number of persons in each group who are *not* CD buyers. The young share of CD buyers may be growing over time, but perhaps the young share of nonbuyers is growing even faster. It is far better to define a base of analysis that includes both buyers and nonbuyers. Dividing CD buyers by this broader-based denominator would define a purchase *rate* and not simply a share of total purchases.

Shares are frequently used in business because they are derived from more convenient samples. Each store or industry has data on its own customers, and so it is easy to break down that *inside sample* into subgroups. Although it is useful to learn what share of total sales come from different sectors, purchase rates for each sector would be preferred for several reasons (Myers 1990c). The great virtue of census data is their comprehensive coverage, thus escaping the limitations of *inside samples* and permitting us to calculate rates instead of shares.

1.4. Preferability of Rates Over Shares

Rates are preferred to shares because they measure the behavior of the subject group whereas shares confound explanations of behavior with the size of the subject group's base. In our poverty example, white persons constitute a much larger share of poverty than blacks, but this is true only because there are many more whites overall. Thus, we see that shares confound the size of the subject group's base with the group's rate of behavior.

We can formally state the relationship between rates and shares as follows:

$$\text{Share}_a = \frac{\text{Rate}_a \times \text{Base}_a}{\sum_i (\text{Rate}_i \times \text{Base}_i)} \quad ,$$

where

Share is the proportion of total subject behavior (e.g. poverty, fertility, retail purchases, migration, etc.) contributed by a particular subgroup *a* among *i* subgroups;

Rate is the likelihood that members of a particular subgroup will engage in a behavior, measured by poverty rates, fertility rates, retail sales per capita, migration rates, etc.; and

Base is the number of members (people, households, or other units) comprising each of the subgroups of actors.

This equation shows that any group's share depends on *other* groups' behaviors and sizes. Not only is a group's share of the total subject behavior determined by its own propensity to spend and by the number of members in its own subgroup (size of the base), but this must be divided by the sum of products between every other group's behavior and base size. This is an analytical quagmire that prevents us from more accurately explaining or forecasting changes over time. For more incisive analysis, we want to distinguish one group from another and also separate the size of its base from its rate of behavior. This is demonstrated in a following section.

2. Disaggregation and Multidimensional Tables

Behavior rates can be defined for subgroups classified by multiple dimensions. The total poverty rate for the nation can be disaggregated in an infinite variety of ways. In Table I we disaggregated poverty by race. Shortly, we will disaggregate it further by adding dimensions of age and sex to race. Alternatively, we might have disaggregated poverty by education level or marital status or state of residence. Any variable recorded by the census questionnaire, and included in PUMS, potentially can be used to disaggregate any other.

In general, we wish to break out the dependent variable by categories of another variable that we think is important. The key decisions in deciding what is appropriate for disaggregation are threefold. First, analysts may hope to *find sizable differences* between categories. Disaggregation of poverty by race is important because the poverty rate is very different between categories of race. Second, analysts often disaggregate by categories of a variable that may *cause* the dependent variable. Thus, in addition to race, we might disaggregate poverty by education level, marital status, sex, or age. Third, analysts often disaggregate by a variable that has *policy significance*. Race is a variable that has policy significance, and it also passes the other two tests of significance as well.

2.1. Mechanics of Disaggregation

Disaggregation involves counting cases in categories of one variable as a subset of each category of another variable. Each variable used in the disaggregation forms one dimension to the resulting matrix. These dimensions are often called *marginals*, about which we will elaborate later. Breaking a population down into two races (black and white or other) is disaggregation by one variable—race. Further disaggregation into male and female persons yields one more dimension, and disaggregation by age adds a third, for a total of three dimensions and 28 defined subgroups (2 races × 2 sexes × 7 ages = 28 subgroups). Fewer and fewer cases will fall into each subgroup as the total is progressively disaggregated by more dimensions.

Poverty can be disaggregated by these population variables. Poverty rates can be defined for each subgroup by appending the poverty variable, with its two categories (below the poverty line; above the line), to a computerized table request. Following the rule that the last variable in the request appears across the top of each subtable, we would write the request as follows: RACE * SEX * AGE * POVERTY. In this format, row percents would define the poverty rate (percent of each subgroup that is in the category of POVERTY identified below the poverty line). Each row would give the poverty rate for each age group. A separate subtable would be generated for each combination of race and sex, yielding in this case four subtables. Table II shows one

Table II. Disaggregation of Poverty Rates—Age Differences Among Black Female Subgroup

Controlling for: RACE = Black SEX = Female		Value of POVERTY		
		Below Poverty Line	Not in Poverty	TOTAL
AGE =				
Under 16	Number Row %	1,903 44.14	2,408 55.86	4,311 100.00
16 to 21	Number Row %	584 36.68	1,008 63.32	1,592 100.00
22 to 44	Number Row %	1,772 30.02	4,130 69.98	5,902 100.00
45 to 54	Number Row %	336 23.50	1,094 76.50	1,430 100.00
55 to 64	Number Row %	346 29.73	818 70.27	1,164 100.00
65 or older	Number Row %	552 37.94	903 62.06	1,455 100.00
Total		5,493 34.65	10,361 65.35	15,854 100.00

Source: Current Population Reports, P-60, no. 166, Table 21. Identical data can be assembled from PUMS tapes for each local area with 100,000 or more population.

Note: Table is formatted to resemble computer printout where column percents and total percents have been suppressed. Table request: RACE • SEX • AGE • POVERTY

of the subtables, that generated for black females, deleting the percentages other than row percents.

Alternative orderings of the variables are possible. We could reverse the last two variables, with age across the top and poverty down the side of the table. In that case column percents would report the poverty rate. We could also move age to an earlier position in the table request statement and replace it on each subtable with race or sex. However, there is a very practical reason for keeping race and sex in their prior position: The combination of those two variables yields only 4 subtables, but age has more categories and would yield 14 subtables if combined with race or sex in a prior position in the statement. Therefore, a practical rule is to keep variables with fewer categories in the prior position unless they are the dependent variable, such as poverty in this case.

2.1.1. Rates Condense Information

How to make use of four pages of output describing the relationship of race, age, and sex to poverty? This information can be distilled to an essential core by reducing it to a series of rates pertaining to the series of population subgroups. The frequency counts for poverty are distributed in two categories: in poverty, not in poverty. By constructing a rate as the ratio between the frequency in poverty and all persons (sum of in and out of poverty), we reduce the information. With knowledge of the poverty rate, information on persons not in poverty becomes redundant. It is simply the complement: 1 − the poverty rate, or in percentage terms, 100 − the poverty rate.

The poverty rates can be extracted from the subtable on each page and combined in a single table of poverty rates, as shown in Table III. Here the poverty rates from each subtable are arranged in parallel columns for inspection. The savings through condensation are considerable. We can use a table of rates to represent a matrix that is at least twice as large. An alternative mode of presentation is to graph the poverty rates, as in Figure 1.[1]

Interpreting the results, we see that blacks have higher poverty rates than whites in every age group. Also, with the exception of children under age 16, females have higher poverty rates than males of the same race. The highest poverty rates of all are found among black females. The pattern by age shows that poverty reaches its lowest rate in the 45 to 54 age group. At younger ages poverty is greater both because earnings are lower among young adults and also because of the financial burden created by children living in the home. This *children effect* is directly represented by the higher poverty rates of children themselves. Black females may be especially prone to poverty because of their child rearing burdens at young ages, whereas black males may escape this poverty burden by living separately.

Table III. Poverty Rates Disaggregated by Age, Sex, and Race

| | Percent Below the Poverty Line | | | |
| | Black Persons | | White* Persons | |
Age of Person	Males	Females	Males	Females
<16	46.6	44.1	16.0	15.6
16–21	31.1	36.7	11.1	14.0
22–44	16.0	30.0	7.2	10.2
45–54	15.4	23.5	5.4	7.2
55–64	20.0	29.7	7.0	9.6
65+	23.6	37.9	6.6	12.7
Total	28.1	34.7	9.3	11.6

Source: Current Population Reports, P-60, no. 166, Table 21. Identical data can be assembled from PUMS tapes for each local area with 100,000 or more population.

Note: *"White" population consists of non-black persons; 3.6% are races other than white.

[1]A line graph is preferred to a bar chart because it maximizes the data-ink ratio and communicates the data pattern much more clearly (Tufte 1983). However, the graph is impaired by the equal spacing of uneven intervals on the X-axis: the 16-21 age group is plotted as equal in width to the 22-44 age group. This bias is judged acceptable for this graph because it does not distort the message of the graph and is outweighed by the benefits of the line-graph format.

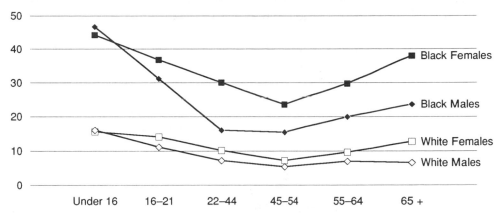

Figure 1. Poverty rates by age, sex, and race in 1989. The rates are the percentage of persons in each subgroup whose incomes fall below the census-measured poverty line. *Source:* Table III.

2.2. Combining Composition and Rates

The preceding review of percentage construction and disaggregation is important for improving the rigor of logic in census data analysis. The basics of analysis can be reduced to two components: *composition and behavior rates.* These two components are derived from tabulation of census data. In the census year the two components form an accounting identity: Number of Actors = Base Group Size × Behavior Rate. Forecasts of future actors can be achieved by projecting the size of each base group and projecting the behavior rate of interest.

Composition describes the makeup of the population or other data group, such as the housing stock or labor force. As we have seen, this is tabulated by disaggregating the total into meaningful subgroups and sorting the cases into their respective categories.

The second dimension, behavior rates, is linked to composition following the rule of symmetry. Behavior rates are measured for each disaggregated subgroup in the population composition. A great many rates might be calculated for the subgroup bases: fertility and mortality rates, poverty rates, unemployment rates, labor force participation rates, migration rates, household formation rates, etc.

Multiplying the behavior rates for each subgroup times the number of persons in that subgroup yields the expected number of persons in each status or behavior group.[2] Total numbers in the category will shift as either the numbers of persons changes in different categories or as the behavior rate changes. Ideally, both factors can be projected. On the rate side we would trace the trends in the particular behavior rate for each separate subgroup and project this trend into the future. On the composition side, we would trace the growth in each of the demographic subgroups

[2]Demographers will recognize this analytical structure as a generalized form of specific procedures such as the cohort-component population projection method. In that application age and sex form the composition, while the rates are fertility, mortality, and migration.

and project the future numbers of persons in each subgroup who are at risk for the behavior.

All demographic change can be expressed as either a change in composition, a change in behavior rates, or a combination of the two. In place of what may seem to be endless variations on changing demographics, a simple *rates × composition analysis structure* provides a much more orderly view.

This *rates × composition* analysis structure can be applied usefully to many questions of change over time. A typical question that arises in local areas is how much poverty (or another behavior or status) will change in the next decade. From the most recent census data, analysts can tabulate a very detailed picture of how much different subgroups contribute to the total behavior. Typically, analysts also have available some forecasts of the changing population composition and even some trends in the subject behavior over time. The task is how to organize this information to shed light on the forecast of interest.

Table IV applies the rates × composition analysis structure to the question of poverty, extending the preceding poverty rate analysis. The baseline condition is the population by subgroup and the poverty rates that are observed at the most recent census. (The rates are the same as those shown in Table III.) We then can project the effects of changes in composition, rates, or both. Section B shows the results of multiplying new population numbers times the poverty rates previously observed for each subgroup. Because of the aging of the population toward age groups with lower poverty rates, we would expect to see a lower total of persons in poverty—2,226 fewer persons than in the baseline period. The number in poverty is reduced among both blacks and whites.

However, the poverty rates for each age-sex-race group may not stay constant. Should rates shift to higher levels, this could offset the gains achieved through a favorable composition change. The lower section of Table IV illustrates a situation where poverty rates rise, more for young blacks than older blacks, and more for blacks than for whites. The end result is that the total number of persons in poverty is greater in section C than in section A. The white numbers in poverty remain lower than in section A, but the black numbers are greater.

More formal analysis can be carried out with these data following a component of change format, such as was illustrated with race data in Chapter 10. Different models of expected change can be constructed by varying the rates and composition inputs to the model. Obviously, great care should be taken in researching trends to support the assumptions of each model.

☐ 3. Manipulating the Marginals of Tables

An important way to use disaggregation in census data applications is based on the concept of *marginals*. Each variable in a table constitutes one dimension that is termed a marginal because the variable appears as a row or column of totals at the margin of the table. Two marginals appeared in Table II: poverty, described by a set of column totals, and age, described by a set of row totals. That table is actually a subtable of a larger table formed by two additional marginals, race and sex. For the moment, let us treat that subtable in isolation.

Table IV. Rates Times Composition Structure of Analysis

A. Baseline Condition

	Composition (Population by Subgroup)					Rates (Poverty Rates by Subgroup)				Subject Behavior or Status (Persons in Poverty)				
	Black Persons		White Persons			Black Persons		White Persons		Black Persons		White Persons		
	Males	Females	Males	Females		Males	Females	Males	Females	Males	Females	Males	Females	TOTAL
<16	4,449	4,311	24,574	23,391		0.466	0.441	0.160	0.156	2,075	1,903	3,933	3,639	
16–21	1,485	1,592	8,993	9,175		0.311	0.367	0.111	0.140	462	584	995	1,287	
22–44	5,003	5,902	39,809	36,801		0.160	0.300	0.072	0.102	802	1,772	2,852	4,045	
45–54	1,135	1,430	10,782	11,275		0.154	0.235	0.054	0.072	175	336	582	807	
55–64	942	1,164	9,147	10,147		0.200	0.297	0.070	0.096	188	346	640	974	
65+	981	1,455	11,097	15,489		0.235	0.379	0.066	0.127	232	552	733	1,966	
Total	13,995	15,854	104,402	109,278						3,934	5,493	9,735	12,718	31,880

B. Different Composition and Same Rates

	Black Persons		White Persons			Black Persons		White Persons		Black Persons		White Persons		
	Males	Females	Males	Females		Males	Females	Males	Females	Males	Females	Males	Females	TOTAL
<16	3,500	3,500	19,000	18,000		0.466	0.441	0.160	0.156	1,632	1,545	3,041	2,800	
16–21	1,000	1,000	7,000	7,000		0.311	0.367	0.111	0.140	311	367	774	982	
22–44	4,500	5,500	36,000	36,000		0.160	0.300	0.072	0.102	721	1,651	2,579	3,659	
45–54	2,000	2,500	13,000	14,000		0.154	0.235	0.054	0.072	308	587	702	1,002	
55–64	1,500	1,800	12,000	14,000		0.200	0.297	0.070	0.096	299	535	840	1,344	
65+	1,200	1,600	14,000	17,000		0.236	0.379	0.066	0.127	284	607	925	2,158	
Total	13,700	15,900	101,000	106,000						3,556	5,293	8,861	11,945	29,654

C. Different Composition and Different Rates

	Black Persons		White Persons			Black Persons		White Persons		Black Persons		White Persons		
	Males	Females	Males	Females		Males	Females	Males	Females	Males	Females	Males	Females	TOTAL
<16	3,500	3,500	19,000	18,000		0.500	0.500	0.165	0.160	1,750	1,750	3,135	2,880	
16–21	1,000	1,000	7,000	7,000		0.330	0.400	0.120	0.140	330	400	840	980	
22–44	4,500	5,500	36,000	36,000		0.200	0.330	0.080	0.110	900	1,815	2,880	3,960	
45–54	2,000	2,500	13,000	14,000		0.170	0.250	0.060	0.080	340	625	780	1,120	
55–64	1,500	1,800	12,000	14,000		0.220	0.320	0.080	0.110	330	576	960	1,540	
65+	1,200	1,600	14,000	17,000		0.250	0.400	0.070	0.130	300	640	980	2,210	
Total	13,700	15,900	101,000	106,000						3,950	5,806	9,575	12,690	32,021

Source: Table III and author's calculations.

265

3.1. Structure of a Table

The entries in a table are described by the intersection of variables that form its dimensions. The cell frequencies found in each jointly defined category of the crosstabulation may be described as $n_{ij...x}$, where i through x identify the different variables defining the table's cells. The sum of all cells is n, the grand total of the table.

Given the entries in a two-dimensional table, n_{ij}, one marginal is found by *summing over*, or *collapsing*, all the categories of the other. The marginal on the i dimension is found by $\Sigma_j n_{ij}$ and is given by a series of row (or column) totals. The marginal for j is found by $\Sigma_i n_{ij}$ and is shown by the other series of column (or row) totals. The grand total, n, is the total marginal, given by $\Sigma_i \Sigma_j n_{ij}$.

These relationships are illustrated in section A of Table V. We see that a total of 15,854 cases is distributed by categories of poverty and age, represented by i and j, respectively, forming cells n_{ij}. With categories of age arrayed in rows, the age marginal appears as a vector of row totals. Conversely, poverty is arrayed in columns and appears as a row of column totals. The grand total appear at the lower right of matrix A.

The opposite procedure to collapsing a dimension, or summing over a marginal, is to disaggregate a table by *adding* another dimension, as discussed in the preceding section. That procedure cannot be conducted within an existing table. While dimensions can always be collapsed by summing across a dimension, the only way to add a new dimension to a table is to retabulate from the raw data or seek some other information from outside the table.

In tables with more than two dimensions, the extra dimensions define separate subtables, as discussed under disaggregation. In such tables, marginals can be formed by pairs or sets of variables. Not only do we have the one-dimensional marginals defined by each separate variable, but there also are two-dimensional, or higher order, marginals. For example, in the case of our crosstabulation of RACE * SEX * AGE * POVERTY, we have the four one-dimensional marginals, then six two-dimensional marginals formed by each conceivable pair of variables, and finally four three-dimensional marginals. The four-dimensional marginal is the complete table. These multidimensional marginals are employed most commonly in statistical models for categorical data, such as hierarchical log-linear models.[3] For census analysis purposes, it often suffices to operate on two-variable tables defined as subsets of larger tables.

3.2. Expected Values in Tables

The marginals for tables can be used in different ways to estimate expected values for the cells within the table. These expected values can be used for statistical tests

---

[3]Log-linear models are a statistical method developed for testing hypotheses and measuring effects in categorical data structured in tables. Two accessible overviews may be found in Davis (1974) and Feinberg (1977).

Table V. Table Manipulations with Marginals

A. Initial Matrix Given

B. Expected Values Under Assumption of Independence (1)

	POVERTY (i)		*Age Marginal*		POVERTY (i)		
	In Poverty	Not	$= \sum_i n_{ij} = n_j$		In Poverty	Not	
AGE (j)							
Under 16	1,903	2,408	4,311		1,494	2,817	4,311
16 to 21	584	1,008	1,592		552	1,040	1,592
22 to 44	1,772	4,130	5,902	"Row Totals"	2,045	3,857	5,902
45 to 54	336	1,094	1,430		495	935	1,430
55 to 64	346	818	1,164		403	761	1,164
65 or older	552	903	1,455		504	951	1,455

Total Marginal

Poverty Marginal	5,493	10,361	15,854	$= \sum_i \sum_j n_{ij} = n$	5,493	10,361	15,854

$= \sum_j n_{ij} = n_i$ "Column Totals"   "Grand Total"

(1) Each cell is estimated as the product of the row marginal and the column marginal, divided by the total marginal.

C. Initial Matrix Scaled by New Total Marginal (2)

Ratio of Desired to Initial Total = 1.262

D. Initial Matrix Scaled by New Age Marginal (3)

	New Age Marginal	Ratio to Marginal of Matrix A	Ratio of Desired to Initial Given at Right

AGE (j)									
Under 16	2,401	3,038	5,438	4,500	1.044	1,986	2,514	4,500
16 to 21	737	1,272	2,008	1,800	1.131	660	1,140	1,800
22 to 44	2,235	5,210	7,445	7,000	1.186	2,102	4,898	7,000
45 to 54	424	1,380	1,804	2,500	1.748	587	1,913	2,500
55 to 64	436	1,032	1,468	2,300	1.976	684	1,616	2,300
65 or older	696	1,139	1,835	1,900	1.306	721	1,179	1,900
	6,929	13,071	20,000	20,000		6,740	13,260	20,000

(2) Each cell in the initial matrix is multiplied by a scalar defined as:
Scalar = Desired Total of 20,000 / Initial Total Marginal
E.g. in the first row, 1.262 is multiplied times 1,903 and 2,408 yielding the results shown in Matrix C.

(3) Cells in each row in the initial matrix A are multiplied by a vector of row scalars defined as:
Scalar = New Age Marginal / Initial Age Marginal
E.g. in the first row, 1.044 is multiplied times 1,903 and 2,408 yielding the results shown in Matrix D.

and for estimations of unknown values. For example, as shown in section B of Table V, given the marginals from matrix A, we may estimate each cell entry as the product of the row and column totals, divided by the grand total of the table. This procedure produces an expected entry of 1,494 cases in the category under 16 and in poverty. The expected value is much lower than actually observed, indicating that there are many more persons in poverty who are young. The expected value for under 16 and in poverty would have been greater, had the number of young persons in the age marginal been greater, or had the number of persons in poverty been greater in that marginal. The fact that the observed values exceed the expected in this age category shows that poor people are disproportionately young (or young people are disproportionately poor).

The estimates produced in matrix B are the basis for the well-known chi-square statistical test that calculates squared deviations of observed from expected values, divided by the expected values. Comparison of matrix A with B by this method yields a chi-square value of 328.24, with 11 degrees of freedom. This highly significant result indicates that the two marginals are not independently distributed.

Other applications of expected values can be constructed in different ways to achieve *estimation* purposes. Rather than testing the significance of known distributions, we can calculate the most likely estimate for unknown distributions. We focus on those applications for the remainder of this chapter.

3.3. Scaling Tables by New Information

Census data provide a detailed portrait of the population for a given place at a given time. However, we frequently need to adjust these data for broader purposes. Either we need to update the data for trends after the census, or we need to *borrow* information from a larger area to use in a smaller area. Although the postcensal application is most common, we often need to make use of detailed microdata tabulations that are otherwise restricted to larger geographic areas. The same set of procedures is required for either use.

The data in a table can be scaled up or down without distorting the underlying relationships between the variables. We can do this because tables have the property that the cross-product ratio between rows and columns is unaffected by the size of the marginals (Mosteller 1968; Fienberg 1977: 16–19). Divide all categories by 10 and you get the same statistical relationship.

3.3.1. Scaling by a New Total

Section C of Table V illustrates the case in which we wish to adjust the table for a larger overall population. If we only know that the population is larger in total, then the most likely estimate for each of the cells is obtained by scaling them all proportionally. The scalar is formed by the ratio of the new total to the grand total for the initial table (i.e., 20,000 / 15,854 = 1.262). This scaling preserves all existing relationships within the table: The age distribution remains the same, the percent in poverty remains the same, and the relationship between age and poverty remains the same. Such an assumption of no change is the conservative assumption.

3.3.2. Scaling by a New Row or Column Marginal

We can improve these estimates if we can obtain more detailed information about the second population. Suppose we know not only its total size but also its age distribution. The new age marginal can be used in combination with the age marginal of the initial table to adjust the cell frequencies. In Table V, a new age marginal is given between matrix C and D. A series of row scalars is then formed by taking the ratio of the new age marginal to the initial age marginal. The cell frequencies in matrix A are then multiplied by their respective row scalars to produce matrix D. The information in matrix D sums to the same new total of 20,000 but its age distribution is altered to conform to the new marginal. Within each age group, the proportion in poverty remains the same. However, the poverty marginal at the bottom of the matrix is changed because it is the sum of the age rows, and these have now been weighted differently by the new age marginal.

3.3.3. Scaling by Multiple Marginals

New information estimated from Table V could be improved another step if we also possessed a new marginal for poverty. We could then scale, or *rake*, the initial table first by the new row marginal, then by the new column marginal. However, each scaling by one marginal will disrupt the totals on the other marginal, requiring a series of successive rakings to achieve a stabilized estimate. The method of *iterative proportional scaling* yields estimates comparable to least-squares estimates of expected values, given the respective row and column totals. That method was originally developed for use with census data (Deming 1948) and is now used for broad statistical purposes with cross-classified categorical data (Fienberg 1977). We demonstrate it fully in the next chapter.

A different example of scaling with multiple marginals is the *rates × composition* analysis structure demonstrated in Table IV. There we used rates to condense the poverty marginal from two categories to one. The baseline table consisted of four dimensions, three of which described population composition and one of which described poverty status. The middle section of Table IV then applied a different composition to the baseline rates. In terms of our marginal concepts, we essentially scaled the initial matrix by a new three-dimensional composition. With the poverty rates held constant, we retained the same poverty marginal as in the initial matrix.

3.4. Generalized Use of Marginals

Thus, the *rates × composition* analysis structure is a specialized form of scaling with marginals, using condensed marginals in the form of rates, and also treating the three composition marginals as a single three-dimensional marginal. A great many such applications can be constructed from the generalized notion of scaling with marginals.

The concept of marginals and their use is both flexible and powerful. For estimation purposes, additional information can be applied to an initial table as available. In this manner, census data from summary tabulations or microdata form the basis for extensive estimates in small areas.

4. Constructing Microdata Files

A firm understanding of appropriate percentage denominators and principles of disaggregation is required for structured analysis of microdata. In the remainder of this chapter we describe how data from different universes can be tabulated and interpreted.

4.1. Data Files for Different Universes

The basic structure of the PUMS files was introduced in Chapter 4. There we described the hierarchical structure of the database, with a housing record, "H," preceding one or more person records, "P" (one for each of the household occupants).[4] This data structure preserves information about who lives with whom in a household and what kind of housing they share.

For specific analysis, the raw data usually need to be extracted into more usable "flat" files with different properties. Four basic types of extracts can be performed from the raw data, creating files of type P, H, PH, or PHH. The first two consist of P or H records in isolation, whereas the latter two combine the information from the P records with the corresponding H records. The PHH file type is merely a subset of PH, with the person records retained only for the householder. We have identified this as a separate file type because of its importance for defining the universe of households. As discussed in Chapter 3, households are the key concept for linking the population and housing data.

One other important subset can be defined from the file types defined above. Labor force data are recorded as person characteristics, including current labor force status, type of job (if employed), and commuting behavior. This subset can be defined relative to a type P file, or it can be defined relative to type PH or type PHH, depending on the intent of analysis.[5]

The necessary programming statements to create each major file type from the raw PUMS data are listed in Appendix B. These examples are written in the SAS syntax, although comparable statements can be executed in SPSSx (Hedderson 1987: Chapter 13) or other major statistical programs.

4.2. Kinds of Questions Addressed with the Data

With the PUMS data, four different universes, can be exploited: population, housing, occupied housing units (or households), and labor force (job holders and job seekers). Each universe or base can be used to address different questions. Two considerations to be kept in mind are what type of file combines the different variables desired and what variable defines the denominator, or base, for the analysis.

---

[4]Persons not living in housing units receive a blank housing record, or one marked group quarters, followed by a single person record.

[5]Records can be selected into the file based on whether the person is presently employed or more generally in the labor force. For more details about the labor force variable, see Chapter 13.

4.2.1. Population Universe

On a population base, we can tabulate the demographic, housing, household, or labor force characteristics of persons. Only portions of these data are recorded directly on the "P" record: all the demographic and labor force variables, plus a variable for relationship in the household. A file type PH or PHH is needed in order to access housing characteristics of persons.

4.2.2. Housing Unit Universe

On a housing base, we can tabulate the characteristics of all housing units (both occupied and vacant) using a type H file. Without information on the occupants, we cannot tabulate demographic, household, or labor force characteristics.

4.2.3. Household (Occupied Housing Unit) Universe

More commonly, analysts use an occupied housing unit base, because this provides information about the household and persons who live in different types of units. With a type PHH file, we can tabulate the demographic characteristics of householders (e.g., race, marital status, and age). These householder characteristics are commonly used by the Census Bureau and others to describe the household. Although this demographic data could be pulled directly from a type P file, without the housing information we cannot distinguish, for example, between the demographic characteristics of owner and renter households.

Similarly, we could distinguish between types of occupied units using a type H file, but we could not relate these different housing types to the different types of households that occupy the units. In general, the household concept is especially rich because it links the housing and its occupants, and the two household type files (PHH or PH) are the most fruitful for analysis.

Microdata users who hesitate at the distinction between analyzing demographic characteristics *of housing* versus the housing characteristics *of persons* may benefit from Fig. 2. Categories of one variable can be analyzed relative to categories of the other. For example, a table of age of person by type of housing can be read in two directions, using either the age marginal or the housing marginal as the denominator (or base) for the analysis. Such a table was presented previously in Table III of Chapter 3.

4.2.4. Employed Worker Universe

Employed workers are a subset of the population universe, restricted to persons 16 years old or more and further restricted to persons who are presently employed. Note that on a population base we could calculate the proportion of persons who are in the labor force or, more specifically, presently employed. (Such a population-based analysis is carried out in the next chapter and details on the labor force concept are given there). Conversely, on an employed labor force base, we can tabulate the job characteristics of workers and the demographic or housing characteristics of workers in different types of jobs.

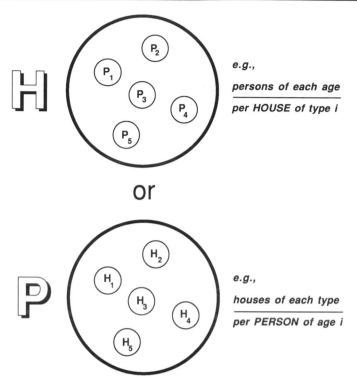

Figure 2. Two alternative bases: housing or population.

4.2.5. Summary

The PUMS raw data contain several overlapping but distinct universes. Not all persons live in housing units and not all housing units are occupied. Also, persons who live in households are a broader set than householders, only one of whom may be identified per household. Finally, employed workers are a subset of all persons, but persons often work in locations different from where they live, and, with multiple workers per household, the employment linkages emanating from the common household become complex.

The hierarchical structure of the PUMS database has rich potential, allowing us to construct tabulations on alternative bases. With these custom tabulations analysts can link population, housing, and employment variables in a fashion required for particular models of community analysis. A review of the basic concepts and linkages defined in Chapter 3 may assist with analysis of these microdata.

□ 5. Conclusion

This chapter has reviewed the basic skills necessary for working with microdata. Deliberate care is needed in constructing ratios and percentages with the appropriate denominators. Similarly, analysts require a firm grasp of the principles of disaggregation and the role that marginals play in tables. The hierarchical sequence of micro-data records and the multiuniverse structure of the database create an especially rich, and challenging, environment for analysis.

The following chapters will develop two concepts in particular. Chapter 13 links labor force behavior to a population base. The chapter carries the added responsibility of demonstrating how tract level data may be augmented by *borrowing* more detailed microdata tabulations. The method of iterative proportional scaling shown there is widely used for many different applications of small area analysis. Chapter 14 then approaches more comprehensively the task of linking variables from different universes. Those linkage analyses are among the most rewarding with microdata, but they are also among the most difficult.

13 □□□

Labor Force Analysis in Small Areas

Economic behavior in the census is analyzed on a population base. A major component of the economic data collected by the decennial census consists of labor force participation and characteristics of persons' jobs. Hundreds of different occupation types and industry types are recorded by the census. These data are recorded for individual persons, describing their employment-related behavior in the week preceding the census. All of the economic data, including labor force participation, are collected from only a sample of the population, with consequent problems of sampling variability in small areas.

Labor force analysis illustrates two general problems faced in local analysis. Solutions proposed in this chapter have broad utility. First, the hierarchy principle of data availability constrains analysis in local areas. Opportunity for detailed analysis of labor force behavior and types of jobs held by different types of persons is provided only by PUMS data for larger areas. Detailed tabulations of labor force behavior by age are not published. Instead, moderately detailed data on labor force status may be retrieved for small areas only from the difficult-to-access STF 4.[1] The fewer cases reported in small areas simply permit fewer categories of labor force and job types than supported by PUMS.

---

[1] The Census Bureau cautions that "STF's 2 and 4 involve rather large logical records and relatively complex structures. STF's 1 and 3 are usually much easier to work with, as well as less expensive to purchase" (U.S. Bureau of the Census 1983a: 3). The STF 4 files needed for detailed labor force analysis also are not produced in CD-ROM format or on microfiche, making them less accessible than other STFs.

Not only is information more sketchy at the local level, but the sample count variables—which include all the economic data—are prone to large sampling error. Consider that a census tract of 3,600 population has only 600 direct observations if sampled at 1-in-6. Of these 600 persons, perhaps only 400 are of labor force age. Incisive analysis of labor force behavior requires that men and women be separately analyzed. Race and ethnicity may also be important, and age is an especially critical factor. However, once divided into so many subgroups, the sample of adults becomes too small for any useful analysis of actual behavior in small areas. Rates of behavior sampled for a specific age group can easily fluctuate by ±15 percentage points. One alternative smooths out the sampling variability of small local estimates and borrows detailed information from higher geographies. The method of *iterative proportional scaling* permits us to estimate detailed behavior in small areas by combining readily available local data with more detailed information drawn from the larger area. The larger area information may be taken from several sources: published detailed characteristics for metropolitan areas, locally available tabulations from PUMS data, or custom tabulations from PUMS data. PUMS data are generally easier to work with than STF 4. In addition, a relatively few tabulations for the larger area are sufficient to serve the needs of analyses in a multitude of small areas. Thus, it may be possible to borrow a PUMS tabulation "off the shelf," whereas labor force analysis of a particular neighborhood would require customized access to STF 4 files for each specific place.

Skill with scaling procedures, as introduced in Chapter 12, provides analysts with much greater flexibility. An iterative proportional scaling method is demonstrated in this chapter that allows us to introduce variables that are not crosstabulated in any STF source. The method has other applications, such as updating census data as they grow older. In the future, this method could take on even broader utility, given current discussions about reducing the detailed coverage afforded small areas in future censuses.

In this chapter we first review the labor force data provided by the census and briefly consider some trends and issues regarding labor force behavior. Next we examine evidence on labor force participation derived from microdata sources. Patterns by age, sex, race, and housing are described. Then we turn to the task of adapting these larger area profiles to the analysis of a small area, using the method of iterative proportional scaling. Results of this estimation are compared to actual data from STF 4 and conclusions drawn.

☐ 1. Labor Force Concepts in the Census

The census collects information on several labor force concepts. Persons age 16 and older who filled out the sample, long-form questionnaire were asked a number of questions relating to their work activity and status in the week prior to filling out the form. This reference week was the last week of March for most persons. A summary of the key concepts is provided below. For more detailed explanations, consult the U.S. Bureau of the Census (1982).

1.1. Labor Force Status

Labor force status is a hierarchical concept, indicated by indented subsets within categories. The 1990 and 1980 censuses report the following categories for labor force, shown here with illustrative numbers:

Persons, 16 years and over	**10,000**
In labor force	7,000
Armed forces	200
Civilian labor force	6,800
Employed	6,500
Unemployed	300
Not in labor force	3,000

Total persons age 16 and over are the universe for labor force analysis, although published sources break this down by sex and sometimes by age. The top category shows 7,000 persons in the labor force and the bottom reports the remainder who are not. The labor force includes both members of the armed forces and of the civilian labor force. The latter is further broken down to include both persons currently employed and persons looking for work. The *unemployment rate* is calculated as the proportion of the civilian labor force that is currently unemployed. Persons not employed and not looking for work are not counted as in the labor force and are not counted as *unemployed*.

The categories shown cover the essentials for labor force analysis. Some additional crosstabulations condense this information by omitting redundant categories. Given the total number of persons, the full set of categories listed above can be reconstructed from three categories alone: employed, unemployed, and not in labor force. Various additional tabulations related to labor force, such as crosstabulation by presence of children of different ages, may be found in different sources.[2]

The concept of employment has certain complexities. Employment is made up primarily of persons *at work* in the reference week. This includes paid employees, persons with their own business or own farm, or persons who worked at least 15 hours as unpaid workers on a family farm or business. Employment also includes persons *with a job but not at work* during the reference week, such as those temporarily absent due to illness, vacation, or another factor. Some workers hold more than one job, but the census collects information only about the worker's primary source of employment. Because of this "moonlighting," the number of employed persons is slightly less than the number of occupied jobs.

Employment and labor force information collected by the decennial census is limited because the census is a residence-based survey and does not visit workplaces.[3]

---

[2]The CP-2 report for each state (including its subareas) lists a variety of different tabulations related to labor force status. For more information about CP-2, see Chapter 4.

[3]The Census Bureau has alternative programs to collect workplace-based information. Detailed economic censuses are conducted every five years (for years ending in 2 or 7), and the annual *County Business Patterns* report job counts by industry type. For information about these alternative sources, see U.S. Bureau of the Census (1991c).

As a result, the census cannot record information about vacant jobs like it does vacant housing. Also, counts of employed persons are recorded by place of residence and not place of work. Thus, for reasons of commuting, the number of workers living in a subarea can be very different from the number of jobs located there.

Persons age 16 and older who are not in the labor force (both not employed and not looking for work) consist mainly of students, homemakers, retired workers, inmates of institutions, and others unable or unwilling to seek employment in the reference week prior to the census. As described below, the number of persons not in the labor force has been declining over time.

The two key measures used to describe a population's labor force status are the percent of population in the labor force (*labor force participation rate*) and the percent of labor force that is unemployed (*unemployment rate*). In this chapter particular attention is directed to describing labor force participation by specific age groups, controlling also for race and sex.

1.2. Industry and Occupation

Respondents to the census long-form questionnaire were asked for details of the job worked during the reference week, or for the most recent job worked in the last five years. Persons with more than one job gave details for the job where they spent the most hours during the reference week. The published census tract tables for 1990 and 1980 list these data as *Occupation and Selected Industries*. Hundreds of different occupations and hundreds more industries are coded by the census. Although all of these may be used to describe individual jobs on the microdata, only broad categories are reported in published tabulations. At the tract level, 13 occupations and 7 industry groups are summarized. Some of these data are addressed in the next chapter.

☐ 2. Trends and Issues

Trends and differentials in labor force behavior are linked to a number of important issues. These include the increase of women working outside the home, the shifting linkage of employment and population growth, and concerns about the proper balance of jobs and houses in local areas. While we possess very good data about these issues at the national level or other larger geographies, much less is known at the local level where individuals play out their lives. Before demonstrating the analytical methods of this chapter, it may prove useful to review the significance of the data.

2.1. Shifting Linkage of Employment and Population Growth

One of the most dramatic social changes of the twentieth century has been the rise in female labor force participation. Masnick and Bane (1980: Chapter 3) offer an historical analysis from 1940 and develop projections of cohorts' trajectories towards the end of the century. Traditionally, young women entered the labor force for only a few years, reaching peak participation around age 20 or 22. After that age, participation rates steadily declined, because women married, had children, and devoted full time to unpaid work inside the home. The 1950 census detected a secondary peak of labor force participation forming around age 40 to 45. Apparently, women were returning to the labor force after their children were older. By 1960, the new middle-aged peak rose even higher, to nearly 50 percent participation, and still higher by 1970. Large baby boom families had more mouths to feed and mothers responded in order to sustain a high standard of living for their families (Oppenheimer 1982).

The pattern of participation observed among women in 1950 through 1970 resembled an M shape: two peaks, separated by a deep valley. By 1970, it was apparent, however, that the valley between the peaks was filling in. More young women were sustaining their labor force participation and not dropping out of the labor force in their late twenties or early thirties. By 1980, a new pattern had nearly eliminated the traditional M shape. Masnick and Bane's analysis of baby boom cohorts showed that these women were entering the labor force with very high participation rates at ages 20 to 24, and they suggested that the upward trajectories of the cohorts would create a new peak of participation when they reached 30, an age range where there was once a valley.

At the same time that women have increased their labor force participation, age structure has shifted toward middle age. During the 1970s, both males and females in the baby boom generation entered age brackets where rates of labor force participation are typically high. The combined effect of rising rates among women and the shift in age structure is that a smaller population growth can absorb more jobs. As shown in Table I, for the six-county Los Angeles region, jobs grew during the 1960s and 1970s at a much faster pace than population, because expanding labor force participation absorbed the added employment.

Table I. Comparison of Employment and Population Growth Over Three Decades[4]

	1960s	1970s	1980s
% employment growth	38.8	34.8	25.1
% population growth	16.4	14.8	26.8

In the 1980s, however, the population could not continue expanding its labor force participation. As a result, the slower increase in jobs could not be satisfied by the resident population and greater migration was induced. Some of the immigrant workers also brought more dependents than characteristic of the established population, further enhancing population growth.

2.2. Variations by Urban Subareas

It is unlikely that this shift in labor force participation has been spread evenly across subareas of the region. Sharp differences between places in racial composition and age composition suggest substantial differences in labor force behavior as well. Patterns of change may differ widely between types of neighborhoods, not only in terms

[4]These data were supplied by Viviane Doche-Boulos from internal files of the Southern California Association of Governments.

of population composition alone, but also in terms of location (suburban areas versus inner city areas), income level, and type of housing (apartments versus single-family subdivisions).

These patterns of change represent interesting research questions, and they also have important significance for policy decisions in local areas. There is growing concern for achieving a reasonable balance between jobs and housing in local areas, motivated in part by a desire to reduce commuting and slow the increase in traffic congestion seen in most metropolitan areas (Cervero 1989). Much of this traffic increase may be due to the movement of the baby boom generation into ages of high car ownership and also to rising labor force participation (Pisarski 1987). However, without detailed knowledge of local labor force patterns, planners cannot fully analyze or project the behavior that is increasing commuting. Also, jobs and houses cannot be balanced unless we know how many workers in an area there are per house in each local area. Old rules of thumb are quickly outdated with regard to labor force behavior. Although the present chapter does not attempt a comprehensive analysis of these issues, a research method is demonstrated that provides some insights into the questions.

☐ 3. Alternative Profiles of Participation Rates

We begin with an examination of county-level data for Los Angeles. Labor force participation is analyzed by focusing on the age profiles of participation, defined for separate sex and race groups, and also by type of housing occupied. These age profiles can be graphically compared to show differences between different subgroups.

Guiding this empirical investigation is our desire to use these county profiles for local area analysis. For improved accuracy we need a profile that is representative of the local population group being modeled. For example, a female county age profile should be used to describe the behavior of local women. Better yet, the profile should be defined for the same race, and even defined by the same housing type as found in the local area. Later we discuss how to evaluate the fit of alternative profiles, but the first step is to collect a set of alternatives for comparison.

3.1. Change Over Time

Figure 1 shows the changing profile of age-specific labor force participation rates for men and women in Los Angeles County.[5] Between 1970 and 1980 a relatively small decrease in labor force participation occurred among men. This drop-off was greatest in the 50 to 64 age range, as males apparently accelerated their departure from the labor force.

---

[5]Data for the profiles in this section may be drawn from PUMS county group data, using a type P extract from the A sample that holds 5 percent of all persons records. Similar data are also reported in published tabulations by race, age, and sex, although in broader age categories. In 1990, county-level data are reported in Series CP-2, Table 154.

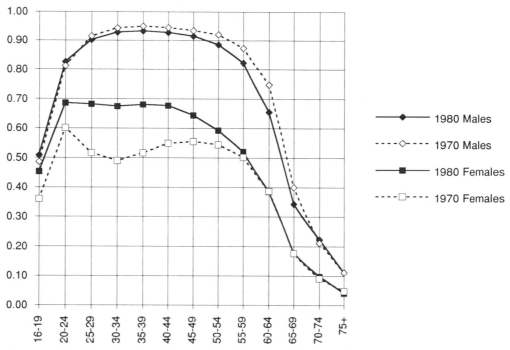

Figure 1. Labor force participation rates by age and sex for Los Angeles County in 1970 and 1980. The rates are the proportion of persons in each age–sex group who are in the labor force (either employed or seeking work).

The changes among women were far more dramatic. In the young adult and middle ages, the female profile closed a third or more of the 1970 gap between male and female labor force participation. Participation rates in the 25 to 39 age range increased by nearly 20 percentage points. Female labor force participation is beginning to assume the same shape as for males. The most significant aspect of the change is that the traditional M shape of the profile has been filled in, creating a high plateau over the 20 to 44 age span. (As will be shown, among black women the valley had already been filled by 1970, and in 1980, participation surged instead to a peak at ages 30 to 39.)

3.2. Differences Between Races and Places

Given the variability of female participation over time, a key question is how much the age profiles of participation might differ between different groups of women in the same census year. Figure 2 compares six different profiles for women in 1980:

1. All women in Los Angeles County;
2. Black women in Los Angeles County;
3. Hispanic women in in Los Angeles County;
4. White women in Los Angeles County;
5. All women in Orange County;
6. All women in San Bernadino and Riverside counties.

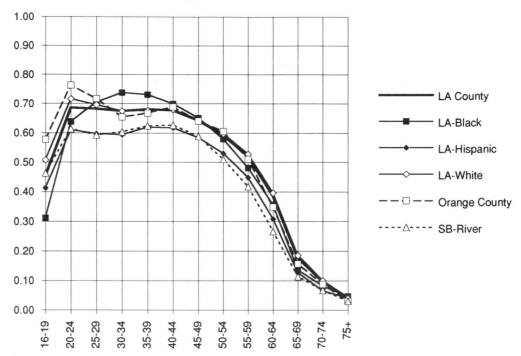

Figure 2. Comparison of 1980 labor force participation rates, by age, for females in different counties or of different races.

The Orange County profile is included to represent an area of higher income suburbs, whereas the profile for San Bernadino and Riverside Counties is included to represent an area of affordable middle-income suburbs and retirement communities.

Substantial differences exist among the profiles, both in their overall *level* of participation and in terms of the *shape* of the age profile. The overall Los Angeles County profile for females is shown as a heavy line. In contrast, the two lowest profiles consist of Los Angeles Hispanics and residents of San Bernadino–Riverside. The two highest profiles are found in Orange County (peak at 20 to 24) and among Los Angeles blacks (peak at 30 to 39). Note the extreme difference in shape between these two profiles: One has a valley at 30 to 39, the other a peak in that range. This difference might reflect the timing of fertility: early childbearing and early return to the labor force for blacks versus delayed childbearing and high labor force participation among young women in Orange County.

3.3. Differences by Type of Occupied Housing

The suburban county profiles may also reflect the effects of housing choices. Homeowners may be pursuing different lifestyles than renters. This could reflect an orientation toward childrearing and away from employment outside the home. Alternatively, the high cost of housing might require greater labor force particpation among younger homeowners (Myers 1985). These housing choices can be measured more directly with PUMS data, allowing us to identify occupants in owned or rented dwellings. The value of bringing tenure into the analysis is that this variable is also available as a complete-count variable in areas as small as census tracts. Thus, an important housing difference discovered in the PUMS data can be measured locally as well.

Type of housing can be evaluated for its additional explanatory value if we retain the previous dimensions of age, sex, and race, and further cross-classify the population members by the type of housing in which they live.[6] We can then compare the labor force participation rates of owner and renter-occupants to see how much housing alters the profiles. The results of this comparison are presented in two graphs, one for males and the other for females, that display the difference in labor force participation as the owners' rate minus the renters' rate in the same race and age category.

The results for males (Fig. 3) show much higher labor force participation rates among owner-occupants. The gap grows especially pronounced in late middle age. (Among the youngest persons, participation is actually greater for renters, probably because young males in owned homes are living at home with their parents.) The greater participation of owner-occupants exceeds 15 percentage points (0.15) among older blacks. Among non-Hispanic whites in this age range, the gap between owners and renters is also at its greatest—about 8 percentage points. Apparently, older males in owner-occupied housing depart the labor force more slowly than those in rented quarters.

The differences among females (Fig. 4) also favor owner-occupants, but the gap between owners and renters does not increase in older ages. More important, females who are non-Hispanic whites reflect the opposite pattern of all other groups analyzed: Their labor force participation is *lower* by 5 to 10 percentage points among owner-occupants. The gap is pronounced in the 25 to 39 age range where the dip in female participation is traditionally found. Apparently the dip exists more strongly among owner-occupants than renters.

3.4. Implications of the Profiles

From the preceding data we reach three conclusions. First, we would expect to find a higher ratio of workers to population in local areas that have their population concentrated in the prime labor force ages. Second, we expect higher or lower participation, depending on the racial composition of the area. Finally, we expect higher participation in areas of owner-occupied housing, with the exception of white, non-Hispanic females, where the reverse is expected.

One way to test the accuracy of these judgments is by applying the designated profiles to the local population composition and comparing the expected number of workers with what is actually observed in the sample data. Further, for local estimation purposes, we can adjust the fit of the larger area profiles until they yield results that are perfectly consistent with other local data that are known.

---

[6]This analysis utilizes a type PH extract, classifying all persons (not just the householder) by the type of housing they occupy. Thus, many persons under age 25 in owned homes are probably children living in their parents' homes.

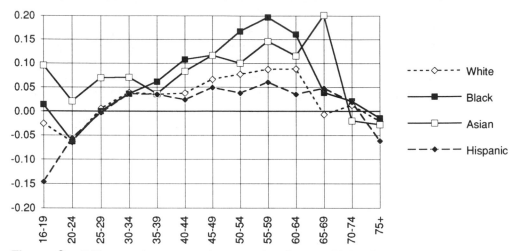

Figure 3.   Difference between male owners' and renters' labor force participation in Los Angeles County in 1980.

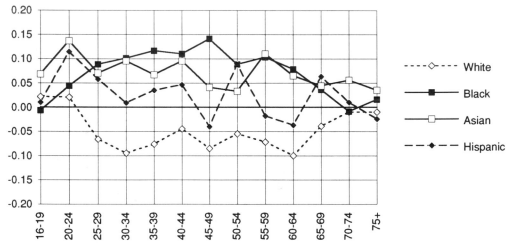

Figure 4.   Difference between female owners' and renters' labor force participation in Los Angeles County in 1980.

☐ 4. Applying Profiles to Local Areas

Profiles of race, sex, and age-specific rates, drawn from the larger area, can be applied to local areas by multiplying these rates times the population numbers observed in each race-sex-age subgroup. This application of the *rates × composition* framework yields expected values that combine larger area behavior information with local area composition.

Two additional steps are required here. After fitting the profiles to the age marginal of a race-sex group, we must then adjust the fit on a second dimension, scaling the labor force numbers by age, up or down, so that the total equals the number of workers of the race-sex group that was reported locally by the census. This procedure applies methods of scaling with marginals presented in the preceding chapter, but the sequential adjustment on two dimensions requires an iterative scaling procedure to achieve a fit on both dimensions.

The final step involves selection of the most appropriate profile to fit to the local area. Although this could be the first step, it is wise to apply a number of alternatives to determine which fits most easily and what differences exist among the alternatives. Thus, we choose the best alternative after trying several on for size.

These procedures will be demonstrated in the case of one population subgroup of one census tract. The example is black females aged 16 and older living in Pasadena area Tract 4602. These women have an overall labor force participation rate of 72.0 percent, well above the average level observed for all women in Los Angeles County (53.5 percent). To complete the detailed labor force analysis for the entire tract, parallel analyses would need to be carried out for both males and females of all racial groups.[7]

4.1. Initial Fitting of Profiles

How closely can we estimate the labor force participation level of this specific group? Three items of data defined for black females are available for use: (1) their larger area labor force behavior profiles by age, (2) their local area age profile, and (3) their total number of labor force participants aged 16 and older. From these data we want to estimate the detailed profile of labor force behavior by age for black females in this small area.

Tract-level data for the black females are presented in Table II. Their age marginal (or breakdown) is given on the right.[8] On the left is their labor force data,[9] and the housing type of black households in the area is given at the very bottom (86 percent owner-occupied).[10] All of these data are 100-percent count variables except for labor force.

4.1.1. Adjusting the Input Data

The fact that the labor force data are *weighted up* by the Census Bureau from a sample creates a slight inconsistency that must be corrected. The labor force numbers do

---

[7]Necessary data may be reported only for racial groups that have at least 400 persons. Groups not separately identified can be grouped into a residual *other* category that equals total persons of each sex minus the identified groups of each sex.

[8] For sources of 100-percent count age data, see Chapter 8.

[9]These sample count data are drawn from Table P-14 of the published census tract books (Series PHC80-2). Similar tables are available for each of the major race and Hispanic origin groups. (See the *short* list of groups provided in Table I of Chapter 10.) Comparable data for blacks in 1990 are found in Table 22 of the published census tract books (Series CPH-3). Later in the chapter, we compare data drawn from detailed crosstabulations of labor force status by age, sex, and race reported in STF 4, Table PB51, in 1980. Comparable data for 1990 may be found in STF 4, Table PB50. See Chapter 4 for details of these data sources.

[10]Using the published census tract sources cited, housing tenure is obtained for each of the major race and Hispanic origin groups in Table H-1 for 1980, and in Table 9 for 1990. Identical data, in computerized form, are found in STF 1, Tables 27 and 28, for 1980, and Tables H9 and H11 for 1990.

Table II. Tract-Level Marginals, 1980, Black Females, Tract 4602

Labor Force Sample Data			Age Marginal	
Labor force age 16+		1207	16 to 19	173
In labor force		869	20 to 24	121
Proportion in labor force		0.720	25 to 29	139
			30 to 34	163
Population 100% Control Data			35 to 39	158
Persons age 16+ (P)		1230	40 to 44	136
Labor force age 16+ (LF)		1207	45 to 49	97
Control ratio = P/LF		1.019	50 to 54	77
			55 to 59	57
Adjusted Labor Force Data			60 to 64	46
Age 16+ (N)		1230.00	65 to 69	27
In labor force (Y)		885.56	70 to 74	18
Proportion in labor force		0.720	75+	18
			Total	1230

Confidence Limits (95%) for Labor Force Proportion

	In LF	Not	Proportion in LF
Adjusted labor force marginal	886	344	0.720
Upper limit In LF*	956	274	0.777
Lower limit In LF*	815	415	0.663

*=Estimate +/− 2*SQRT(5*Y*(1−(Y/N))).

Housing Type
Proportion owner-occupied   0.858

not exactly correspond to the 100-percent count population data. Although they are very close, it is important to have an exact match. Accordingly, we must adjust the labor force data, controlling these to the complete count data by multiplying all the labor force numbers by the ratio of complete count to sample count, age 16 and over.[11] The numerator of the control ratio is taken from the 100-percent count table where the age marginal is listed. The denominator (labor force age total) is taken from the sample count table where labor force data are reported, as described previously. The control ratio is 1.019, meaning that all categories of labor force participation are scaled up by about two percent. This has no effect on the overall labor force participation rate, since both numerator and denominator are adjusted by the same proportion, but the adjustment assures that the labor force data will match up correctly with the population data.

A second complication with the sample count labor force data arises due to sampling variance. As discussed in Chapter 4, we must recognize that the labor force numbers are estimated from a sample and carry out confidence interval calculations shown there. Table I indicates that the actual labor force participation in this area

---

[11]The distinction between 100-percent count and sample count data is explained in Chapter 4. It is not unusual for small discrepancies to exist between totals prepared from one source or the other. In general, we take the 100-percent count total as more accurate, but we make use of the distribution measured within the sample count data. That distribution can then be applied to the 100-percent count control total.

could be greater or less by 70 persons, raising or lowering the participation rate by 5.7 percentage points. Initially, we will work only with the observed labor force numbers, as adjusted, but we will return to the sampling error issue later in the chapter.

4.1.2. Fitting Alternative Profiles

The alternative regional profiles to be fit to the tract data are presented in Table III.[12] All profiles are for females in Los Angeles County. The alternatives consist of all races combined, black females, blacks living in rented units, blacks living in owned units, or a blended profile that combines the rented and owned profiles in proportion to the tenure of black households in the tract. In this case, the blended tenure profile is weighted to 86 percent owner profile and 14 percent renter profile.

We now proceed to fit each of these profiles to the local tract data. Table IV shows the required calculations. Column 1 lists the tract's age marginal (from Table II), while column 2 lists the profile that is being applied (from Table III). Each of the alternatives can be tried in succession, saving the results after each trial. The first alternative—all females in the county—is shown in the example. The expected labor force numbers are calculated in columns 3 and 4, as shown in the simple formulas at the head of the columns. This procedure is simply a scaling of the regional labor force data (condensed into the form of labor force participation rates) by a local age marginal, following the general method described in Chapter 12. Readers may also recognize this as an example of the *age-expected* calculations first introduced with marital status in Chapter 7. More generally, this is an application of the *rates* × *composition* analytical framework.

The totals at the bottom of the columns summarize the results of the calculations. Expected labor force participants, based on this regional profile, are 728 in number. This is well short of the target number of 886 that was actually observed, a deviation

Table III. Alternative Profiles of Labor Force Participation, 1980, Females, Los Angeles County

	Total All Races	Total Blacks	Black Renters	Black Owners	Blended Tenure
16 to 19	0.453	0.312	0.305	0.312	0.306
20 to 24	0.685	0.638	0.663	0.619	0.657
25 to 29	0.682	0.704	0.773	0.685	0.760
30 to 34	0.674	0.738	0.795	0.694	0.781
35 to 39	0.680	0.731	0.787	0.671	0.770
40 to 44	0.677	0.699	0.739	0.629	0.723
45 to 49	0.643	0.651	0.715	0.574	0.695
50 to 54	0.592	0.579	0.605	0.517	0.592
55 to 59	0.520	0.480	0.515	0.412	0.500
60 to 64	0.385	0.356	0.374	0.296	0.363
65 to 69	0.177	0.136	0.146	0.110	0.141
70 to 74	0.097	0.081	0.074	0.083	0.075
75+	0.041	0.047	0.056	0.040	0.054

[12]The profiles were derived from PUMS sources, as described previously.

of some 17.8 percent. (It is also well short of the lower bound to the confidence interval around the observed sample estimate.) The bottom of Table IV reports these summary statistics for the estimated number *in* or *out* of the labor force. Because the two labor force categories are complements, forced to add to the total population, we need evaluate only one number—those *in* the labor force.

The degree of fit for all five alternative profiles is summarized in Table V. The two worst fitting alternatives are that for all black females and that for black renters. The error here is 19 to 25 percent. The two best fitting alternatives are that for black owners and the blended tenure profile. The error here is considerably smaller: 14 to 15 percent. All of the profiles yield estimates that are at least 121 persons short of the actual labor force, implying that the local women are working to a higher degree.[13]

Two options now exist to achieve a better fit. The first is to construct still more carefully tailored alternative profiles from the PUMS data. Use of data on owner-occupants helped considerably, but perhaps we should further narrow this to occupants in houses valued as high as those in this neighborhood. We might also seek profiles for the same educational attainment class as found in the neighborhood, or the same marital status mix.

Table IV. Calculation of Expected Labor Force by Age

| | Tract Population | Profile to Apply | Expected Labor Force Status | |
| | | | In LF | Not in LF |
	[1]	[2]	[3 = 1 × 2]	[4 = 1 − 3]
16 to 19	173	0.453	78	95
20 to 24	121	0.685	83	38
25 to 29	139	0.682	95	44
30 to 34	163	0.674	110	53
35 to 39	158	0.680	107	51
40 to 44	136	0.677	92	44
45 to 49	97	0.643	62	35
50 to 54	77	0.592	46	31
55 to 59	57	0.520	30	28
60 to 64	46	0.385	18	28
65 to 69	27	0.177	5	22
70 to 74	18	0.097	2	16
75+	18	0.041	1	17
Total	1230		728.06	501.94
Target marginal (adjusted sample data)			885.56	344.44
Deviation (expected − observed)			−157.50	157.50
% absolute deviation			17.79	45.73

[13]The best fitting estimate is significantly different from the observed, because it is also 51 workers less than the lower limit of the confidence interval reported in Table II for the observed sample count.

Table V. Summary of Trial Fit for the Alternative Profiles

	Total All Races	Total Blacks	Black Renters	Black Owners	Blended Tenure
Total expected in LF	728.06	717.39	666.59	763.71	749.90
Target marginal (adjusted sample data)	885.56	885.56	885.56	885.56	885.56
Deviation (expected − observed)	−157.50	−168.17	−218.97	−121.85	−135.66
% absolute deviation	17.79	18.99	24.73	13.76	15.32

Rather than pursue additional profiles, the second option is to force a fit by scaling the best fitting initial profile to match the observed tract marginal exactly. Essentially this method raises (or lowers) the expected labor force participation rate in all age groups until the sum total of participation matches the total that has been locally observed. As demonstrated below, the differences between the alternative profiles are substantially reduced by this method.

4.2. Iterative Proportional Fitting or Scaling

In the previous stage of analysis, we fit a regional labor force profile to the local area's age marginal. This procedure scales the regional profile to match the local area on one dimension—its observed age structure. However, the regional profile being scaled has two dimensions: Age × Labor Force Status. The profiles initially fitted to the local age structure may be further controlled to match the locally observed labor force status.

The difficulty encountered, when controlling first to one marginal and then the other, is that the adjustment process on one dimension disrupts the balance on the other. To solve this problem, demographers at the Census Bureau and elsewhere have long used a method known as *iterative proportional fitting, or scaling,* (also known as the *Deming method* or the method of *biproportional adjustment*).[14] This procedure is extremely powerful but must also be used wth care.

Table VI illustrates a spreadsheet layout for carrying out the iterative scaling. Although the exact formulas are not shown, they are very simple, merely involving addition, multiplication, and division. The key is to follow the prescribed sequence, and that logic is carefully portrayed in the spreadsheet layout.

Each cycle has two stages, matrix A and matrix B, where the table of Age × Labor Force Status is adjusted first to fit the row totals and then the column totals. Cycle 1 begins with the expected table produced in Table IV. The regional profile has been fit to the local age marginal, controlling that profile to the local age totals. That result is entered here as matrix A in cycle 1.

Observe that the expected total number of persons in the labor force is only 728 in matrix A, whereas the observed target labor force total is 886. We can form a ratio of the target total to the column total, yielding a value of 1.216 for the *In LF* column and 0.686 for the *Not* column. Matrix B is then computed as the numbers in the two

---

[14]This procedure was briefly introduced in Chapter 12. We now provide a full demonstration of the method. Further details can be found in Mosteller (1968) or Shyrock and Siegel (1976: 547–549).

Table VI. Iterative Proportional Scaling of Age by Labor Force Status Matrix to Fit Observed Marginals (Targets) in Local Area

Cycle 1	A Initial Expected Labor Force Status		B Controlled to Target Column Totals-1				
Age	In LF	Not	In LF	Not	Row Total-1	Target Age Total	Row Ratios-1
16 to 19	78	95	95	65	160.26	173	1.0795
20 to 24	83	38	101	26	126.97	121	0.9530
25 to 29	95	44	115	30	145.41	139	0.9544
30 to 34	110	53	134	37	170.32	163	0.9583
35 to 39	107	51	131	35	165.45	158	0.9554
40 to 44	92	44	112	30	142.06	136	0.9568
45 to 49	62	35	76	24	99.45	97	0.9736
50 to 54	46	31	56	22	77.17	77	1.0000
55 to 59	30	28	36	19	55.23	57	1.0396
60 to 64	18	28	21	19	40.58	46	1.1232
65 to 69	5	22	6	15	21.23	27	1.2820
70 to 74	2	16	2	11	13.12	18	1.3557
75+	1	17	1	12	12.74	18	1.4125
Column Total-1	728.06	501.94	885.56	344.44			
Target LF Total	886	344	886	344			
Column Ratios-1	1.2163	0.6862					

Explanation of Calculations
Column Ratios = Target LF Total/Column Total
Row Ratios = Target Age Total/Row Total
Controlled to Target Column Totals:
    Matrix B = Matrix A $\times$ Column Ratios
Controlled to Target Row Totals:
    Matrix A = Matrix B (of previous cycle) $\times$ Row Ratios (of previous cycle)
    (The initial Matrix A is calculated as in Table IV)
Note: Matrix A or B is the table of Age by Labor Force Status

Cycle 2	A Controlled to Target Row Totals-1		B Controlled to Target Column Totals-2				
Age	In LF	Not	In LF	Not	Row Total-2	Target Age Total	Row Ratios-2
16 to 19	103	70	105	67	171.93	173	1.0062
20 to 24	96	25	98	24	121.56	121	0.9954
25 to 29	110	29	112	28	139.41	139	0.9955
30 to 34	128	35	130	34	163.90	163	0.9958
35 to 39	125	33	127	32	158.77	158	0.9956
40 to 44	107	29	109	28	136.52	136	0.9957
45 to 49	74	23	75	22	97.11	97	0.9971
50 to 54	56	22	56	21	77.22	77	0.9994
55 to 59	38	20	38	19	57.26	57	1.0028
60 to 64	24	22	24	21	45.13	46	1.0100
65 to 69	8	20	8	19	26.57	27	1.0241
70 to 74	3	15	3	14	17.25	18	1.0308
75+	1	17	1	16	17.37	18	1.0360
Column Total-2	871.74	358.26	885.56	344.44			
Target LF Total	886	344					
Column Ratios-2	1.0158	0.9614					

Table VI. (Cont.)

Cycle 3	A Controlled to Target Row Totals-2		B Controlled to Target Column Totals-3				
Age	In LF	Not	In LF	Not	Row Total-3	Target Age Total	Row Ratios-3
16 to 19	105	68	105	68	172.89	173	1.0006
20 to 24	97	24	97	24	121.06	121	0.9995
25 to 29	111	28	111	28	138.85	139	0.9995
30 to 34	130	34	130	33	163.29	163	0.9996
35 to 39	126	32	127	32	158.14	158	0.9995
40 to 44	108	28	109	27	135.99	136	0.9995
45 to 49	75	22	75	22	96.86	97	0.9997
50 to 54	56	21	57	21	77.18	77	0.9999
55 to 59	38	19	39	19	57.40	57	1.0003
60 to 64	25	21	25	21	45.54	46	1.0010
65 to 69	8	19	8	19	27.15	27	1.0025
70 to 74	3	15	3	15	17.73	18	1.0032
75+	1	17	1	17	17.93	18	1.0038
Column Total-3	884.11	345.89	885.56	344.44			
Target LF Total Column	886	344					
Ratios-3	1.0016	0.9958					

columns of matrix A times these two respective scaling ratios. For example, the 16 to 19 age group has 78 expected persons in the labor force in matrix A, but when multiplied by 1.216, the expected result in matrix B is 95.

Matrix B is now controlled to match the total labor force numbers in the local areas, but it has lost its previous fit on the row dimension. The row totals no longer match the local, target age marginal. Instead, we see that the row ratios (target/row total) diverge substantially from 1.000, especially at the older ages.

The second part of Table VI continues the iterative *raking* of the rows and columns through two more cycles. By the third cycle we see substantial convergence of the two alternative adjustments. After row raking has produced matrix A, the column ratios only moderately depart from 1.0000. Similarly, after column raking has produced matrix B, the row ratios are close to 1.0000, except at the oldest ages. Two more cycles would smooth the row and column ratios to the fourth decimal place, but the marginal benefit is slight and we do not show those cycles here.

The alternative to the expansive spreadsheet format presented here is to write a computer program where, for a number of cycles, first the column ratios from matrix A would determine matrix B and then the row ratios from matrix B would determine matrix A. All cycles could then be carried out within the same space, saving the need to stretch the calculations over multiple pages. This program for sequential raking could be written either as a spreadsheet macro or in any programming language.[15]

[15]Some suggestions are found in Fienberg (1977: 34). An iterative fitting routine is also available in SAS, and the author has written a simple spreadsheet macro in Microsoft Excel.

4.2.1. Results of Alternative Trials

The result of fitting the regional profile to local marginals is an estimated, detailed labor force profile for the local area. This profile is tailored to the local area's age structure and its observed labor force totals. The estimated local age profile of labor force participation has the same general shape as the regional profile, but the level of the curve is raised or lowered in order to yield the observed number of local workers.

A number of alternative regional profiles may be fit to the local marginals and then compared. The greater the differences among the resulting estimates, the greater the importance of choosing the most approporiate profile. The wrong input profile may yield an estimated local age curve that is of the wrong shape.

Figure 5 compares the estimated results fitted from the alternative input profiles. Although a fair amount of variation is seen among the input profiles, the fitted results converge more tightly because they have been scaled up or down by the same local area marginals. Only the flat curve based on all Los Angeles County females departs noticably from the consensus profile. Of course, that curve is a less appropriate model for this local population of black females than the alternatives based on blacks alone. The curve based on females living in rented homes also drops off more than others at the older ages. Given that this is a neighborhood composed largely of homeowners, the model based on black females living in owned homes, or blended tenure, seems more appropriate.

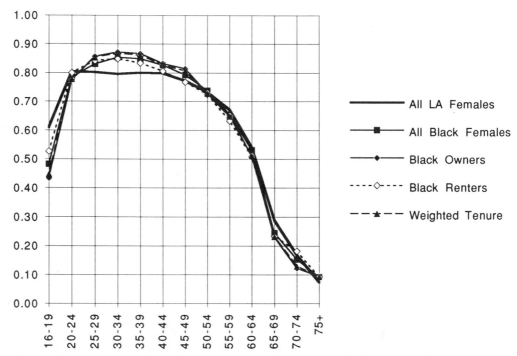

Figure 5. Fitted profiles of labor force participation by age for the case of black females in suburban Pasadena Tract 4602 in 1980. The profiles are derived from alternative regional models such as those shown in Fig. 2.

These observations are buttressed by our earlier finding that the regional profile for black owner-occupants gave the closest initial fit to the local data (see Table V). The other profiles have been forced to sum to the observed local total by elevating the level of their curves substantially. A conservative choice among alternatives would give preference to the regional profile that had the closest initial fit and required less scaling.

4.2.2. Statistical Significance of the Estimates

These estimates are subject to some range of error. Two sources of error are prominent. First, we incur some sampling error in the PUMS estimates used for the alternative profiles. As discussed in Chapter 4, these data are based on a sample and we should inspect them for their degree of variability. Fortunately, a five percent sample of the very large Los Angeles County population yields a sizable number of cases, reducing sampling error greatly. But a sample of residents defined by race, sex, age, and tenure status can become fairly small. For practical purposes, we may assume that the sampling variability creates estimates in adjoining age groups whose errors are offsetting. An additional safeguard is to inspect the curves for excessive irregularity, smoothing the age distribution to the profile if necessary.

A second source of error proves more important. The local labor force data used for fitting the regional profile has substantial sampling error of its own. As demonstrated in Table II, our local example of black females has an observed labor force participation rate of 72.0 percent, plus or minus 5.7 percentage points (95-percent confidence interval). This variability should be factored into our estimation of the detailed local age profile of labor force participation. A simple means of evaluating this error is to use the high and low estimates of total labor force participation in successive trials to fit the regional profile. Figure 6 illustrates the results, based on the regional profile for black female owner-occupants and the three alternative estimates of local labor force: observed, high estimate, and low estimate.

As can be seen, the high degree of variability due to the alternative local estimates is much greater than that shown in Fig. 5, which is due to the alternative regional profiles. One means of smoothing this local sampling error (not demonstrated here) is to estimate the expected level of labor force participation for each locality using multiple regression on a set of census tracts and employing relevant characteristics of the local areas as predictor variables. The expected value for a locality could be used in place of the observed value, or the expected and observed values could be averaged.

4.3. Comparison with Estimates Obtained Directly from STF 4

The question remains whether or not more accurate data on local labor force behavior, by age, can be obtained directly from the STF 4 files. Would these data be worth accessing, now that we see how much variation accompanies our estimates?

Those direct age-specific observations in STF 4 are among the most difficult to access data from the census, as discussed earlier in the chapter. We should evaluate the effort required. Less detailed age categories are reported in the STF 4 table than the 5-year groupings we have emphasized. In addition, even with the wider, 10-year

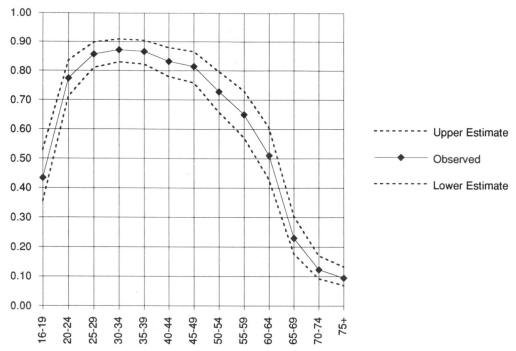

Figure 6. Confidence interval surrounding a fitted labor force participation profile selected from Fig. 5. The 95-percent confidence interval is estimated from sampling error for the observed local labor force marginal in suburban Pasadena Tract 4602 in 1980.

categories, sample sizes in small areas become very small when broken down by sex and race. This leads to potentially large sampling error.

Comparison of our best previous estimate with the reported data is instructive. The sample sizes reported in the STF 4 tabulation (Table PB51) are small. For example, at age 45 to 54, a weighted sample count of 181 black females, or approximately 30 sample observations, is reported. Of this sample, 76.2 percent are in the labor force. According to statistical significance calculations described in Chapter 4, this sample estimate has a 95-percent confidence interval of ±12.5 percentage points. The next reported age group is only five years wide, with a much smaller sample, and the confidence interval grows to ±20 percentage points.

An overview of sampling error across the age profile is gained by plotting the observed values and high/low estimates at each age. For ease of comparison, Fig. 7 superimposes the PUMS-derived age profile of labor force participation on the STF-derived age profile. Note how tightly bounded is the PUMS-derived confidence interval in contrast to the wider band around the STF-derived estimates. The STF data are also much more erratic, forming a sawtooth pattern that reflects the instability of the small sample sizes.

The PUMS-derived age profile generally tracks within the confidence bounds of the STF estimates. The one exception is at age 60 to 64, a sample estimate that is anomolously high and based on 31 reported observations (only five unweighted cases). In fact, reflecting the sampling instability at the older ages, zero observations

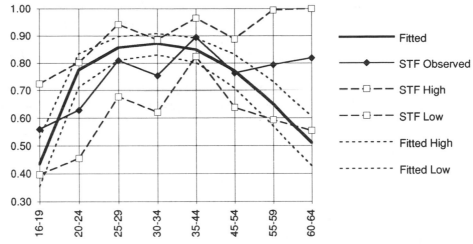

Figure 7. Comparison of the fitted, PUMS-based estimates with the sampling error surrounding actual local observations reported in STF 4.

are reported in the STF sample data at age 65 and older, whereas the 100-percent age marginal shows 61 black females in that age group.

It may seem counterintuitive that the detailed characteristics estimated from regional data for this locality should be more accurate than local data directly observed from census samples. When the samples are small, sampling error overwhelms the local detail. Remember that the regionally based estimates are scaled to key items of local data—the overall numbers in the labor force and the 100-percent age marginal. Those items ensure that the regional data are closely tailored to the local area.

4.4. Closer Geographic Targeting of Regional Profiles

Some might argue, with good reason, that Los Angeles County is too large and diverse a region to yield useful regional profiles for fitting to local areas. The PUMS-A file for this area does not define a specific area centered around Pasadena. That is only available on PUMS-B, but that 1-percent file affords much smaller samples. For example, in 1980, the Pasadena area held only about 22,000 black females, about 14,000 between ages 20 and 64, and only about 2,000 between ages 45 and 54. When sampled at 1-percent, we would expect to find only about 20 black females of ages 45 to 54 in the Pasadena portion of the file. Sample sizes this small in specific age groups are not adequate to estimate a regional profile reliably. If PUMS-A had defined the same small area, the sample size would be five times larger. The specificity of geographic areas in the different PUMS files varies from state to state because these are locally determined.

An alternative is to make fuller use of the STF 4 data, extracting information for the City of Pasadena and surrounding jurisdictions. With a sampling rate of 1-in-6, data reported in STF 4 would be collected from around 330 of the 2,000 black females age 45 to 54 in the area. To refine the profile, the 10-year wide age groups in that data could be interpolated into 5-year groups using methods described in Chapter 8.[16] This locally drawn profile then could be used with the tract-level marginals in the same manner as the regional PUMS-derived profiles. A major drawback is that, unlike with PUMS, it is impossible to further define the STF-derived race-age-sex labor force profiles by additional variables such as tenure, price range, marital status, or education level. If the profiles are restricted solely to race, age, and sex, then STF 4 may serve well as a source of more localized profiles.

☐ 5. Conclusion

This chapter has pursued two objectives. The topic of labor force participation is an important one, and changes among women have been pronounced. We also have seen some noteworthy differences between race and Hispanic origin groups, and also between places. Because of its social and economic implications, this variable deserves close monitoring in every local area.

In addition, we have used this chapter to delve into methods for bringing PUMS data into more localized service. The techniques of fitting regional profiles and scaling with marginals have broad utility. We have selected an example that is relatively clear-cut. We hope readers will recognize how this method can be carried out with other variables, as well, and how it might be adapted to other uses.

---

[16]More detailed age groupings are offered in 1990 than in 1980. The two categories above age 60 have been expanded to five categories. However, the two 10-year groups, 35 to 44 and 45 to 54, have been retained.

14 □□□

Tabulating and Linking Different Universes

The preceding chapters have demonstrated analysis within a single universe at a time, using the example of the population universe. Tabulation of alternative universes requires two differences from the analysis of the previous chapter. Instead of persons, we may wish to tabulate houses, households or jobs. Each of these constitutes a different universe of cases, even though all derive from the same PUMS database.

A second difference from the preceding chapter is more complex: We often wish to link tabulations from one universe with those from another. For example, we can construct ratios of persons per house or houses per job. In the first case, the housing universe forms the base for analysis, whereas in the latter case employment forms the base for analysis.

The present chapter has four sections. First, we seek a fuller understanding of the three major alternative universes and how they may be tabulated. We then proceed to the general question of how different universes can be linked. The third section provides detailed treatment of the linkage between population and housing universes. The final section addresses the employment universe and some of its linkages.

□ 1. Basing Analysis on Alternative Universes

The PUMS data allow for remarkable flexibility of analysis. Direct access to microdata records permits an infinite variety of creative tabulations. In particular, the multiuniverse structure of the PUMS data provides opportunity to link variables from different bases in many useful ways. Three major universes are explored in this chapter: (1) population, (2) occupied housing, and (3) labor force or employment. Custom tabulation requires careful and deliberate treatment of these alternative bases.

A *universe* is defined as "the set of entities of which the characteristics are studied, or about which an inference is made" (U.S. Bureau of the Census 1982: 51). The general definition is very flexible: The population of the United States, the population of California, and the elderly population of California each form a universe. However, the latter two are really subsets of the former. Recall that in Chapter 3 we defined the concept of universe more restrictively to signify sets of elements that are each of a different type. Persons, households, housing units, and jobs are separate universes of major interest. A synonym often used for universe is *base*, as when a universe serves as the base for a tabulation or base for a ratio.

1.1. Linkages Within the PUMS Database

The hierarchical file structure of the PUMS database, described previously, provides for several linkages among the different universes. For every occupied housing unit, defining a household, there is at least one person resident. Each of the persons in the unit may hold one or more jobs, although only the primary job for each person is actually recorded. If a household has only one member, the relationships among the three variables are straightforward: The one person is matched with one housing unit of a given type and one (or no) job of a given type.

In situations where more than one person occupies the same housing unit, the association among variables is less clear-cut, yet our tabulation procedures must accommodate this common occurrence. If household size is greater than one, more than one person is matched with each housing unit, and more than one job may be matched with the unit as well. Therefore, houses may have more than one set of person characteristics and more than one set of associated job characteristics. One simple convention for reducing this complexity is to analyze characteristics of only one person per household—the householder. That method only serves limited purposes.

From the perspective of the employment base, there is even more variability. Jobs are occupied by only one person who lives in only one housing unit. However, one job may support other persons in the same unit, and multiple jobs may support the same group of people living in the same housing unit. Also, unlike the necessary association that every occupied housing unit must have a person resident, not all occupied housing units have a job associated. Without the built-in association found between population and housing, analysis of the employment base is more complex. A final, unique feature of employment is that it introduces spatial linkages: persons live in houses at one point in space, but they usually work at different locations. This has particular importance for transportation planners, but it also is a concern of employment planners and others who strive to improve conditions for an area's residents based on jobs in that area.

1.2. Analytical Questions that Exploit the Data

To utilize the PUMS data fully, three different universes must be exploited: population, occupied housing units (or households), and labor force (job holders and job seekers). The necessary programming statements to create each universe from the raw PUMS data are listed in Appendix B. Each universe can be used to address different questions:

On a population base, we can tabulate the demographic, housing, or labor force and job characteristics of persons.

On an occupied housing unit base, we can tabulate the persons per household, the type of housing occupied, or the jobs-housing ratio.

On an employed labor force base, we can tabulate the persons supported per job, type of housing occupied per job, or type of job that is held.

1.2.1. Single-Universe Analysis

A subset of these questions is limited to single universe analysis. Examples include inquiring about the demographic composition of the population (its age, race, or marital status), or about the characteristics of the housing stock (its age, type, and size of units), or finally about the types of job positions held by workers (by occupation and industry type). These analyses all remain within a single universe, and all of the tabulation methods discussed in Chapter 12, using the example of poverty, can be applied equally within the population, housing, or employment universes.

However, as soon as we inquire about the type of jobs held by different types of people, or the types of people living in different types of houses, we have entered a multiuniverse analysis. Those questions are certainly richer, but they are also more difficult to address.

1.2.2. Linkages in Multiuniverse Analysis

Figure 1 diagrams the major linkages that may be measured between the three main variables. The arrows in the diagram pass from the universe that forms the

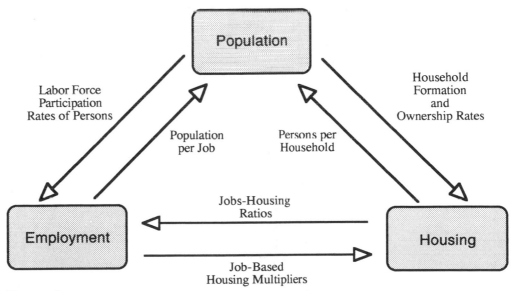

Figure 1. Bidirectional linkages among population, housing, and employment.

base, or the denominator, for an analysis to the universe that is linked by its position in the numerator. Note that linkages can be formed in either direction, depending on theory of appropriate causal order, questions being asked, or on opportunity afforded by known variables. Readers may wish to review the discussion in Chapter 3 regarding causal direction among the variables and its relation to level of geography.

Population-based analysis is linked to housing most directly via *household formation rates*, and it links to employment via *labor force participation rates*. When each of these rates is multiplied by the number of persons in each population subgroup, the product is the number of households or the number of labor force participants. Most analysts prefer population-based inquiries to the others. On that base we can construct detailed profiles of rates of behavior and combine these with projected changes in the population base.

However, many instances occur in urban analysis for which housing- or employment-based analyses are valuable. Most often those instances involve cases where housing or employment is a prior variable and we wish to use that base to predict another. Examples include where a new housing subdivision is planned or where employment growth is projected by industry. The population consequences of those events can only be modeled in a multiuniverse structure with data constructed on a housing or employment base.

Housing-based analysis links to population most directly via *persons per household*, and it links to employment via a *jobs-housing ratio*. When each of these ratios is multiplied by the number of houses of each type, the product is the number of persons or the number of jobs. If the number of houses is given, the method can estimate the likely number of persons or jobs.

Employment-based analysis links to population via *persons per job*, and it links to housing via *housing units per job*. When multiplied by the number of jobs, an estimate of population and housing associated with the jobs can be created.

What makes the alternative tabulations of PUMS data especially fruitful is that we can disaggregate the alternative bases into types of persons, houses, or jobs, respectively, and then compute tabulations of disaggregated characteristics on this disaggregated base. The alternative is to assume that crude ratios, such as persons per household or persons per job, apply equally in all cases. Yet research with PUMS data will show that wide variation occurs in both the numerator and denominator of these ratios.

1.3. Constructing Multiuniverse Ratios

We consider three examples of tabulation linking the different bases—one for each of the three alternatives. Labor force participation is selected as our example of population-based tabulation. Persons per household will be used to illustrate a housing-based tabulation. Finally, housing per job ratios will be used to illustrate employment-based analysis. In the examples that follow, each variable is assigned a subscript that is unique across all equations, aiding comparison between the alternatives.

1.3.1. Population-Based Tabulation

We begin with a tabulation of persons in the labor force, expressed as labor force participation rates. These rates can be disaggregated by age, sex, and race. The numerator of the rate is an employment variable, whereas the denominator is formed by population variables.

The denominator is disaggregated by three variables: age has 13 categories, sex has 2, and race has 4 categories (black, Hispanic, non-Hispanic white, and other). Thus, for this analysis, the population denominator is disaggregated into 104 subgroups. The numerator has one additional variable, labor force participation, with two categories: in the labor force or not. The percentage in the labor force is calculated for each of the subgroups formed by the denominator:

$$\text{LFP}_{ijkl} = \frac{P_{ijkl}}{P_{ijk}} \ , \tag{1}$$

where

LFP = labor force participation rate,
P = persons,
i = age,
j = sex,
k = race, and
l = labor force status.

1.3.2. Housing-Based Tabulation

Let us tabulate the number of persons by age and sex living in each type of housing unit. The numerator is formed by population variables, whereas the denominator is formed by housing variables. The denominator of housing units may be disaggregated into one (e.g., tenure) or more characteristics (e.g., tenure, structure type, and number of bedrooms).

The numerator of person characteristics is disaggregated along the same dimensions as the denominator but further disaggregated by the two person characteristics of age and sex. Tabulation of housing characteristics alone would require this statement: TENURE * TYPE * BEDRMS. Tabulation of population characteristics alone would require: AGE * SEX. The combined tabulation would be requested as: TENURE * TYPE * BEDRMS * AGE * SEX. This would generate an age/sex profile (in the form of a separate table) for each combination of tenure, type, and bedrooms. If tenure has two categories, and type and bedrooms each have four, a total of 32 housing types is defined and each would have its own resident population disaggregated by age and sex.

One way of imagining the linkage of the two sets of variables is of one matrix resting on another. The base matrix is the TENURE * TYPE * BEDRMS disaggregation of the housing stock. The numerator matrix is connected to this by the same three housing dimensions but adds a population profile disaggregated by the two additional population dimensions. Because there may be more than one person per housing unit, we need to read the raw PUMS data twice: once to construct the matrix of housing unit counts by tenure and type, and again to construct the matrix of person counts by housing subgroup and by person characteristics.[1]

The ratio of persons per household, for each age and sex group, disaggregated by each type of housing unit, is given by the expression:

$$PPH_{tsbij} = \frac{P_{tsbij}}{H_{tsb}} ,$$ (2)

where

PPH = persons per household,
 P = persons,
 H = occupied housing units,
 t = tenure (rent or own),
 s = structure type (single-family,
        multifamily, or mobile home),
 b = number of bedrooms,
 i = age, and
 j = sex.

The total number of persons per household in each kind of housing unit is given by summing across all the ages and sexes:

$$PPH_{tsb} = \sum_i \sum_j \frac{P_{tsbij}}{H_{tsb}} .$$ (3)

Equation (3) describes calculation of average household size for different types of housing units. This total household size is a more familiar measure, but the formulation makes clear that it is really the sum of individual persons of each age and sex who live in the units. Equation (2) disaggregates not only the denominator but also the numerator, so that total household size in each type of unit is broken down into age and sex components.

1.3.3. Employment-Based Tabulation

The most difficult tabulations are based on employment for several reasons. One problem is that there are so many more categories of job types that it becomes difficult to manage. More serious is the problem that there is no necessary match between jobs and houses. Some households have no workers and some have two or more. Therefore, greater care is required when constructing employment-based multipliers.

We can tabulate the number of housing units or persons that accompany the number of workers in each type of job. We focus on housing units per job. The numerator is formed by housing variables, whereas the denominator is formed by employment

---

[1]Once tabulated the two matrices can be entered into spreadsheets for element-wise division to create the desired ratios. Alternatively, the tabulated matrices can be output as datasets with a record for each unique combination of cells and an added variable that counts the frequency of cases in each cell. The two datasets derived from the matrices can be linked by renaming the "count" variables to two different names and then merging the output datasets on the three housing dimensions that they share. Division of the person and housing unit count variables will yield the desired ratio for each subgroup of the housing unit base.

variables. The denominator of occupied jobs may be disaggregated by industry, occupation type, or both.

Unlike other cases we have examined, note that a spatial dimension enters into employment analysis. We need to specify separately the person's residence from the place of work. Residence location is an attribute of the housing unit, like structure type, whereas work place is an attribute of the job.

The numerator of housing characteristics is disaggregated along the same dimensions as the denominator but further disaggregated by the desired housing characteristics. Tabulation of job characteristics alone would require: INDUSTRY * WORKPLACE. Tabulation of housing characteristics alone would require this statement: RESIDENCE * TENURE * TYPE. The combined tabulation would be requested as: INDUSTRY * WORKPLACE * RESIDENCE * TENURE * TYPE. This would generate a tenure/type profile (in the form of a separate table) for each residence location and for each combination of industry/workplace.

Again imagine the linkage of the two sets of variables as one matrix resting on another. The base matrix is the INDUSTRY * WORKPLACE disaggregation of the job types held by persons. The numerator matrix is connected to this by the same two job dimensions but adds a housing profile disaggregated by the additional housing dimensions. The ratio of housing units per job, for each tenure and structure type, disaggregated by industry and workplace, is given by the expression:

$$HPJ_{ewrls} = \frac{H_{ewrts}}{P_{ew}} , \tag{4}$$

Where

HPJ = housing units per job,
P = persons,
H = occupied housing units,
e = employee's industry type,
w = employee's place of work,
r = employee's place of residence,
l = housing tenure, and
s = structure type.

Since the housing unit is tabulated for each worker who resides there, in effect the unit is double counted if there are two workers. Accordingly, the houses per jobs should be divided by the number of workers in the house. However, that method has a major drawback. All workers in a housing unit are counted equally, regardless of earnings relative to other household members. Also, it is difficult to assign credit for the same house to the different industries whose paychecks support it. A solution is outlined for these problems near the end of this chapter.

☐ 2. Procedures for Linking Two Different Universes

The best procedure for linking the different universes is to construct separate tabulations on each base, building these tabulations so that each shares a common set of classifying variables. For small-scale tabulations these tables can be manually divided; larger tabulations require a programming solution. We first demonstrate a simple example that could be manually executed. After that a structured programming solution is outlined.

2.1. Example of Ratios of Population per Household

A common example of linking universes is between persons and households. A tabulation of households by tenure (owner- versus renter-occupied), can be related to a tabulation of persons living in each type of occupied housing. One application of linking these two universes is to calculate persons of each age per household (the APH method discussed in Chapter 8). The numbers of persons of each age and housing type are simply divided by the number of housing units of each type.

This example is presented in Table I. The two-cell tabulation at the top is of occupied housing units by tenure. The tabulation below that is of persons living in occupied units, broken out by age and tenure. The two tabulations are drawn from the

Table I. Linking Two Bases to Form Persons per Household Ratios

	Occupied Housing Count by Tenure (a)	
	Owned	Rented
	66,916	69,838

	Population Count by Age and Tenure (b)			Persons per Household by Age and Tenure (c)	
	Owned	Rented		Owned	Rented
0–4	12,080	16,230	0–4	0.181	0.232
5–9	13,958	12,923	5–9	0.209	0.185
10–14	17,763	10,930	10–14	0.265	0.157
15–19	19,497	12,435	15–19	0.291	0.178
20–24	13,220	22,178	20–24	0.198	0.318
25–29	12,230	22,638	25–29	0.183	0.324
30–34	15,122	15,944	30–34	0.226	0.228
35–39	14,424	9,996	35–39	0.216	0.143
40–44	12,802	7,139	40–44	0.191	0.102
45–49	12,484	5,680	45–49	0.187	0.081
50–54	13,511	5,732	50–54	0.202	0.082
55–59	13,526	5,541	55–59	0.202	0.079
60–64	10,542	4,782	60–64	0.158	0.068
65–69	7,999	4,429	65–69	0.120	0.063
70–74	5,640	3,688	70–74	0.084	0.053
75+	6,942	5,697	75+	0.104	0.082
Total	201,740	165,962	Total	3.015	2.376

Notes: (a) Tabulation of "Tenure" from a householder universe (PHH extract of PUMS).
(b) Tabulation of "Age × Tenure" from a persons in households universe (PH extract from PUMS).
(c) Ratio of number of persons in each age group to the total number of owned or rented units.
Source: Los Angeles County, 1980; PUMS-A (5% sample of households).

same PUMS database and share a common variable—tenure.[2] We then calculate the resulting persons per household, by age and tenure, by dividing the age number by the number of total owners or renters. Persons defined by age and tenure are divided by the total number of households of the respective tenure.

The results show that the overall household size of owner-occupied units is greater than that in renter-occupied units. We also find a pronounced difference in the age profile, as graphed in Fig. 2. Renter-occupied units are heavily loaded toward adults in their twenties and babies, whereas owner-occupied units have much higher ratios of middle-aged adults and teenagers.

2.1.1. Contrast to a Single-Universe Analysis

The method of linkage chosen here addresses persons and housing from a housing base (ratios to housing). To understand the process entirely, consider the alternatives *not* chosen. Instead of the multiuniverse, housing-based analysis above, a population-based analysis could approach the same issue within a single universe, asking different questions: either, what percentage *of persons in each type of housing unit* are in each age group, or what percentage *of each age group* lives in each type of housing unit. For these questions, the population counts by age and tenure in Table I are all that are required.[3] Those counts could simply be percentaged down the col-

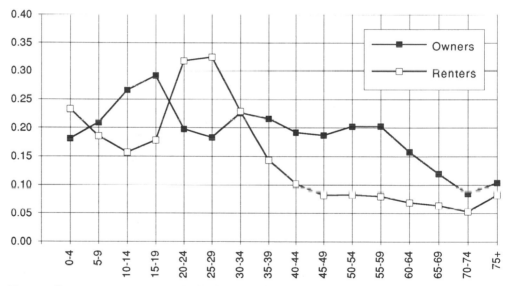

Figure 2. Persons per household, by age of person and tenure of household.

[2]The persons in households are derived from a type PH extract that combines the housing record with the population record, filtering out persons not living in a household (occupied housing unit). The count of housing units that are owner- or renter-occupied is derived from either a type H extract or from a PHH extract that filters out all persons from a type PH extract that are not householders. The count of householders by tenure equals the count of households by tenure. For more details of these extracts, refer to Chapter 12 and Appendix B.

[3]The type PH extract loads information on the occupied housing unit into the person record, thereby making housing a characteristic of persons and making it part of the population base.

umns or across the rows. In the former case, we would find, for example, that 6.6 percent of persons in owner-occupied housing are age 20 to 24; in the latter case, 37.3 percent of persons age 20 to 24 live in owner-occupied units.[4] This analysis makes use of a single universe—persons in households—so the entire calculation can be carried out within the single tabulation.

In contrast, the method demonstrated in Table I relates the person numbers to total household numbers that are separately tabulated for each universe. (Note that the totals at the bottom of the columns in the population count are totals of *persons* in owned or rented homes, not totals of housing units of that type.) The results of the population-based analysis (what percent *of persons*) are very different from the housing-based results shown (how many persons *per housing unit*).[5]

Recapping, single-universe analyses typically are expressed as percentage breakdowns of the total universe or of subgroups within the universe. However, multiuniverse analysis is necessary if we want to construct ratios between numbers in different universes (such as persons per household or houses per job).

2.2. Linking Multidimensional Tabulations

The basic strategy illustrated above can be carried out with much larger sets of variables. For example, we might calculate persons per household for very specific types of persons in very specific types of dwellings. The housing base might be defined by tenure, structure type, and number of bedrooms, or a total of three dimensions. The population characteristics to be related to this base might be defined by age and sex, or a total of two dimensions. The housing totals forming the denominators in the person per household ratios are expressed as a three-dimensional tabulation; however, the population numbers require a five-dimensional tabulation (the two population dimensions plus the three housing dimensions). In effect, the person characteristics must include not only age and sex, but also the housing characteristics. It is the overlap between the housing characteristics in the numerator and denominator that permits us to link the two different tabulations.

One way to view the structure of the linkage is through an expression that shows the dimensions employed in numerator and denominator. The necessary expression was given previously as Equation (2).

A second way to view the linkage is through the actual programming solution. Table II shows an example represented in the programming syntax of SAS. The three basic program steps are: (1) construct the two tabulations, outputting them as computer files, (2) arrange the two computer files for merger, and (3) calculate the ratio and print. The program generates a data file from each tabulation by outputting one record for each cell in the tabulation. The record contains the values for each variable that defines that cell, plus one additional variable, "Count," which records the number of cases in the cell. The heart of the programming procedure is to sort the

---

[4]Note that this is the percentage of persons, not of householders. The homeownership rate is always calculated as a percentage of households, or householders, of each age. Many of the persons who are owner-occupants in their early twenties are probably living in their parents' homes.

[5]Should our explanations seem redundant, it is deliberate. The basic distinctions between simple calculations on one base or another are conceptually very slippery and difficult to grasp at once. Readers also may wish to consult Fig. 2 of Chapter 12.

Table II. Program for Linking Tabulations in Two Different Universes, as Written in the SAS Language

```
/* STEP ONE FOLLOWS--CREATE TABULATIONS */
Proc Freq Data=Persons;                              /* Uses 'Persons' file previously created */
                                                     /* through PH extract */
    Tables Tenure*Structure*Bedrms*Age*Sex*Race /
                Noprint Out = Pop;                   /* 'Out = ' Creates table as output file -- see Note */

Proc Freq Data=OccUnits;                             /* Uses occupied units file previously */
                                                     /* created through PHH extract */
    Tables Tenure*Structure*Bedrms / Noprint Out = Units;
```

```
/* STEP TWO FOLLOWS--ARRANGE DATA FOR MERGER */
Data Pop2; Set Pop;
    Rename Count = Popct;
Proc Sort; By Tenure Structure Bedrms;               /* Sort two files in identical order */
Data Units2; Set Units;
    Rename Count = Unitct;
Proc Sort; By Tenure Structure Bedrms;               /* Sort two files in identical order */
```

```
/* STEP THREE FOLLOWS--CALCULATE RATIOS*/
Data PopUnits;
    Merge Pop2  Units2;
    By Tenure Structure Bedrms;                      /* This 'By Group' is shared by both tabulations.  */
Data Divide;                                         /* The merger appends the housing unit count  */
    AgeRatio = Popct / Unitct;                       /*     to each population type, repeating the  */
Proc Print;                                          /*     housing count across records for all population */
Title 'Persons per Household Calculations';          /*     types where it applies. */
```

Note: /* This signifies a comment the computer will ignore. */

        Output files from tabulations are an efficient method for compressing large PUMS files.
A separate data record is created for each cell in the tabulation and a "Count" variable
is added to record the number of cases in each cell. Thus, a five-variable tabulation is output
with six variables, and a 20,000-record PUMS file is compressed to the same number of records
as there are cells in the table.

two output tables by the variables they share in common, merging them on those shared variables, and placing the population and housing counts on the same data record. They can be divided then to form a ratio. Results of applying this basic procedure are presented in the next section.

☐ 3. Housing-Based Population Analysis

In Chapter 3 we introduced the housing unit method of population estimation. That is the most common procedure used in local area analysis to update population counts after the decennial census data are no longer current. Administrative data on building permits describe changes in the local housing stock, but no comparable register of population changes exists. Therefore, analysts have learned to use the housing data to help estimate population changes.

3.1. Extending the Housing Unit Method

We now illustrate how this basic idea can be easily extended to more detailed analysis of local population characteristics. Pursuing methods developed in Myers and Doyle (1990), as well as those of Smith and Lewis (1983) and Burchell and Listokin (1978), we demonstrate the rich information to be extracted from local PUMS data files. So great is the detail that can be produced, we will only sample some of the more interesting examples. Local users will want to tailor their analysis to specific requirements for such applications as school enrollment, parks or health care use, or more general estimation.

The housing unit method multiplies persons per household times the number of occupied housing units to estimate residential population.[6] This method can be improved on two fronts. On the housing side, we can disaggregate the housing base into more specific dwelling types, by tenure, structure type, number of bedrooms, or other characteristics. On the population side, we also can add much more detail than assumed previously. Persons in households can be disaggregated by five-year age brackets from 0 to 4 to 75 and older, by sex, and by race. The extension to race has particular value because, as seen in Chapter 10, sharp differences in household size and age often occur when one race or Hispanic group replaces another. The methods demonstrated here, along with those in Myers and Doyle (1990), also fulfill the final step in the two-tiered linkage of population and housing that was discussed in Chapter 3.

3.2. Age Profiles in Different Types of Housing

PUMS data for Los Angeles County in 1980 are analyzed for information about the occupants of different types of housing. The results presented here should be viewed as illustrative. Different patterns may be found in different locations and over time.[7] Our aim is to show what can be gained by this type of analysis.

Previously we identified the age profile found in rented versus owner-occupied housing (Fig. 2). An alternative way to describe type of housing is by structure type—whether the unit is in a one-unit structure (single-family), part of a multiunit structure (multifamily), or a mobile home (see Chapter 6). The advantage of struc-

---

[6]The method also requires an estimate of the proportion of housing units that is not vacant, a task that must be accomplished outside the census database (often using local utility records or postal vacancy surveys). To complete an estimate of the total population, a separate estimate of nonresidential population (persons in group quarters, institutions, or homeless) must also be added in.

[7]Locations that have a housing stock that is skewed toward one type or another will find a higher representation of all age groups in the housing type that is more prevalent. The age profiles for multifamily housing in New York City, for example, will certainly differ from Los Angeles. Also rural areas with many mobile homes will find more young families living in them than in this example. Changes over time also may be expected as the baby boom generation works its way up the age scale. Profiles in all housing types will be altered as the baby boom exits its twenties and moves into its forties.

ture type is that it is recorded on building permits, whereas tenure is not (Smith and Lewis 1983).

Figure 3 displays very different age profiles (persons per household of each age) of occupants in three major structure types. Occupants of multifamily housing have an age profile that is strongly concentrated in their twenties, accompanied by a smaller concentration of young children, much like that found for renters in Fig. 2. In contrast, single-family occupants have much higher concentrations in the ages of older children and in middle-aged adults. Mobile homes are strongly skewed toward late middle age and elderly persons.

Comparison of Figs. 2 and 3 suggests that tenure and structure type are roughly approximate, with renters matched to multifamily housing and owners matched to single-family housing. (The small number of mobile homes makes little impact on the owner distribution.) However, tenure and structure type are not fully equivalent. Many single-family housing units are renter-occupied (24 percent) and many multifamily units are owner-occupied (11 percent). Accordingly, housing units can be defined by both tenure and structure type. Whether or not this is useful depends on how much difference tenure makes for the age profiles within categories of structure type.

One difference is shown in Fig. 4. Single-family renters have many more small children than single-family owners. The rest of the renters' profile resembles that for renters as a whole. Because three-quarters of all single-family occupants are owners, the profile for owners closely matches that for single-family as a whole.

3.3. Age Profiles by Number of Bedrooms

The age profiles in different structure types might merely reflect differences in the size of units in those structures. Small apartments may be less attractive to large

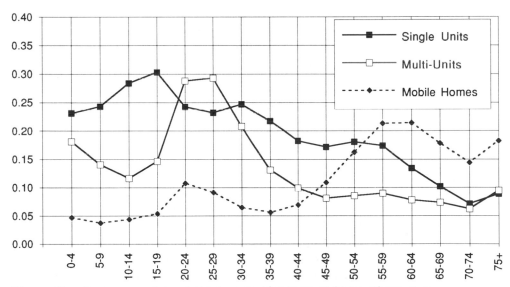

Figure 3. Persons per household, by age of person and type of structure.

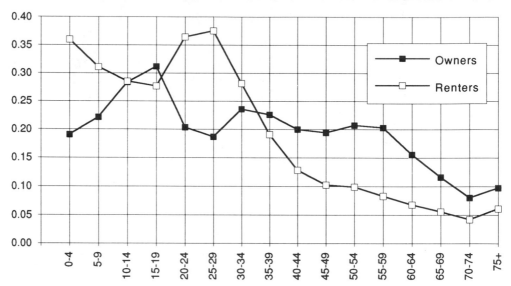

Figure 4. Contrast between age profiles of persons per household in owned and rented one-unit houses.

families than single-family houses, which are typically larger. A direct way to measure this size effect is to construct age profiles for units with different numbers of bedrooms.

In the interest of brevity, we will look in-depth at only one structure and tenure type. Different age profiles can be constructed for single-family units with different numbers of bedrooms. The effect of tenure is also controlled by examining profiles only within owner-occupied units. Figure 5 shows the difference made by number of bedrooms.

Household sizes are much greater in houses with more bedrooms. The overall household size rises from 2.42 (two bedrooms), to 3.20 (three bedrooms), to 3.92 (four bedrooms). As shown in Fig. 5, the differences occur largely through increases in teenage children or middle-age adults.

3.4. Differences by Sex in Age Profiles

Thus far, we have ignored any differences by sex; males and females have been combined. However, for some purposes we need separate counts of males and females. Are there any substantial differences in the age profiles of the two sexes according to their type of housing?

In this analysis we control for previous factors investigated above. In all the different types of housing, defined by tenure, structure, and number of bedrooms, only a couple of cases of sex differences stand out, and these are small. Those differences are concentrated in limited portions of the age span.

Within two-bedroom rented apartments, previous results suggest a higher ratio of persons in their twenties. Figure 6 shows that females are slightly more prevalent than males in these units. Within two-bedroom owned houses, females again pre-

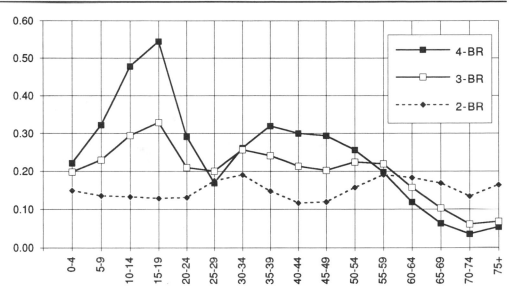

Figure 5. Persons per household, by age, in owner-occupied houses, by number of bedrooms.

dominate, but this time at ages 55 and older (Fig. 7). As seen previously in Fig. 5, these small houses have much higher levels of elderly than larger ones. Females predominate because mortality has reduced the number of males in these older age groups.

Finally, in four-bedroom owned houses, we find a deficit of persons in their twenties, as shown previously in Fig. 5. This age range is a transitional stage in the housing life cycle, as people leave their parents' homes and begin to establish independent living quarters. Figure 8 shows that females exit the large four-bedroom houses more quickly than males, but after age 25 they return more rapidly than males. After age 40, males and females balance out (not shown). This pattern

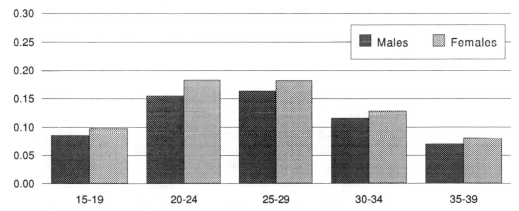

Figure 6. Persons per household, by age and sex, in two-bedroom rented apartments.

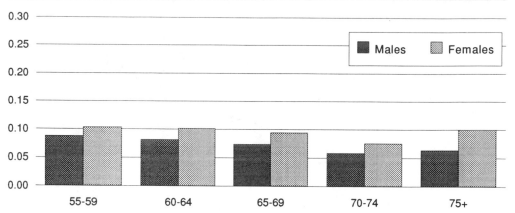

Figure 7. Persons per household, by age and sex, in two-bedroom owned houses.

reflects life-cycle differences between males and females. While females leave their parents' homes more quickly for marriage or other events (see Chapter 7), males are left in greater numbers. Conversely, 10 years later, as women begin to move their families into larger houses, wives tend to be two years younger than their husbands, with the result that females outnumber males around age 30.

Overall, the differences in age profiles by sex are slight. These differences are better explained by mortality and family life-cycle behavior than characteristics of the housing involved. For most purposes, it seems safe to collapse the sex dimension and look at age profiles for males and females combined.

3.5. Differences by Race and Hispanic Origin

In Chapter 10 we learned how racial change in local areas can be viewed as occupancy of the housing stock by different races with different household sizes. We now expand this insight on two dimensions: on the population side, we ask how the age profiles differ among racial groups; on the housing side, we ask how much the age

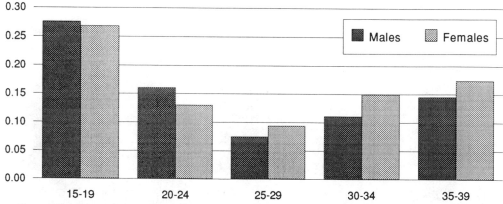

Figure 8. Persons per household, by age and sex, in four-bedroom owned houses.

profile of each race differs by type of housing. Answers to these questions can usefully extend the housing unit method to questions of racial transition by age group in local areas.

Household sizes differ markedly by race and type of housing. (For convenience, the analysis that follows is confined to the three largest race and Hispanic groups, and is also limited to the three dominant house types.) As summarized in Table III, owned, single-unit houses have the largest household sizes for each race or Hispanic origin group, and Hispanics have the largest household sizes overall. In fact, the racial differences appear to be described by a simple additive model. In every housing type, relative to white, non-Hispanics, the household sizes of Hispanics are about 1.6 greater and household sizes among blacks are about 0.6 greater.

This pattern indicates that Hispanic household size is simply elevated by a fixed amount in all house types relative to the other groups. Is this additive effect present in every age group in all house types, or are some age groups relatively more frequent in certain house types? Answers to these questions are sought here.

Before comparing the races, readers may wish to consult Fig. 4 (shown previously) because that figure compares owned and rented single-family houses for the entire population. An additional question is how much the age profiles by race differ from those that lack race detail.

The first housing type is single-unit structures that are owner-occupied. Figure 9 shows an age profile for white, non-Hispanics that is relatively flat, with roughly 0.2 persons in every age group. In contrast, among blacks, the number of teenagers is about double that of whites. Hispanic children are still more common in this house type, and the number of persons in their twenties and thirties is also elevated over that of blacks and whites. Beyond age 40, the age profiles for all three groups converge.

Rented, single-family houses have a substantially younger age profile for all three races (Fig. 10). There are many more young children, especially among Hispanics. There also are many more adults in their twenties and fewer persons over 40. These differences are consistent with those seen earlier in comparisons of

Table III. Ratio of Persons per Household by Race and Housing Type

| | Race and Hispanic Origin of Household Occupants | | |
	White, Non-Hispanic	Black	Hispanic
Owned, one-unit	2.75	3.34	4.33
Rented, one-unit	2.49	2.97	4.07
Rented apartment	1.65	2.23	3.20

Note: Computed by linking separate tabulations of persons, by race and housing type, and householders, by race and housing type. In rare instances, persons might not live in households headed by persons of the same race.
Source: Los Angeles County, 1980; persons living in households PUMS-A (5% sample of households).

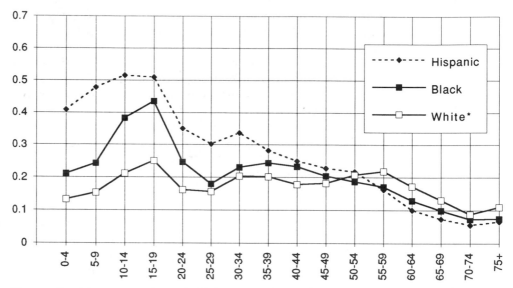

Figure 9.   Persons per household, by age and race, in one-unit owner-occupied housing units. ("White" population is non-Hispanic.)

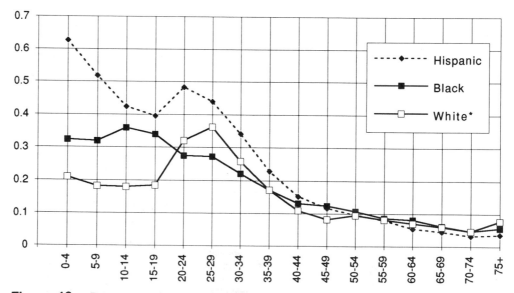

Figure 10.   Persons per household, by age and race, in one-unit renter-occupied housing units. ("White" population is non-Hispanic.)

owned and rented single-family houses (Fig. 4). In terms of differences between the races, it is striking that the number of persons in their twenties, per unit, is greater among whites than blacks despite their lower overall household size in this type of unit.

Racial differenes are greatest in the final housing type—rented apartments (Fig. 11). The age profile of white, non-Hispanic occupants has many fewer children per unit—one-quarter that of blacks and one-sixth that of Hispanics. A shift from white, non-Hispanic occupancy to black or Hispanic occupancy in this type of unit would produce even more dramatic changes in children's service needs (schools, parks, etc.) than in the other housing types.

3.6. Localizing the PUMS-derived Profiles

The preceding analysis of detailed age profiles by type of housing is based on PUMS data available only for areas of at least 100,000 population. These data can be made available for smaller areas through an estimation process similar to that shown in Chapter 13. An advantage in this case is that, unlike labor force, all the variables are 100-percent count, so we avoid the substantial sampling error that may occur at the local level.

The analysis requires linking two tabulations—one of population and one of occupied housing units. Thus, we need to develop a local version of both these tabulations. The counts of occupied units, by tenure, structure, and number of bedrooms,

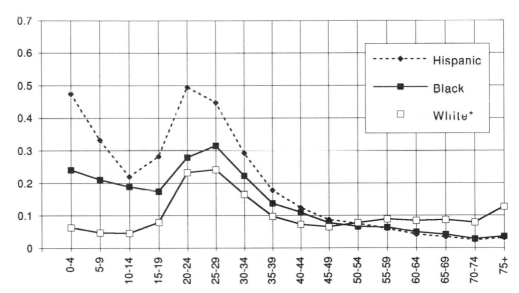

Figure 11. Persons per household, by age and race, in multiunit renter-occupied housing units. ("White" population is non-Hispanic.)

may be obtained directly from local sources, either the published books[8] or the summary tape files.[9] However, the tabulation of persons by age and housing type must be constructed through iterative proportional scaling.

At the local level, a 1990 tabulation of race by age of persons in households is provided in STF 2, Table PB6 (Table B10 in 1980). We also have a count of the number of persons living in housing units defined by race and tenure [STF 2, Table HB4 (Table B21 in 1980)], or in a more detailed version, offered for 1990 only, defined by race, tenure, and structure type (STF 2, Table HB9). The task is to allocate persons by age into the different housing types given knowledge of the total count per type and the overall age profile.

In brief, the necessary procedure requires three steps. First, a tabulation is prepared from PUMS of age by housing type occupied. This can be prepared for the total population or specific to different race groups. Second, the PUMS matrix is arranged for scaling by the two locally available marginals. The age marginal is the count of all persons, by age, who live in households. The housing type marginal consists of a series of totals of persons in each housing type. If a PUMS matrix for specific race groups is to be scaled, then the marginals should be selected for that race group. Third, the initial, PUMS-derived matrix is scaled by the marginals to find the best fitting matrix that matches the local marginals. Each race group is scaled independently in parallel fashion.

The resulting matrix estimated for the local area can then be linked to the appropriate counts of occupied units, and ratios can be calculated following the methods described above. These estimates are the best that can be achieved with any publicly available census data.

☐ 4. Employment-Based Analysis

Labor force analysis has already been investigated as the major topic of Chapter 13. That approach to employment analysis is population based—the percentage of persons who are in the labor force. In this section we use employment as the basis for analyzing other variables.

A great many specific industry or occupation categories can be defined. The PUMS Technical Documentation lists 7 pages of industrial codes and 13 pages of occupation codes (U.S. Bureau of the Census 1983b). Useful practice may dictate examining broad groupings of categories such as the one-digit categories of the standard industrial classification (SIC codes). Alternatively, researchers might define some frequent job types of particular salience, such as legal secretaries and attorneys, or police officers, construction workers, and school teachers.

---

[8]Census tract books in 1990 contain counts of occupied units by race and tenure, or by race and structure type, or by race and number of bedrooms (Series CPH-3). Different tables report the data for each race. For places of 10,000 or more population, counts by race and tenure are crosstabulated by structure or bedrooms (Series CH-2).

[9]The count of occupied units by race, tenure, and structure type is given in STF 2, Table HB8 (in 1980, Table B28). Number of bedrooms is a sample variable and is reported by race and tenure in STF 4, Table HB2 (in 1980, Table HB10).

4.1. Need for Employment-Based Analysis

Analysts often need to base their census use on employment counts. Expected employment growth will yield an increase in population, housing, and various service needs. Different types of jobs may have different levels of impact.

A highly contentious policy based on employment growth is that of office-housing linkage fees. Commercial real estate developers are assessed a fee based on the presumed impacts of new employees brought into the area (Hausrath 1988). Although some analyses have found that most employment moving into new commercial buildings is not new to the area (Patton 1988), to date the link between jobs and type of housing has not been well-researched. The availability of new microdata from the 1990 census may provide the data that are required for this type of detailed analysis of linkage.

4.2. Greater Complexity of Analysis

The employment universe is more difficult to work with than any other. Employment data do not nest within the hierarchical PUMS household record in the way that population and housing data do. Although employment is a property of persons, that activity usually takes place at a different location from the residence. Also, as discussed earlier in the chapter, employment is not confined to matching one-to-one with housing unit or household data. There may be no jobs held by household members or there may be several. Previously demonstrated methods must be extended even further for purposes of employment-based analysis. A full treatment of the subject is beyond the scope of this book, but a basic orientation can be provided.

4.3. Multiuniverse Employment Analysis

Within the single employment universe we might explore the characteristics of jobs and how they differ from place to place. However, many interesting questions link characteristics of the jobs to those of the job holders (persons) and to their housing units. In this section we will explore several issues. First, we inquire into the spatial link between place of work and place of residence. Next, we adopt the perspective of one place, the City of Long Beach, and inquire how many of the workers employed there, by type of industry, also live in that city. Following that question, we explore what type of housing units workers occupy. Substantial variation is found, depending both on place of residence and place of work.

Much of the following analysis is focused on the City of Long Beach, located in the southern portion of Los Angeles County, bordering Orange County, and one of the two central cities for the Los Angeles–Long Beach PMSA. Long Beach is a city of under half a million tucked within a Los Angeles and Orange counties region of about 10 million. It also is an older port city with a less advantaged population. Traditionally, Long Beach has had a strong manufacturing base of shipyards and aircraft manufacturing, but that base is now shifting with the loss of large manufacturing installations and the addition of more offices. Based on the 1980 PUMS data, we can examine the linkages between employment and other factors. With 1990 PUMS data, analysts could proceed to study changes over the decade in more depth.

4.3.1. Linking Work Place and Residence

One of the central analytic tools of transportation planners is the origin-destination matrix. Those data are required for hundreds of small traffic analysis zones and the table maps the number of trips from one zone to the next. PUMS data cannot deliver any kind of detail like that, but data can be organized in a similar structure for several useful purposes other than traffic analysis.

Places of residence and work are identified in the PUMS-A files based on their location in counties or county groups, and also distinguishing central cities of metropolitan areas from the remainder of their regions. Because workers may commute long distances, it is frequently necessary to define a laborshed with a wide range. The following areas within a 40-minute drive of Long Beach can be identified on PUMS-A:

Los Angeles County (Los Angeles–Long Beach PMSA)
    Long Beach City
    Los Angeles City
    Remainder of the county
Orange County (Anaheim–Santa Ana–Garden Grove PMSA)
    Anaheim City
    Garden Grove City
    Santa Ana City
    Remainder of the county
Other portions of southern California.

Simplifying the analysis still further for this illustration, we will treat Orange County as a whole and also group Los Angeles City with the remainder of Los Angeles County. Table IV portrays the resulting origin-destination matrix formed by these broad areas. The superiority of this grouping over that afforded in the transportation planning package from the census (COMSIS 1983), with its fine spatial grain, is that we can carry out very detailed analysis of characteristics and their linkages in the PUMS database.

Roughly 3,400 workers are identified as working in Long Beach in 1980. PUMS-A is a 5-percent sample, but only half of the work place and migration data were coded (for budgetary reasons), making these variables a 2.5 percent sample. Therefore, we would need to weight the worker counts by 40 to approximate their true level in 1980.

4.3.1.1. Distinguishing Two Employment Universes.
Two different percentage bases are identified in Table IV. The row totals represent resident workforce and column totals represent local employment. This shows how place of residence and place of work form *two different universes* within the employment universe. These may overlap substantially in very large areas, but certainly not in the case of smaller areas

Table IV. Origin-Destination Tabulation of Workers, by Place of Residence and Place of Work

Residence Origin	Work Place Destination				
	Long Beach	Rest of LA County	Orange County	Elsewhere in Southern California	TOTAL
Long Beach	1,628	1,507	313	14	3,462
Percent of residents	47.02	43.53	9.04	0.40	100.00
Percent of local workers	47.48	2.48	1.57	0.05	
Rest of LA County	995	54,470	1,651	812	57,928
Percent of residents	1.72	94.03	2.85	1.40	100.00
Percent of local workers	29.02	89.51	8.28	2.75	
Orange County	748	3,297	17,115	305	21,465
Percent of residents	3.48	15.36	79.73	1.42	100.00
Percent of local workers	21.81	5.42	85.84	1.03	
Other Southern California	58	1,579	859	28,395	30,891
Percent of residents	0.19	5.11	2.78	91.92	100.00
Percent of local workers	1.69	2.59	4.31	96.17	
TOTAL	3,429	60,853	19,938	29,526	113,746
	100.00	100.00	100.00	100.00	

Source: Southern seven counties of California; persons "at work" in week prior to census and part of migration-place of work subsample coded in PUMS-A database.

such as Long Beach. Analysts must decide which base is more appropriate for their analyses.

Of the locally employed workers, just under half (47.5 percent) live in Long Beach as well. The proportion of local workers in other areas who also reside there is much higher—89.5 percent in the Los Angeles County remainder, 85.8 percent in Orange County, and 96.2 percent in the remainder of southern California. This lower residential capture of workers is not surprising in view of the smaller size of Long Beach, or given its position midway between Los Angeles and Orange counties. In general, the smaller a subarea within a region, the less likely local employees are to live there.

The alternative way of reading this table is to ask what share *of local residents* also works in the same area. Asked this way, we find the Long Beach number is almost identical to the percentage running the other direction. The percentages are the same because the row and column totals for Long Beach are almost identical: The number of jobs in Long Beach is perfectly balanced with the number of workers who reside in Long Beach.

The jobs-housing balance is more even in Long Beach than elsewhere. The rest of Los Angeles County shows a surplus of jobs over resident workers, and so it must import. Note that Los Angeles County sends 995 workers to Long Beach, but Long Beach sends 1,507 workers to its big neighbor to the north. Conversely, Orange County and the rest of southern California have a shortfall of jobs relative to their workforce. For example, Orange County sends twice as many workers to Los Angeles County as it receives in return.

4.3.2. Capture of Residents from Workers in Different Industries

For the remainder of the chapter we adopt the perspective of a work place employment universe. The key questions facing local decision makers usually involve the population or housing impacts of locally situated employment. How large are these impacts and how many will fall inside the jurisdiction? A full methodology to address those questions is beyond our scope; however, we can offer a few suggestions about how PUMS data may play a role.

High-paying jobs may have greater or less benefit to local residents, depending on the likelihood that local persons will fill them. We find sharp variation exists in Long Beach in the local residency of workers in different industries. Figure 12 shows that in industries at the top of the graph more of the workers are traveling from out of town to go to work. The least likely workers to reside in Long Beach are those in the high-paying blue collar industries such as manufacturing and construction. The *most* likely residents are those in lower paying industries such as retail, entertainment, and personal services. The latter industries also employ higher proportions of women, although we have not analyzed that question in the PUMS data (see Rosenbloom 1989). Clearly, much more work deserves to be carried out with these PUMS data, especially comparing changes from 1980 to 1990.

4.3.2.1. Significance of Estimates from Small Samples.
A note is warranted here on statistical significance. Some of the job categories in Fig. 12 have fewer than 200 sample observations, causing concern about substantial sampling error. Accordingly we have represented the 95-percent confidence interval by a white error bar that shows the upper range of the interval. Groups with the largest error bars are those

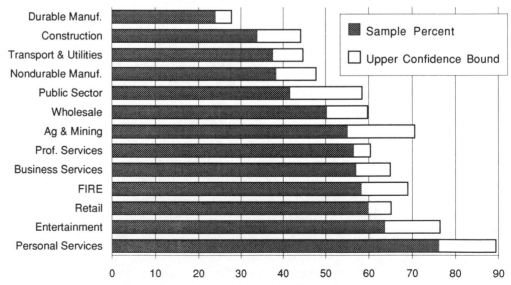

Figure 12. Percent of workers employed in Long Beach who live in Long Beach, by industry. The white bar indicates the upper tail of the 95-percent confidence interval.

that have the smallest sample size. For example, only 118 construction workers appear in the PUMS sample as employed in Long Beach. Of those, 33.9 percent also live in Long Beach, but the confidence interval ranges from 20.6 percent up to 44.2 percent. (The portion above 33.9 percent is represented by the white bar.)

The confidence interval is calculated as plus or minus two standard deviations from the observed value. The exact formula is that given in U.S. Bureau of the Census (1983b: 20) as adjusted for the reduced (one-half of normal) sampling rate for the place of work data. The formula is explained in Chapter 4.

4.3.3. Joint Decision of Housing and Residential Location

Locally employed workers choose residential locations in part on the type of housing offered there. A prime reason workers commute longer distances is that they can purchase more housing with their salary at some distance from job centers. A very simple analysis with our data is to compare the housing circumstances of locally employed residents with those who travel further.

We have measured type of housing on two dimensions—tenure and structure type. Combining the two, we find that one joint category—owned, single-family houses—accounts for most of the variation in the data. Therefore, we will focus on the percentage of workers living in that homeownership category as a useful measure of housing choices.

Figure 13 graphs this percentage for workers employed in the four locations of Table IV, conditioned on where they reside. Long Beach workers show the lowest homeownership, regardless of where they live. Compared to workers in different locations, the Long Beach residents have significantly lower homeownership.[10] Conversely, when compared to other Long Beach residents, those who work in Long Beach have lower homeownership than those who work elsewhere. The differences with Orange County and the rest of Los Angeles County are each statistically significant.

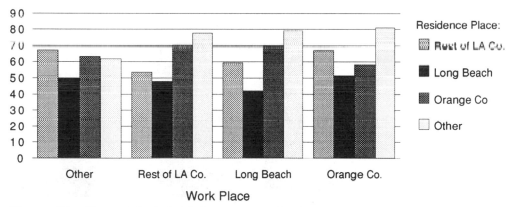

Figure 13. Percent single-family homeowners, by place of work and place of residence.

[10]The calculation of significance of a difference in percentages is based on the formula explained in Chapter 4.

Similarly, we note that residents of Orange County or Los Angeles County who work in that same location have significantly lower homeownership than those workers who travel outside. It is not certain whether this pattern reflects the higher incomes of longer commuters or other factors. Further study would be recommended.

4.3.4. Housing Occupied by Different Workers

These housing differences could arise because the workers are in different industries. With the PUMS data we can control for job type while focusing on housing type and location. Figure 14 presents the data for three of the larger job groupings defined by industry. (Alternatively, we could carry out the analysis with groupings based on occupation.) The subsample represented here is employed in Long Beach, and we differentiate them based on where they live.

Clearly, the Long Beach residents in all these industries have lower homeownership than residents commuting from Orange or Los Angeles counties. Orange County residents have higher homeownership than all others, reflecting the more suburban, low-density housing offered in that area. In contrast, some might assume that the higher housing prices in Orange County would discourage homeownership there.

4.3.5. Summarizing the Long Beach Evidence

Certain types of jobs appear to benefit Long Beach residents more than others. The evidence in Fig. 12 is startling in its portrayal of Long Beach residents' capture of lower paying service jobs. We also learned from Fig. 13 that the Long Beach workforce is much less likely to be homeowners than others, but their chances are increased somewhat if they work outside the city. Figure 14 then showed that, among persons employed in Long Beach, those who reside outside of Long Beach are much more likely to be homeowners.

Overall, these inquiries are merely illustrative of the type of analysis that can be performed within the employment universe of PUMS data. Much more detailed

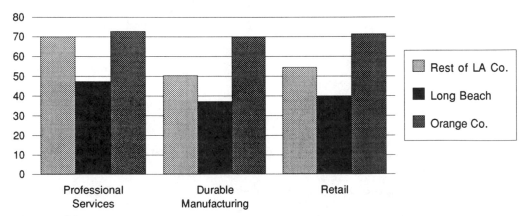

Figure 14.   Percent of Long Beach workers living in owned, one-unit structures, by type of job and place of residence.

analysis, covering a broader set of factors, is warranted for understanding the case of Long Beach.

4.4. Constructing True Employment-Based Multipliers

The preceding analysis has been of workers. In that analysis some of the housing units have been tabulated more than once. A house in Long Beach with two workers, for example, might be counted once with a manufacturing worker in Orange County and again with a retail worker in Long Beach. To estimate the impacts of employment growth, we need a method for linking the employment universe that does not double-count houses. Here we offer only a brief design of the necessary method; the complete programming solution is beyond our present scope, although the author has executed such a program in a number of cities.

The hierarchical structure of the PUMS database contains the necessary information, but it is not as easily extracted as the examples in Appendix B. Basically, we need to create a type PH extract, but at the same time record the number of workers who share the housing unit on each person's record. With that information, a weight can be applied to the case as it is tabulated. For example, with three workers in a house, each would receive one-third credit for the unit. Alternatively, credit might be assigned based on the percentage of household income contributed by each worker. The effect of these weighting schemes is to assign portions of the shared housing unit to each worker's industry of employment. With those multipliers we can then estimate, for example, how many housing units of each type will be demanded inside the city for every 100 jobs in a particular industry.

☐ 5. Conclusion

PUMS data provide a remarkably rich database for local analysis. Variables can be combined in unlimited fashion. In this chapter we have emphasized the linkages between different universes that are possible within the hierarchical structure of the database. These are among the most conceptually difficult analyses one is likely to undertake. Yet linkages between universes answer questions that are commonly raised in the local context. Analysts with a good basic knowledge of microdata analysis should be prepared to address these questions. The examples given here may help to illustrate the types of procedures required.

15 □□□

Conclusion

The decennial census is a massive enterprise that is difficult to carry out but it yields a cornucopia of essential information on our nation and its many subareas. The census has such urgent importance and the data are so widely available, in both quantity and depth, that data users are wise to maximize their understanding. A great many judgments are required for effective use of the data.

The preceding chapters have provided a broad introduction to the census and its uses. Our central focus has been to describe what data can be used to understand local communities. This focus has been expressed in two ways—explanation about the data themselves and demonstration about how to use the data. So many issues have been surveyed that it may prove helpful to review the major lessons in this conclusion.

□ 1. What Are the Data?

Before planning any analysis, users need to know what data are available for use. We have offered guidelines for selecting the most appropriate data source and have described all of the offerings available (see Chapter 4, particularly Table V). The informed user will have acquired a sophisticated understanding of those data from the explanations of Chapters 3 and 4.

We began with a summary of the purposes behind the census and the politics that shape its form, turning later to an explanation of the degree of error associated with

325

sampling and undercoverage of the population, or *census undercount*. The level of accuracy required of the data and the content of data to be collected are political decisions based on public needs and purposes for the census. Within the limits of these publicly provided data, individual users must exercise further decisions about accuracy and detail appropriate to their purposes.

One of the most vital issues about the census is the *content of the questionnaire* from which all census information stems. The questionnaire, along with most census reports, is divided into two halves constituting the decennial census of population and housing. Rather than hold to that rigid division we have defined four major categories of census items—housing, household, population, and economic—and questions in each category were explained. The discussion in Chapter 3 emphasized how the concepts measured in the different parts of the questionnaire fit together. The household and householder concepts are the crux of the census database, because they form the bridge between population and housing.

Not all questions are asked of all persons: the distinction between the *100-percent component* and the much larger *sample component* is essential to acknowledge (Table I of Chapter 3). The sample questions—including all of the economic data— are less accurate, due to the associated sampling error, as explained in Chapter 4. The sample or 100-percent count data also are reported in different data products (Tables III and IV of Chapter 4).

Users require a familiarity with *census geography* if they are to select the best data source for their particular area. In Chapter 4, we emphasized the *hierarchy principle of data availability*, showing how more detailed data are reported for higher geographies. Protection of confidentiality, small sample sizes, and limitations of reporting space all discourage the release of detailed data for smaller local areas. Sometimes users will need to combine more detailed data from higher geographies with more geographically specific, but limited, counts from lower level, small areas, as demonstrated in Chapter 13.

A key decision in data selection is whether to use *printed reports* or *computerized data files*. The advantages of each have been described. In general, printed reports are more easily accessed, especially when relatively few places are being investigated, but computerized files provide access to more detailed data. The differences have been explained among the various printed report series, as have differences among the different computer files.

The book has given attention to both *summary tabulations* and *microdata records*. Summary tabulations are tables reported in either printed reports or in summary tape files (some of which are available in CD-ROM form). In contrast, microdata are summaries of individual household and person questionnaires, with names and addresses removed and some additional variables added (see Fig. 1 of Chapter 4). Microdata are more difficult to use but allow for very flexible, customized analysis, as shown in the later chapters of the book.

Many of the issues just described are interlocking. The basic structure of content in the questionnaire, combined with the distinction between 100-percent and sample questions, is expressed in the various data products from the census. More detailed versions of the data—such as the microdata—are only available for larger areas and only in computerized form. The most widely available data are summary tabulations for the small areas of census tracts and block numbering areas, and the most often used versions of those data are issued in published reports. That data source is the common denominator of census data and has served as the basis for many of the book's examples.

☐ ## 2. How to Use the Data

Emphasis on how the different parts of the questionnaire fit together leads into models for analyzing relationships in the data. A major theme of the book has been to emphasize the larger story described by separate bits of data. Integrated portraits of change in communities can be constructed through analysis and linkage of different tables of census data.

Throughout the book we have emphasized an opportunistic, data-oriented strategy of analysis. Beginning with the data available for a local area, our goal has been to extract the maximum insight possible. That inductive strategy has been directed by prior knowledge of the questionnaire and database structure and by a modest amount of theory for deductively guiding the investigation.

2.1. Theory Directing Local Analysis

Chapter 3 introduced a number of causal models useful for thinking about how change can be described in local areas. We contrasted *top-down* and *bottom-up* models, the former appealing to national researchers and the latter appealing to local analysts. The difference in perspective is directly related to a *reversal in causal direction*. At the level of the region or larger areas, it is clear that employment growth induces migration and population growth, creating household formation and the need for housing. However, at the local level, housing takes precedence. Available housing units are filled by households, whose members have different ages, races or other characteristics, including their employment behavior.

This bottom-up housing-initiated view is adopted by the census taker who begins data collection with a list of housing addresses. It also is the basis for the well-known housing unit method of population estimation. For these reasons, we might say that housing serves as the essential substructure for small-area demography. The sequence of Chapters 6 through 13 follows that logic, proceeding from housing to households, then to age structure, migration, race, income and poverty, and labor force participation. This is only a *general* causal order, because it is obvious that other important causal currents are at work, such as from race to labor force to income, and then to housing types that can be afforded.

Some causal forces work at higher geographies and others at lower levels. Housing takes precedence at lower levels because, once built, it is relatively immobile and unchangeable, lasting for 80 years or more. People flow in and out of housing, and their family characteristics change dramatically if they remain for even 20 years. Meanwhile, the neighborhood housing stock remains relatively constant, playing a causally prior role, although dramatic events such as gentrification or redevelopment may infrequently alter the picture.

A more sophisticated analysis seeks to interplay causal dynamics at the regional level with those at the local level. Three general models were outlined in Chapter 3. The employment-based transportation planning model is most highly developed and requires very complex computer programming, but it tends to slight local characteristics. A population-based social services planning model follows the classic demographic accounting framework, but is baffled by local migration and must adjust local parameters until they mesh with a separate model for the larger region. Finally, a two-tiered model was suggested that integrated elements of the others, using one basis of causality—economic and demographic—for the larger region and another, housing-based model at the local level. Readers should take these models as conceptual organizing devices for thinking about how local area analysis fits into analysis of the broader region. Other formulations are also possible. An analyst's objective should be to find an approach that is *suitable to the particular problem at hand* and that is *workable within existing constraints* of time, data, and ability.

2.2. General Analytical Strategies

Despite a general orientation toward bottom-up analysis, presentations in the book have avoided a single-minded theory, because there is no single model that neatly addresses all needs. Instead, a variety of specific methods have been portrayed in different chapters, and their relative advantages and disadvantages have been discussed. Overall, several general analytical strategies have been demonstrated, and these may be usefully summarized as follows:

1. make presentation count;
2. analyze data based on linkages in the questionnaire;
3. conduct parallel analysis of population subgroups;
4. ransack the data opportunities;
5. borrow information from higher geographies;
6. employ a variety of analytical formats for the same data; and
7. consider a range of technical solutions.

Early in the book, we stressed an overall belief that *presentation counts*. No matter how accurate and insightful the census data analysis, the work may be valueless if the findings cannot be communicated clearly to an audience of decision makers or interested observers. Chapter 5 emphasized skills in constructing tables and graphs, the two most common devices for displaying quantitative information. Mapmaking is another essential skill for this purpose, but that practice is so complex that it is the subject of the specialized field known as cartography. Recently, microcomputer-based mapping and GIS programs have begun to bring that form of data display within the reach of average users (Lucerna and Landis 1989).

A fundamental strategy has been to analyze census data in terms of *linkages* built into the census questionnaire. This was demonstrated most clearly in Chapter 7, which analyzed marital status, families, and household relationships. The direct relationships among these variables define one category as a subset of another, guiding our construction of ratios and percentages. The most unusual outcome of the chapter was to break down total household size in terms of its relationship components, i.e., how many married persons per household, children per household, etc. With that accounting framework, changes in household size between censuses can be attributed to losses and gains in particular marital statuses or family relationships. For example, we found that one suburban neighborhood lost one whole child per household in just a decade, while a Hollywood gay neighborhood gained nonrelative household members almost as fast as it lost spouses.

A third strategy has been *parallel analysis* of population subgroups, demonstrated most clearly in the analysis of racial groups in Chapter 10, but also in the analysis with microdata in Chapters 12 through 14. Identical tables are repeated for each race or Hispanic-origin group with a minimum of 400 persons, permitting a replication of detailed analysis for each separate group. The separate analyses can then be combined to represent overall dynamics for the whole population and to learn what share of the detailed changes is accounted for by each subgroup. For example, in one of the Pasadena tracts we found that white population of all ages had been moving out, but replacement by black in-movers was highly concentrated among early middle-aged adults, accompanied by their children. We also found that blacks were occupying a growing share of the housing stock, but their share of the *newer* stock was far greater than their share of the pre-1940 housing. Apparently, the white population retained greater preference for that older vintage.

In a fourth strategy, we have strived to *ransack the data opportunities* available for analysis. As an example, Chapter 9 on migration identified five different local tabulations that could be used for mobility analysis: state and county residence five years before the census, metropolitan residency status five years before the census, length of occupancy in the current home, lifetime migration (place of birth), and year of immigration. We also called attention to gross migration data by age available for counties in a special report from the census. Given the complexity of migration—a spatial process over time—no single tabulation is adequate for description. All of these data opportunities deserve to be exploited for the different perspectives they offer.

Readers will have observed that this ransacking strategy has an important conservative restraint. Rather than go directly to the most detailed summary tape file information available, our emphasis has been to first extract maximum use of the published data. The limitations of those published sources have been noted where relevant, and users who require extra information may then proceed to the computer files. Even then, our emphasis has been to rely on the simpler STF 1 or 3 before delving into the more detailed STF 2 or 4. Microdata files are a last resort used for customized tabulation, but they sacrifice small-area detail. In Chapter 14 we demonstrated how microdata can be usefully combined in multiuniverse analysis.

A fifth strategy has been to *borrow information from higher geographies* to supplement data for small-area analysis. This strategy responds to the hierarchy principle of data availability—more detailed data are published for larger places, counties and states, and the most detailed computerized data are only available for places of 100,000 or more population. This strategy was the centerpiece of Chapter 13 on labor force analysis, carried out through the technical procedure of *iterative propor-*

tional scaling, but we also employed it at points in other chapters where we needed to augment published data for census tracts. The *rates × composition* analysis structure, fully described in Chapter 12, was also used in Chapter 7 to calculate age-expected marital status by using the small area's composition (age structure by sex) and the larger area's rates (percent never-married by age and sex).

A sixth strategy has been to explore a *variety of analytical formats* with the same raw data. This emphasis on alternatives was demonstrated most clearly in Chapter 8 on age structure where we surveyed a number of alternative formats for analyzing age. The traditional population pyramid is based on percentages of total population. Two alternatives were recommended for specific purposes. One approach divides the number in an age group by the number of households instead of by total population, creating *age-specific persons per household*. These age ratios have the advantage that they do not depend on changes in other age groups or in the total population and that they can be used as multipliers. Another advantage of this method is that it breaks down household size into age components, as shown at length in Chapter 14, extending Chapter 7's analysis where household size was broken down into family relationships that make up households.

A major alternative with age data emphasizes local *cohort* analysis instead of *age group* comparisons. By linking the absolute counts of two age groups in two different census years, we demonstrated how *cohort trajectories* can be traced as the size of cohorts changes over the decade. Net increases or decreases, termed cohort retention, reflect the net migration attractions of a neighborhood for persons in particular age ranges and provide sharper insight into behavior that changes age structure. This cohort retention method was employed in striking fashion for the analysis of racial change in Chapter 10.

A final strategy of our analytical presentations has been to emphasize a *range of technical solutions* to problems. Some techniques require greater skill than others, or demand access to mainframe computers and tape drives, or require more difficult data management, or simply demand more time-consuming analysis. However, in the *real world* use of census data, it is often true that the best solution is the simplest, cheapest, and quickest. Our objective has been to portray a menu of different solutions to choose from, including not only more technical methods, but also showing how analysts can get by with a more limited data source and a simpler analysis if needed. This strategy was most clearly demonstrated in Chapter 8 on age structure in which a variety of methods were described for handling data grouped in 10-year categories. We also encountered the strategy in Chapter 11 on income where tabulations are grouped into wide dollar ranges that must be interpolated for many purposes. Users should choose the level of technique that is most appropriate to their needs and abilities.

Throughout the book our emphasis has been on a common denominator of available data and simpler methods. These are not only more accessible, but they also make more transparent the structure of census data and the strategy of analysis. Minimizing complexity, without discounting it, has greatly facilitated the instructional objectives of the book. Experts will recognize readily how they can extend the basic techniques demonstrated here, and, as other users grow experienced through practice with the 1990 data, they also will be able to advance to more elaborate solutions when needed.

☐ 3. Continuing Progress in Census Analysis

Despite the relative continuum of the census from decade to decade, minor changes are always being made in the conduct of the census and in its content. However, more dramatic changes have occurred in the past decade in the *analysis* of census data, largely because of the rapid advances in computing technology. The 1980 census was the first to bring microdata analysis within range of many academic and professional users. Data files were prepared for many more local county groups, and the lower cost of mainframe computer time, combined with the ready availability of mainframe programs such as SAS and SPSS, made analysis more practical. Computerized mapping with summary tape files also became more feasible in 1980.

The 1990 census is the first to be accessed by microcomputer users. Many times more analysts will have access to those computers than the mainframes used with 1980 census data. We are witnessing a democratization of census computing, some of it based on micro versions of mainframe statistical packages, or on GIS mapping analysis programs, but more often using spreadsheets for analysis of tables and preparation of graphs.

It is too soon to tell what progress will be made in the analysis of 1990 census data. We hope this book provides users with the foundation they require for informed and insightful analysis. Future editions will aim to document the progress of local census data analysis. If analysts will help to keep the author informed of local successes, we will be pleased to showcase that work—with proper credit—for the benefit of others. With so much data, and so little time, it is imperative that we share experiences. May your ventures be successful.

Appendix A ⬜⬜⬜

1990 Census Questionnaire

CENSUS '90

OFFICIAL 1990 U.S. CENSUS FORM

Thank you for taking time to complete and return this census questionnaire. It's important to you, your community, and the Nation.

The law requires answers but guarantees privacy.

By law (Title 13, U.S. Code), you're required to answer the census questions to the best of your knowledge. However, the same law guarantees that your census form remains confidential. For 72 years--or until the year 2062--only Census Bureau employees can see your form. No one else--no other government body, no police department, no court system or welfare agency--is permitted to see this confidential information under any circumstances.

How to get started--and get help.

Start by listing on the next page the names of all the people who live in your home. Please answer all questions with a black lead pencil. You'll find detailed instructions for answering the census in the enclosed guide. If you need additional help, call the toll-free telephone number to the left, near your address.

Please answer and return your form promptly.

Complete your form and return it by April 1, 1990 in the postage-paid envelope provided. Avoid the inconvenience of having a census taker visit your home.

Again, thank you for answering the 1990 Census.
Remember: Return the completed form by April 1, 1990.

Para personas de habla hispana --
(For Spanish-speaking persons)

Si usted desea un cuestionario del censo en español, llame sin cargo alguno al siguiente número: 1-800-XXX-XXXX
                    (o sea 1-800-XXX-XXXX)

U.S. Department of Commerce
BUREAU OF THE CENSUS

FORM D-2

OMB No. 0607-0628
Approval Expires 07/31/91

Page 1

The 1990 census must count every person at his or her "usual residence." This means the place where the person lives and sleeps most of the time.

1a. List on the numbered lines below the name of each person living here on Sunday, April 1, including all persons staying here who have no other home. If EVERYONE at this address is staying here temporarily and usually lives somewhere else, follow the instructions given in question 1b below.

Include

- Everyone who usually lives here such as family members, housemates and roommates, foster children, roomers, boarders, and live-in employees
- Persons who are temporarily away on a business trip, on vacation, or in a general hospital
- College students who stay here while attending college
- Persons in the Armed Forces who live here
- Newborn babies still in the hospital
- Children in boarding schools below the college level
- Persons who stay here most of the week while working even if they have a home somewhere else
- Persons with no other home who are staying here on April 1

Do NOT include

- Persons who usually live somewhere else

- Persons who are away in an institution such as a prison, mental hospital, or a nursing home
- College students who live somewhere else while attending college
- Persons in the Armed Forces who live somewhere else

- Persons who stay somewhere else most of the week while working

Print last name, first name, and middle initial for each person. Begin on line 1 with the household member (or one of the household members) in whose name this house or apartment is owned, being bought, or rented. If there is no such person, start on line 1 with any adult household member.

	LAST	FIRST	INITIAL		LAST	FIRST	INITIAL
1				7			
2				8			
3				9			
4				10			
5				11			
6				12			

1b. If EVERYONE is staying here only temporarily and usually lives somewhere else, list the name of each person on the numbered lines above, fill this circle ⟶ ○ and print their usual address below. DO NOT PRINT THE ADDRESS LISTED ON THE FRONT COVER.

House number	Street or road/Rural route and box number	Apartment number
City	State	ZIP Code
County or foreign country	Names of nearest intersecting streets or roads	

NOW PLEASE OPEN THE FLAP TO PAGE 2 AND ANSWER ALL QUESTIONS FOR THE FIRST 7 PEOPLE LISTED. USE A BLACK LEAD PENCIL ONLY.

Page 2   *PLEASE ALSO ANSWER HOUSING QUESTIONS ON PAGE 3* ➤

	PERSON 1	**PERSON 2**
Please fill one column ➤ for each person listed in Question 1a on page 1.	Last name First name Middle initial	Last name First name Middle init.

2. How is this person related to PERSON 1?

Fill ONE circle for each person.

If **Other relative** of person in column 1, fill circle and print exact relationship, such as mother-in-law, grandparent, son-in-law, niece, cousin, and so on.

PERSON 1: START in this column with the household member (or one of the members) in whose name the home is owned, being bought, or rented.

If there is no such person, start in this column with any adult household member.

PERSON 2: If a RELATIVE of Person 1:
- ○ Husband/wife
- ○ Natural-born or adopted son/daughter
- ○ Stepson/ stepdaughter
- ○ Brother/sister
- ○ Father/mother
- ○ Grandchild
- ○ Other relative �”

If NOT RELATED to Person 1:
- ○ Roomer, boarder, or foster child
- ○ Housemate, roommate ■
- ○ Unmarried partner
- ○ Other nonrelative

3. Sex

Fill ONE circle for each person.

PERSON 1: ○ Male ○ Female

PERSON 2: ○ Male ○ Female

4. Race

Fill ONE circle for the race that the person considers himself/herself to be.

If **Indian (Amer.)**, print the name of the enrolled or principal tribe.

If **Other Asian or Pacific Islander (API)**, print one group, for example: Hmong, Fijian, Laotian, Thai, Tongan, Pakistani, Cambodian, and so on.

If **Other race**, print race.

PERSON 1:
- ○ White
- ○ Black or Negro
- ○ Indian (Amer.) (Print the name of the enrolled or principal tribe.) �”
- ○ Eskimo
- ○ Aleut

Asian or Pacific Islander (API)
- ○ Chinese
- ○ Filipino ■
- ○ Hawaiian
- ○ Korean
- ○ Vietnamese
- ○ Japanese
- ○ Asian Indian
- ○ Samoan
- ○ Guamanian
- ○ Other API �”
- ○ Other race (Print race) ➚

PERSON 2:
- ○ White
- ○ Black or Negro
- ○ Indian (Amer.) (Print the name of the enrolled or principal tribe.) �”
- ○ Eskimo
- ○ Aleut

Asian or Pacific Islander (API)
- ○ Chinese
- ○ Filipino ■
- ○ Hawaiian
- ○ Korean
- ○ Vietnamese
- ○ Japanese
- ○ Asian Indian
- ○ Samoan
- ○ Guamanian
- ○ Other API ➚
- ○ Other race (Print race) ➚

5. Age and year of birth

a. Print each person's age at last birthday. Fill in the matching circle below each box.

b. Print each person's year of birth and fill the matching circle below each box.

PERSON 1:
a. Age
b. Year of birth   1

a. Age			b. Year of birth		
0 ○ 0 ○	0 ○	0 ○	1 ● 8 ○	0 ○ 0 ○	0 ○
1 ○ 1 ○	1 ○	1 ○	9 ○ 1 ○	1 ○ 1 ○	1 ○
2 ○	2 ○			2 ○ 2 ○	2 ○
3 ○	3 ○			3 ○ 3 ○	3 ○
4 ○	4 ○	■		4 ○ 4 ○	4 ○
5 ○	5 ○			5 ○ 5 ○	5 ○
6 ○	6 ○			6 ○ 6 ○	6 ○
7 ○	7 ○			7 ○ 7 ○	7 ○
8 ○	8 ○			8 ○ 8 ○	8 ○
9 ○	9 ○			9 ○ 9 ○	9 ○

PERSON 2:
a. Age
b. Year of birth   1

a. Age			b. Year of birth		
0 ○ 0 ○	0 ○	0 ○	1 ● 8 ○	0 ○ 0 ○	0 ○
1 ○ 1 ○	1 ○	1 ○	9 ○ 1 ○	1 ○ 1 ○	1 ○
2 ○	2 ○			2 ○ 2 ○	2 ○
3 ○	3 ○			3 ○ 3 ○	3 ○
4 ○	4 ○	■		4 ○ 4 ○	4 ○
5 ○	5 ○			5 ○ 5 ○	5 ○
6 ○	6 ○			6 ○ 6 ○	6 ○
7 ○	7 ○			7 ○ 7 ○	7 ○
8 ○	8 ○			8 ○ 8 ○	8 ○
9 ○	9 ○			9 ○ 9 ○	9 ○

6. Marital status

Fill ONE circle for each person.

PERSON 1:
- ○ Now married
- ○ Widowed
- ○ Divorced
- ○ Separated
- ○ Never married

PERSON 2:
- ○ Now married
- ○ Widowed
- ○ Divorced
- ○ Separated
- ○ Never married

7. Is this person of Spanish/Hispanic origin?

Fill ONE circle for each person.

If **Yes, other Spanish/Hispanic**, print one group.

PERSON 1:
- ○ No (not Spanish/Hispanic)
- ○ Yes, Mexican, Mexican-Am., Chicano
- ○ Yes, Puerto Rican ■
- ○ Yes, Cuban
- ○ Yes, other Spanish/Hispanic (Print one group, for example: Argentinean, Colombian, Dominican, Nicaraguan, Salvadoran, Spaniard, and so on.) ➚

PERSON 2:
- ○ No (not Spanish/Hispanic)
- ○ Yes, Mexican, Mexican-Am., Chicano
- ○ Yes, Puerto Rican
- ○ Yes, Cuban
- ○ Yes, other Spanish/Hispanic (Print one group, for example: Argentinean, Colombian, Dominican, Nicaraguan, Salvadoran, Spaniard, and so on.) ➚

FOR CENSUS USE ➤

PERSON 1: ○   ○

PERSON 2: ○   ○

Page 3

PERSON 7

Last name

First name — Middle initial

If a RELATIVE of Person 1:
- ○ Husband/wife
- ○ Brother/sister
- ○ Natural-born or adopted son/daughter
- ○ Father/mother
- ○ Grandchild
- ○ Other relative ⤷
- ○ Stepson/ stepdaughter

If NOT RELATED to Person 1:
- ○ Roomer, boarder, or foster child
- ○ Unmarried partner
- ● Housemate, roommate
- ○ Other nonrelative

- ○ Male
- ○ Female

- ○ White
- ○ Black or Negro
- ○ Indian (Amer.) (Print the name of the enrolled or principal tribe.) ⤷
- ○ Eskimo
- ○ Aleut

Asian or Pacific Islander (API)
- ○ Chinese
- ○ Japanese
- ● Filipino
- ○ Asian Indian
- ○ Hawaiian
- ○ Samoan
- ○ Korean
- ○ Guamanian
- ○ Vietnamese
- ○ Other API ⤷

- ○ Other race (Print race) ⤶

a. Age

0	0	0	0	0	
1	0	1	0	1	0
2	0	2	0		
3	○	3	0		
4	0	4	●		
5	0	5	0		
6	0	6	0		
7	0	7	○		
8	0	8	0		
9	0	9	0		

b. Year of birth

1

1	●	8	0	0	0
		9	0	1	0
		2	0	2	0
		3	○	3	0
		4	0	4	0
		5	0	5	0
		6	0	6	0
		7	0	7	0
		8	0	8	0
		9	0	9	0

- ○ Now married
- ○ Separated
- ○ Widowed
- ○ Never married
- ○ Divorced

- ○ No (not Spanish/Hispanic)
- ○ Yes, Mexican, Mexican Am., Chicano
- ○ Yes, Puerto Rican
- ○ Yes, Cuban
- ○ Yes, other Spanish/Hispanic (Print one group, for example: Argentinean, Colombian, Dominican, Nicaraguan, Salvadoran, Spaniard, and so on.) ⤷

- ○
- ○

NOW PLEASE ANSWER QUESTIONS H1a—H26 FOR YOUR HOUSEHOLD

H1a. Did you leave anyone out of your list of persons for Question 1a on page 1 because you were not sure if the person should be listed — for example, someone temporarily away on a business trip or vacation, a newborn baby still in the hospital, or a person who stays here once in a while and has no other home?
- ○ Yes, please print the name(s) and reason(s). ⤷
- ○ No

b. Did you include anyone in your list of persons for Question 1a on page 1 even though you were not sure that the person should be listed — for example, a visitor who is staying here temporarily or a person who usually lives somewhere else?
- ○ Yes, please print the name(s) and reason(s). ⤷
- ○ No

H2. Which best describes this building? Include all apartments, flats, etc., even if vacant.
- ○ A mobile home or trailer
- ○ A one-family house detached from any other house
- ○ A one-family house attached to one or more houses
- ○ A building with 2 apartments
- ○ A building with 3 or 4 apartments
- ○ A building with 5 to 9 apartments
- ○ A building with 10 to 19 apartments
- ○ A building with 20 to 49 apartments
- ○ A building with 50 or more apartments
- ○ Other

H3. How many rooms do you have in this house or apartment? Do NOT count bathrooms, porches, balconies, foyers, halls, or half-rooms.
- ● 1 room
- ○ 4 rooms
- ○ 7 rooms
- ○ 2 rooms
- ○ 5 rooms
- ○ 8 rooms
- ○ 3 rooms
- ○ 6 rooms
- ○ 9 or more rooms

H4. Is this house or apartment —
- ○ Owned by you or someone in this household with a mortgage or loan?
- ○ Owned by you or someone in this household free and clear (without a mortgage)?
- ○ Rented for cash rent?
- ○ Occupied without payment of cash rent?

If this is a ONE-FAMILY HOUSE —

H5a. Is this house on ten or more acres?
- ○ Yes
- ○ No

b. Is there a business (such as a store or barber shop) or a medical office on this property?
- ○ Yes
- ○ No

Answer only if you or someone in this household OWNS OR IS BUYING this house or apartment —

H6. What is the value of this property; that is, how much do you think this house and lot or condominium unit would sell for if it were for sale?
- ○ Less than $10,000
- ○ $70,000 to $74,999
- ○ $10,000 to $14,999
- ○ $75,000 to $79,999
- ○ $15,000 to $19,999
- ○ $80,000 to $89,999
- ○ $20,000 to $24,999
- ○ $90,000 to $99,999
- ○ $25,000 to $29,999
- ○ $100,000 to $124,999
- ○ $30,000 to $34,999
- ○ $125,000 to $149,999
- ○ $35,000 to $39,999
- ○ $150,000 to $174,999
- ○ $40,000 to $44,999
- ○ $175,000 to $199,999
- ○ $45,000 to $49,999
- ○ $200,000 to $249,999
- ○ $50,000 to $54,999
- ○ $250,000 to $299,999
- ○ $55,000 to $59,999
- ○ $300,000 to $399,999
- ○ $60,000 to $64,999
- ○ $400,000 to $499,999
- ○ $65,000 to $69,999
- ○ $500,000 or more

Answer only if you PAY RENT for this house or apartment —

H7a. What is the monthly rent?
- ○ Less than $80
- ○ $375 to $399
- ○ $80 to $99
- ○ $400 to $424
- ○ $100 to $124
- ○ $425 to $449
- ○ $125 to $149
- ○ $450 to $474
- ○ $150 to $174
- ○ $475 to $499
- ○ $175 to $199
- ○ $500 to $524
- ● $200 to $224
- ○ $525 to $549
- ○ $225 to $249
- ○ $550 to $599
- ○ $250 to $274
- ○ $600 to $649
- ○ $275 to $299
- ○ $650 to $699
- ○ $300 to $324
- ○ $700 to $749
- ○ $325 to $349
- ○ $750 to $999
- ○ $350 to $374
- ○ $1,000 or more

b. Does the monthly rent include any meals?
- ○ Yes
- ○ No

FOR CENSUS USE

A. Total persons

B. Type of unit

Occupied	Vacant
○ First form	○ Regular
○ Cont'n	○ Usual home elsewhere

C1. Vacancy status
- ○ For rent
- ○ For seas/rec/occ
- ○ For sale only
- ○ Rented or sold, not occupied
- ○ For migrant workers
- ○ Other vacant

C2. Is this unit boarded up?
- ○ Yes
- ○ No

○	○
1	1
2	2
3	3
● 4	4
5	5
6	6
7	7
8	8
9	9

D. Months vacant
- ○ Less than 1
- ○ 6 up to 12
- ○ 1 up to 2
- ○ 12 up to 24
- ○ 2 up to 6
- ○ 24 or more

E. Complete after
- ○ LR ○ TC ○ QA JIC 1
- ○ P/F ○ RE ○ I/T ○
- ○ MV ○ FD ○ EN
- ○ P0 ○ P3 ○ P6
- ○ P1 ○ P4 ○ IA JIC 2
- ○ P2 ○ P5 ○ SM ○

F. Cov.
- ○ 1b ○ 1a ○ 7 ○ H1

G. DO

ID

0	0	0	0	0	0	0	0	0	0	0
1	1	1	1	1	1	1	1	1	1	1
2	2	2	2	2	2	2	2	2	2	2
3	3	3	3	3	3	3	3	3	3	3
4	4	4	4	4	4	4	4	4	4	4
5	5	5	5	5	5	5	5	5	5	5
6	6	6	6	6	6	6	6	6	6	6
7	7	7	7	7	7	7	7	7	7	7
8	8	8	8	8	8	8	8	8	8	8
9	9	9	9	9	9	9	9	9	9	9

H8. When did the person listed in column 1 on page 2 move into this house or apartment?

- ○ 1989 or 1990
- ○ 1985 to 1988
- ○ 1980 to 1984
- ○ 1970 to 1979
- ○ 1960 to 1969
- ○ 1959 or earlier

H9. How many bedrooms do you have; that is, how many bedrooms would you list if this house or apartment were on the market for sale or rent?

- ○ No bedroom
- ○ 1 bedroom
- ○ 2 bedrooms
- ○ 3 bedrooms
- ○ 4 bedrooms
- ○ 5 or more bedrooms

H10. Do you have COMPLETE plumbing facilities in this house or apartment; that is, 1) hot and cold piped water, 2) a flush toilet, and 3) a bathtub or shower?

- ○ Yes, have all three facilities
- ○ No

H11. Do you have COMPLETE kitchen facilities; that is, 1) a sink with piped water, 2) a range or cookstove, and 3) a refrigerator?

- ○ Yes
- ○ No

H12. Do you have a telephone in this house or apartment?

- ○ Yes
- ○ No

H13. How many automobiles, vans, and trucks of one-ton capacity or less are kept at home for use by members of your household?

- ○ None
- ○ 1
- ○ 2
- ○ 3
- ○ 4
- ○ 5
- ○ 6
- ○ 7 or more

H14. Which FUEL is used MOST for heating this house or apartment?

- ○ Gas: from underground pipes serving the neighborhood
- ○ Gas: bottled, tank, or LP
- ○ Electricity
- ○ Fuel oil, kerosene, etc.
- ○ Coal or coke
- ○ Wood
- ○ Solar energy
- ○ Other fuel
- ○ No fuel used

H15. Do you get water from —

- ○ A public system such as a city water department, or private company?
- ○ An individual drilled well?
- ○ An individual dug well?
- ○ Some other source such as a spring, creek, river, cistern, etc.?

H16. Is this building connected to a public sewer?

- ○ Yes, connected to public sewer
- ○ No, connected to septic tank or cesspool
- ○ No, use other means

H17. About when was this building first built?

- ○ 1989 or 1990
- ○ 1985 to 1988
- ○ 1980 to 1984
- ○ 1970 to 1979
- ○ 1960 to 1969
- ○ 1950 to 1959
- ○ 1940 to 1949
- ○ 1939 or earlier
- ○ Don't know

H18. Is this house or apartment part of a condominium?

- ○ Yes
- ○ No

If you live in an apartment building, skip to H20.

H19a. Is this house on less than 1 acre?

- ○ Yes — Skip to H20
- ○ No

b. In 1989, what were the actual sales of all agricultural products from this property?

- ○ None
- ○ $1 to $999
- ○ $1,000 to $2,499
- ○ $2,500 to $4,999
- ○ $5,000 to $9,999
- ○ $10,000 or more

H20. What are the yearly costs of utilities and fuels for this house or apartment?

If you have lived here less than 1 year, estimate the yearly cost.

a. Electricity

$ [   ] .00
Yearly cost — Dollars

OR

- ○ Included in rent or in condominium fee
- ○ No charge or electricity not used

b. Gas

$ [   ] .00
Yearly cost — Dollars

OR

- ○ Included in rent or in condominium fee
- ○ No charge or gas not used

c. Water

$ [   ] .00
Yearly cost — Dollars

OR

- ○ Included in rent or in condominium fee
- ○ No charge

d. Oil, coal, kerosene, wood, etc.

$ [   ] .00
Yearly cost — Dollars

OR

- ○ Included in rent or in condominium fee
- ○ No charge or these fuels not used

The sample questionnaire contains housing questions H8 to H26 shown here on pages 4 and 5.

QUESTIONS FOR YOUR HOUSEHOLD

INSTRUCTION:

Answer questions H21 TO H26, if this is a one-family house, a condominium, or a mobile home that someone in this household OWNS OR IS BUYING; otherwise, go to page 6.

H21. What were the real estate taxes on THIS property last year?

| $ | .00 |

Yearly amount — Dollars

OR

○ None

H22. What was the annual payment for fire, hazard, and flood insurance on THIS property?

| $ | .00 |

Yearly amount — Dollars

OR

○ None

H23a. Do you have a mortgage, deed of trust, contract to purchase, or similar debt on THIS property?

○ Yes, mortgage, deed of trust, or similar debt

○ Yes, contract to purchase

} Go to H23b

○ No — *Skip to H24a*

b. How much is your regular monthly mortgage payment on THIS property? Include payment only on first mortgage or contract to purchase.

| $ | .00 |

Monthly amount — Dollars

OR

○ No regular payment required — *Skip to H24a*

c. Does your regular monthly mortgage payment include payments for real estate taxes on THIS property?

○ Yes, taxes included in payment

○ No, taxes paid separately or taxes not required

d. Does your regular monthly mortgage payment include payments for fire, hazard, or flood insurance on THIS property?

○ Yes, insurance included in payment

○ No, insurance paid separately or no insurance

H24a. Do you have a second or junior mortgage or a home equity loan on THIS property?

○ Yes

○ No — *Skip to H25*

b. How much is your regular monthly payment on all second or junior mortgages and all home equity loans?

| $ | .00 |

Monthly amount — Dollars

OR

○ No regular payment required

Answer ONLY if this is a CONDOMINIUM —

H25. What is the monthly condominium fee?

| $ | .00 |

Monthly amount Dollars

Answer ONLY if this is a MOBILE HOME —

H26. What was the total cost for personal property taxes, site rent, registration fees, and license fees on this mobile home and its site last year? Exclude real estate taxes.

| $ | .00 |

Yearly amount — Dollars

Please turn to page 6. ➛

Page 6

PERSON 1

Last name   First name   Middle initial

8. In what U.S. State or foreign country was this person born? ⌐

(Name of State or foreign country; or Puerto Rico, Guam, etc.)

9. Is this person a CITIZEN of the United States?
- ○ Yes, born in the United States — *Skip to 11*
- ○ Yes, born in Puerto Rico, Guam, the U.S. Virgin Islands, or Northern Marianas
- ○ Yes, born abroad of American parent or parents
- ○ Yes, U.S. citizen by naturalization
- ○ No, not a citizen of the United States

10. When did this person come to the United States to stay?
- ○ 1987 to 1990
- ○ 1985 or 1986
- ○ 1982 to 1984
- ○ 1980 or 1981
- ○ 1975 to 1979
- ○ 1970 to 1974
- ○ 1965 to 1969
- ○ 1960 to 1964
- ○ 1950 to 1959
- ○ Before 1950

11. At any time since February 1, 1990, has this person attended regular school or college? Include only nursery school, kindergarten, elementary school, and schooling which leads to a high school diploma or a college degree.
- ○ No, has not attended since February 1
- ○ Yes, public school, public college
- ○ Yes, private school, private college

12. How much school has this person COMPLETED? Fill ONE circle for the highest level COMPLETED or degree RECEIVED. If currently enrolled, mark the level of previous grade attended or highest degree received.
- ○ No school completed
- ○ Nursery school
- ○ Kindergarten
- ○ 1st, 2nd, 3rd, or 4th grade
- ○ 5th, 6th, 7th, or 8th grade
- ○ 9th grade
- ○ 10th grade
- ○ 11th grade
- ○ 12th grade, **NO DIPLOMA**
- ○ **HIGH SCHOOL GRADUATE** - high school DIPLOMA or the equivalent (For example: GED)
- ○ Some college but no degree
- ○ Associate degree in college - Occupational program
- ○ Associate degree in college - Academic program
- ○ Bachelor's degree (For example: BA, AB, BS)
- ○ Master's degree (For example: MA, MS, MEng, MEd, MSW, MBA)
- ○ Professional school degree (For example: MD, DDS, DVM, LLB, JD)
- ○ Doctorate degree (For example: PhD, EdD)

13. What is this person's ancestry or ethnic origin? ⌐
(See instruction guide for further information.)

(For example: German, Italian, Afro-Amer., Croatian, Cape Verdean, Dominican, Ecuadoran, Haitian, Cajun, French Canadian, Jamaican, Korean, Lebanese, Mexican, Nigerian, Irish, Polish, Slovak, Taiwanese, Thai, Ukrainian, etc.)

14a. Did this person live in this house or apartment 5 years ago (on April 1, 1985)?
- ○ Born after April 1, 1985 — *Go to questions for the next person*
- ○ Yes — *Skip to 15a*
- ○ No

b. Where did this person live 5 years ago (on April 1, 1985)?

(1) Name of U.S. State or foreign country ⌐

(If outside U.S., print answer above and skip to 15a.)

(2) Name of county in the U.S. ⌐

(3) Name of city or town in the U.S. ⌐

(4) Did this person live inside the city or town limits?
- ○ Yes
- ○ No, lived outside the city/town limits

15a. Does this person speak a language other than English at home?
- ○ Yes
- ○ No — *Skip to 16*

b. What is this language? ⌐

(For example: Chinese, Italian, Spanish, Vietnamese)

c. How well does this person speak English?
- ○ Very well   ○ Not well
- ○ Well   ○ Not at all

16. When was this person born?
- ○ Born before April 1, 1975 — *Go to 17a*
- ○ Born April 1, 1975 or later — *Go to questions for the next person*

17a. Has this person ever been on active-duty military service in the Armed Forces of the United States or ever been in the United States military Reserves or the National Guard? If service was in Reserves or National Guard only, see instruction guide.
- ○ Yes, now on active duty
- ○ Yes, on active duty in past, but not now
- ○ Yes, service in Reserves or National Guard only — *Skip to 18*
- ○ No — *Skip to 18*

b. Was active-duty military service during — Fill a circle for each period in which this person served.
- ○ September 1980 or later
- ○ May 1975 to August 1980
- ○ Vietnam era (August 1964—April 1975)
- ○ February 1955—July 1964
- ○ Korean conflict (June 1950—January 1955)
- ○ World War II (September 1940—July 1947)
- ○ World War I (April 1917—November 1918)
- ○ Any other time

c. In total, how many years of active-duty military service has this person had?

Years

18. Does this person have a physical, mental, or other health condition that has lasted for 6 or more months and which —

a. Limits the kind or amount of work this person can do at a job?
- ○ Yes   ○ No

b. Prevents this person from working at a job?
- ○ Yes   ○ No

19. Because of a health condition that has lasted for 6 or more months, does this person have any difficulty —

a. Going outside the home alone, for example, to shop or visit a doctor's office?
- ○ Yes   ○ No

b. Taking care of his or her own personal needs, such as bathing, dressing, or getting around inside the home?
- ○ Yes   ○ No

If this person is a female —
20. How many babies has she ever had, not counting stillbirths? Do not count her stepchildren or children she has adopted.

None 1 2 3 4 5 6 7 8 9 10 11 12 or more
○ ○ ○ ○ ○ ○ ○ ○ ○ ○ ○ ○ ○

21a. Did this person work at any time LAST WEEK?
- ○ Yes — Fill this circle if this person worked full time or part time. (Count part-time work such as delivering papers, or helping without pay in a family business or farm. Also count active duty in the Armed Forces.)
- ○ No — Fill this circle if this person did not work, or did only own housework, school work, or volunteer work. — *Skip to 25*

b. How many hours did this person work LAST WEEK (at all jobs)? Subtract any time off; add overtime or extra hours worked.

Hours

22. At what location did this person work LAST WEEK? If this person worked at more than one location, print where he or she worked most last week.

a. Address (Number and street) ⌐

(If the exact address is not known, give a description of the location such as the building name or the nearest street or intersection.)

b. Name of city, town, or post office ⌐

c. Is the work location inside the limits of that city or town?
- ○ Yes   ○ No, outside the city/town limits

d. County ⌐

e. State ⌐   **f. ZIP Code** ⌐

FOR PERSON 1 ON PAGE 2

23a. How did this person usually get to work LAST WEEK? If this person usually used more than one method of transportation during the trip, fill the circle of the one used for most of the distance.

- ○ Car, truck, or van
- ○ Bus or trolley bus
- ○ Streetcar or trolley car
- ○ Subway or elevated
- ○ Railroad
- ○ Ferryboat
- ○ Taxicab
- ○ Motorcycle
- ○ Bicycle
- ○ Walked
- ○ Worked at home — *Skip to 28*
- ○ Other method

If "car, truck, or van" is marked in 23a, go to 23b. Otherwise, skip to 24a.

b. How many people, including this person, usually rode to work in the car, truck, or van LAST WEEK?

- ○ Drove alone
- ○ 2 people
- ○ 3 people
- ○ 4 people
- ○ 5 people
- ○ 6 people
- ○ 7 to 9 people
- ○ 10 or more people

24a. What time did this person usually leave home to go to work LAST WEEK?

- ○ a.m.
- ○ p.m.

b. How many minutes did it usually take this person to get from home to work LAST WEEK?

Minutes — *Skip to 28*

25. Was this person TEMPORARILY absent or on layoff from a job or business LAST WEEK?

- ○ Yes, on layoff
- ○ Yes, on vacation, temporary illness, labor dispute, etc.
- ○ No

26a. Has this person been looking for work during the last 4 weeks?

- ○ Yes
- ○ No — *Skip to 27*

b. Could this person have taken a job LAST WEEK if one had been offered?

- ○ No, already has a job
- ○ No, temporarily ill
- ○ No, other reasons (in school, etc.)
- ○ Yes, could have taken a job

27. When did this person last work, even for a few days?

- ○ 1990
- ○ 1989
- ○ 1988
- ○ 1985 to 1987
} *Go to 28*
- ○ 1980 to 1984
- ○ 1979 or earlier
- ○ Never worked
} *Skip to 32*

28-30. CURRENT OR MOST RECENT JOB ACTIVITY. Describe clearly this person's chief job activity or business last week. If this person had more than one job, describe the one at which this person worked the most hours. If this person had no job or business last week, give information for his/her last job or business since 1985.

28. Industry or Employer

a. For whom did this person work?
If now on active duty in the Armed Forces, fill this circle ──► ○ and print the branch of the Armed Forces.

(Name of company, business, or other employer)

b. What kind of business or industry was this?
Describe the activity at location where employed.

(For example: hospital, newspaper publishing, mail order house, auto engine manufacturing, retail bakery)

c. Is this mainly — Fill ONE circle

- ○ Manufacturing
- ○ Wholesale trade
- ○ Retail trade
- ○ Other (agriculture, construction, service, government, etc.)

29. Occupation

a. What kind of work was this person doing?

(For example: registered nurse, personnel manager, supervisor of order department, gasoline engine assembler, cake icer)

b. What were this person's most important activities or duties?

(For example: patient care, directing hiring policies, supervising order clerks, assembling engines, icing cakes)

30. Was this person — Fill ONE circle

- ○ Employee of a PRIVATE FOR PROFIT company or business or of an individual, for wages, salary, or commissions
- ○ Employee of a PRIVATE NOT-FOR-PROFIT, tax-exempt, or charitable organization
- ○ Local GOVERNMENT employee (city, county, etc.)
- ○ State GOVERNMENT employee
- ○ Federal GOVERNMENT employee
- ○ SELF-EMPLOYED in own NOT INCORPORATED business, professional practice, or farm
- ○ SELF-EMPLOYED in own INCORPORATED business, professional practice, or farm
- ○ Working WITHOUT PAY in family business or farm

31a. Last year (1989), did this person work, even for a few days, at a paid job or in a business or farm?

- ○ Yes
- ○ No — *Skip to 32*

b. How many weeks did this person work in 1989?
Count paid vacation, paid sick leave, and military service.

Weeks

c. During the weeks WORKED in 1989, how many hours did this person usually work each week?

Hours

32. INCOME IN 1989 —
Fill the "Yes" circle below for each income source received during 1989. Otherwise, fill the "No" circle. If "Yes," enter the total amount received during 1989.

For income received jointly, see instruction guide.
If exact amount is not known, please give best estimate.
If net income was a loss, write "Loss" above the dollar amount.

a. Wages, salary, commissions, bonuses, or tips from all jobs — Report amount before deductions for taxes, bonds, dues, or other items.

- ○ Yes ──►
- ○ No

$ .00
Annual amount — Dollars

b. Self-employment income from own nonfarm business, including proprietorship and partnership — Report NET income after business expenses.

- ○ Yes ──►
- ○ No

$ .00
Annual amount — Dollars

c. Farm self-employment income — Report NET income after operating expenses. Include earnings as a tenant farmer or sharecropper.

- ○ Yes ──►
- ○ No

$ .00
Annual amount — Dollars

d. Interest, dividends, net rental income or royalty income, or income from estates and trusts — Report even small amounts credited to an account.

- ○ Yes ──►
- ○ No

$ .00
Annual amount — Dollars

e. Social Security or Railroad Retirement

- ○ Yes ──►
- ○ No

$ .00
Annual amount — Dollars

f. Supplemental Security Income (SSI), Aid to Families with Dependent Children (AFDC), or other public assistance or public welfare payments.

- ○ Yes ──►
- ○ No

$ .00
Annual amount — Dollars

g. Retirement, survivor, or disability pensions — Do NOT include Social Security.

- ○ Yes ──►
- ○ No

$ .00
Annual amount — Dollars

h. Any other sources of income received regularly such as Veterans' (VA) payments, unemployment compensation, child support, or alimony — Do NOT include lump-sum payments such as money from an inheritance or the sale of a home.

- ○ Yes ──►
- ○ No

$ .00
Annual amount — Dollars

33. What was this person's total income in 1989?
Add entries in questions 32a through 32h; subtract any losses. If total amount was a loss, write "Loss" above amount.

- ○ None   OR

$ .00
Annual amount — Dollars

Please turn to the next page and answer questions for Person 2 on page 2. If this is the last person listed in question 1a on page 1, go to the back of the form.

Appendix B □□□

Example Programs for Extracting Summary Tape Files and Microdata

Chapter 4 introduced the two major forms of census data that are made available in computer files: the summary tape files (STF) and public use microdata samples (PUMS). This Appendix lists computer programs (in the SAS syntax) for extracting census data from these two types of data files.

<u>Neither the authors of these example programs, Dowell Myers and Gregory Lipton, nor their publisher, Academic Press, are responsible for the success or failure of any analysis conducted with the programs listed here</u>. The following programs are intended merely to illustrate the structure of the required programming statements. These programs are not warranted to execute properly and they should not be run "as is." Working programs would require additional job control language (JCL) specific to the operating system and computer hardware of the user's own facility. Working programs would also require careful testing and debugging by the individual user.

Different census data files have variables assigned in different column locations, and therefore the user must consult carefully the technical documentation for each file. The example format for an STF extract program assumes use of the 1990 STF 1A file (California), and is provided courtesy of Gregory Lipton, City of Los Angeles. In contrast, the example PUMS programs (by Dowell Myers) are based on the 1980 PUMS A file (California). At this writing, none of the 1990 PUMS files have been released, but their format is expected to be very similar to the 1980 versions, as described in Chapter 4.

Part 1: Example Extract Program for STF Data

General Comment: Summary tabulations from the 1990 Census are available in computer readable form—the summary tape files (STF). Unlike printed reports, great care must be used in obtaining information.

343

Titles are not provided to describe the variables, the level of data aggregation, or the geography. Data on a tape is usually for more than one level of geography, including data for census block, block group, tract, city or higher level of geography. Conceptually, selecting the proper variable and level is easy. Operationally, care is required.

When obtaining census data, check city and county totals against printed reports or the work of others to insure that the proper variable and level of geography is obtained. Especially troublesome is that census tracts may be split between political boundaries. For example, part of Tract 9302 is in Los Angeles while the majority is outside the City. The Census Bureau provides totals for the tract as well as totals for the portion of the tract within the City of Los Angeles (or outside). Analysts must deliberately choose the census tract portion (inside, outside, or whole) that is most appropriate to the research purpose.

```
*   STF1A90—See Documentation for Variable Descriptions
*   Program Prepared by:
*       Gregory Lipton   (213) 485-8946
*       Community Development Department
*       City of Los Angeles
*       215 W 6th St.
*       Los Angeles, CA 90014
*  All Errors or Suggestions Should Be Sent to the Above


************************************************************************************.
                                                                                   ,
Data Temp;
        Infile In;
        Input
        SUMLEV $ 11-13 /* Key variable for identifying level of geography */

/* Some of the Possible Geographic Codes Follow */
        BLOCK  $ 47-50
        BLCKGRP $ 51
        @ 52 TRACT  BZ6.2
        COUNTY $ 72-74
        PLACECEN $ 112-115
        PLACEFIP $ 116-120
        PMSA   $ 126-129
        AREANAME $ 192-231  /* First 40 Characters of Area Name */

/* TABLE P1 */
        TPERSON 301-309 /* Total Number of Persons */

/* TABLE P8 */
        THISPANI 652-660 /* Total Number of Persons of Hispanic Origin */

/* TABLE P9 */
        TNHISP 661-669           /* Total NOT of Hispanic Origin */
        MEXICAN 670-678          /* MEXICAN */
        PUERTORI 679-687         /* PUERTO RICAN */
        CUBAN 688-696            /* CUBAN */
        OTHERHIS 697-705         /* OTHER HISPANIC */


*********************************************************************************.
                                                                                ,
```

```
*   Comments on Selecting Gegraphies by
*   Use of Summary Level Codes (SUMLEV)
*
*   For Areas Desired Within County Use SUMLEV Codes:
*   140 = Tract  /  150 = Block Group  /  155 = Place  /  050 = County

*   For Areas Desired Within City Use SUMLEV codes:
*   080 = Census Tract  /  091 = Block Group
*   Note: areas identified within cities must also be accompanied by a
*          FIPS or census place code
**********************************************************************************;
/* Example Selection of  Summary Level Follows */
/* Selected Area is the City of Los Angeles */

IF SUMLEV='091' AND PLACEFIP='44000' ;

/* The Output File Consists of 7 Cells of Data Describing Total Population */
/*  and Hispanic Origin for Each Block Group in the City of Los Angeles    */
**********************************************************************************;
```

Part 2: Example Extract Programs for PUMS Data

Four different programs are listed below, one for each of the major types of extracts: P, H, PH, and PHH. For explanation of these extract types, see Chapter 4 and Chapter 12.

Type H Extract

General Comment: This is the simplest of the four, requiring merely a selection of the H record type.

```
Data New.TypeH;
Infile Raw;
Length Default = 2;

Input TYPE $ 1 COGRP 6–8 @;

If COGRP EQ 41; /* Select Cases for Long Beach City */

If TYPE = 'H' then Do; /* Only read the H records */
        Input  PERSONS 26–27 TENURE 29 UNITS 36–37;
        Output;
End;
```

Type P Extract

General Comment: This program requires selection of only the P records. However, the location of each person is given on the H record that precedes each set of P records. Therefore, we need to read the H records also, extracting the COGRP location from there and putting this information on each P record. A further complication is that columns 6–8 record different information on the P and H records. Therefore, the COGRP from the H record is renamed and "retained" as PLACE before inputting the P records.

```
Data New.TypeP;
Retain PLACE;          /* This command holds retained variables from previous */
                       /* records and puts them on each record as it is output. */
                       /* In this manner, the contents of H records can be passed */
                       /* to the P records that follow them. */
Infile Raw;
Length Income8 = 5.;
Length Default = 2;

Input TYPE $ 1 COGRP 6-8 @;

If TYPE = 'H' then PLACE = COGRP;
If PLACE EQ 41;   /* Select Cases for Long Beach City */

If TYPE = 'P' then Do;   /* Only read the P records */
        Input  SEX 7 AGE 8-9 INCOME8 134-138 ;
        Output;
End;
```

Type PH Extract

General Comment: This program requires selection of both P and H records, combining these for output. Observe that the same column numbers on the two different records contain different variables; thus separate Input statements are required for each. Location is determined as in the Type P extract. All housing variables that are to be output for each person must be listed in the "Retain" statement.

```
Data New.TypePH;
Retain PLACE PERSONS TENURE UNITS;
                       /* This command holds retained variables from previous */
                       /* records and puts them on each record as it is output. */
                       /* In this manner, the contents of H records can be passed */
                       /* to the P records that follow them. */
Infile Raw;
Length Income8 = 5.;
Length Default = 2;

Input TYPE $ 1 COGRP 6-8 @;

If TYPE = 'H' then PLACE = COGRP;
If PLACE EQ 41;   /* Select Cases for Long Beach City */

If TYPE = 'H' then Do;   /* Assigns variables to specific columns on the H record*/
        Input  PERSONS 26-27 TENURE 29 UNITS 36-37;
        Return;
End;

If TYPE = 'P' then Do;   /* Assigns variables to specific columns on the P record*/
        Input  SEX 7 AGE 8-9 INCOME8 134-138 ;
        Output;
End;
```

Type PHH Extract

General Comment: This program is identical to a PH extract except that only the householder record is output. One way to achieve that is to select the householder *after* the PH file is complete. A more efficient method is shown here, selecting the householder *before* inputting person records.

```
Data New.TypePHH;
Retain PLACE PERSONS TENURE UNITS;
Infile Raw;
Length Income8 = 5.;
Length Default = 2;

Input TYPE $ 1 RELAT 2–3 COGRP 6–8 @;

If TYPE = 'H' then PLACE = COGRP;
If PLACE EQ 41;  /* Select Cases for Long Beach City */

If TYPE = 'H' then Do;  /* Assigns variables to specific columns on the H record*/
        Input  PERSONS 26–27 TENURE 29 UNITS 36–37;
        Return;
End;

If TYPE = 'P' and RELAT = 0 then Do;  /* Select only the Householder's record*/
        Input  SEX 7 AGE 8–9 INCOME8 134–138 ;
        Output;
End;
```

Extensions to the Labor Force and Employment Universes

The Type P, PH, and PHH extracts can all be extended to labor force and employment universes. Once the appropriate variables have been read into the person file, cases can be selected for persons currently in the labor force or currently employed. A more specific employment universe can be defined by workplace if that variable is read into the file. Cases of employed persons can then be selected if they work in a particular place, creating a workplace-based employment universe.

Appendix C □□□

List of State Data Centers and Key Contacts

Alabama

Center for Business and Economic Research
University of Alabama
Box 870221
Tuscaloosa, AL 35487-0221

Ms. Annette Watters
205-348-2953

Alaska

Alaska State Data Center
Research and Analysis
Department of Labor
P.O. Box 25504
Juneau, AK 99802-5504

Ms. Kathryn Lizik
907-465-4500

Arizona

Arizona Department of Economic Security
Mail Code 045Z
1789 West Jefferson St.
Phoenix, AZ 85007

Ms. Betty Jeffries
602-542-5984

Arkansas

State Data Center
University of Arkansas–Little Rock
2801 South University
Little Rock, AR 72204

Ms. Sarah Breshears
501-569-8530

California

State Census Data Center
Department of Finance
915 L Street
Sacramento, CA 95814

Ms. Linda Gage, Director
916-322-4651
Mr. Richard Lovelady
916-323-2201

Colorado

Division of Local Government
Colorado Department of Local Affairs
1313 Sherman Street, Room 521
Denver, CO 80203

Ms. Rebecca Picaso
303-866-2156

Connecticut

Comprehensive Planning Division
Connecticut Office of Policy and
Management
80 Washington Street
Hartford, CT 06106-4459

Mr. Theron Schnure
203-566-8285

Delaware

Delaware Development Office
99 Kings Highway
P.O. Box 1401
Dover, DE 19903

Ms. Judy McKinney-Cherry
302-739-4271

District of Columbia

Data Services Division
Mayor's Office of Planning
Room 570, Presidential Bldg.
415 12th Street, N.W.
Washington, DC 20004

Mr. Gan Ahuja
202-727-6533

Florida

Florida State Data Center
Executive Office of the Governor
Office of Planning and Budgeting
The Capitol
Tallahassee, FL 32399-0001

Mr. Steven Kimble
904-487-2814

Georgia

Division of Demographic & Statistical
Services
Georgia Office of Planning and Budget
254 Washington Street, S.W., Room 640
Atlanta, GA 30334

Ms. Marty Sik
404-656-0911

Guam

Guam Department of Commerce
590 South Marine Drive
Suite 601, 6th Floor GITC Building
Tamuning, Guam 96911

Mr. Peter R. Barcinas
671-646-5841

Hawaii

Hawaii State Data Center
Department of Business and Economic
Development
Kamamalu Building, Room 602A
220 S. King Street, Suite 400
Honolulu, HI 96813
(mailing address)
P.O. Box 2359
Honolulu, HI 96804

Ms. Emogene Estores
808-586-2482

Idaho

Idaho Department of Commerce
700 West State Street
Boise, ID 83720

Mr. Alan Porter
208-334-2470

Illinois

Division of Planning and Financial Analysis
Illinois Bureau of the Budget
William Stratton Building, Room 605
Springfield, IL 62706

Ms. Suzanne Ebetsch
217-782-1381

Indiana

Indiana State Library
Indiana State Data Center
140 North Senate Avenue
Indianapolis, IN 46204

Ms. Roberta Eads
317-232-3733

Iowa

State Library of Iowa
East 12th and Grand
Des Moines, IA 50319

Ms. Beth Henning
515-281-4350

Kansas

State Library
Room 343-N
State Capitol Building
Topeka, KS 66612

Mr. Marc Galbraith
913-296-3296

Kentucky

Urban Studies Center
College of Urban & Public Affairs
Univesity of Louisville
Louisville, KY 40292

Mr. Ron Crouch
502-588-7990

Louisiana

Office of Planning and Budget
Division of Administration
P.O. Box 94095
900 Riverside
Baton Rouge, LA 70804

Ms. Karen Paterson
504-342-7410

Maine

Division of Economic Analysis and Research
Maine Department of Labor
20 Union Street
Augusta, ME 04330

Ms. Jean Martin
207-289-2271

Maryland

Maryland Department of State Planning
301 West Preston Street
Baltimore, MD 21201

Mr. Robert Dadd
301-225-4450

Massachusetts

Massachusetts Institute for Social and
Economic Research
128 Thompson Hall
University of Massachusetts
Amherst, MA 01003

Dr. Stephen Coelen, Director
413-545-3460
Ms. Nora Groves
413-545-0176

Michigan

Michigan Information Center
Department of Management and Budget
Office of Revenue and Tax Analysis
P.O. Box 30026
Lansing, MI 48909

Mr. Eric Swanson
517-373-7910

Minnesota

State Demographer's Office
Minnesota State Planning Agency
300 Centennial Office Building
658 Cedar Street
St. Paul, MN 55155

Mr. David Birkholz
612-297-2360
Mr. David Rademacher
612-297-3255

Mississippi

Center for Population Studies
The University of Mississippi
Bondurant Bldg., Room 3W
University, MS 38677

Ms. Pattie Byrd, Manager
601-232-7288

Missouri

Missouri State Library
2002 Missouri Boulevard
P.O. Box 387
Jefferson City, MO 65102

Ms. Marlys Davis
314-751-3615

Montana

Census of Economic Information Center
Montana Department of Commerce
1424 9th Avenue
Capitol Station
Helena, MT 59620-0401

Ms. Patricia Roberts
406-444-4393

Nebraska

Center for Applied Urban Research
The University of Nebraska–Omaha
Peter Kiewit Conference Center
1313 Farnam-on-the-Mall
Omaha, NE 68182

Mr. Tim Himberger
402-595-2311

Nevada

Nevada State Library
Capitol Complex
401 North Carson
Carson City, NV 89710

Ms. Betty McNeal
702-687-5160

New Hampshire

Office of State Planning
2 1/2 Beacon Street
Concord, NH 03301

Mr. Tom Duffy
603-271-2155

New Jersey

New Jersey Department of Labor
Division of Labor Market and Demographic
Research
CN 388-John Fitch Plaza
Trenton, NJ 08625-0388

Ms. Connie O. Hughes, Asst. Dir.
609-984-2593

New Mexico

Economic Development and Tourism
Department
1100 St. Francis Drive
Santa Fe, NM 87503

Mr. John Beasley
505-827-0272

New York

Division of Policy and Research
Department of Economic Development
1 Commerce Plaza, Room 905
99 Washington Avenue
Albany NY 12245

Mr. Robert Scardamalia
518-474-6005

North Carolina

North Carolina Office of State Budget and
Management
116 West Jones Street
Raleigh, NC 27603-8005

Ms. Francine Stephenson, Dir.
919-733-7061

North Dakota

Department of Agricultural Economics
North Dakota State University
Morrill Hall, Room 224
P.O. Box 5636
Fargo, ND 58105

Dr. Richard Rathge
701-237-8621

Ohio

Ohio Data Users Center
Ohio Department of Development
P.O. Box 1001
Columbus, OH 43266-0101

Mr. Barry Bennett
614-466-2115

Oklahoma

Oklahoma State Data Center
Oklahoma Department of Commerce
6601 Broadway Extension
(mailing address)
P.O. Box 26980
Oklahoma City, OK 73126-0980

Ms. Karen Selland
405-841-5184

Oregon

Center for Population Research and Census
Portland State University
P.O. Box 751
Portland, OR 97207-0751

Ms. Maria Wilson-Figueroa
503 725 3922

Pennsylvania

Pennsylvania State Data Center
Institute of State and Regional Affairs
Pennsylvania State University at Harrisburg
Middletown, PA 17057-4898

Mr. Michael Behney
717-948-6336

Puerto Rico

Puerto Rico Planning Board
Minillas Government Center
North Bldg., Avendia De Diego
P.O. Box 41119
San Juan, PR 00940-9985

Sra. Lillian Torres Aguirre
809-728-4430

Rhode Island

Department of Administration
Office of Municipal Affairs
One Capitol Hill
Providence, RI 02908-5873

Mr. Paul Egan
401-277-6493

South Carolina

Division of Research and Statistical Services
South Carolina Budget and Control Board
Rembert Dennis Bldg., Room 425
Columbia, SC 29201

Mr. Mike Macfarlane
803-734-3780

South Dakota

Business Research Bureau
School of Business
University of South Dakota
414 East Clark
Vermillion, SD 57069

Ms. DeVee Dykstra
605-677-5287

Tennessee

Tennessee State Planning Office
John Sevier State Office Bldg.
500 Charlotte Ave. Suite 307
Nashville, TN 37243-0001

Mr. Charles Brown
615-741-1676

Texas

State Data Center
Texas Department of Commerce
9th And Congress Streets
(mailing address)
P.O. Box 12728
Capitol Station
Austin, TX 78711

Ms. Susan Tully
512-472-9667

Utah

Office of Planning and Budget
State Capitol, Room 116
Salt Lake City, UT 84114

Ms. Linda Smith
801-538-1036

Vermont

Office of Policy Research and Coordination
Pavilion Office Building
109 State Street
Montpelier, VT 05602

Mr. Bernie Johnson
802-828-3326

Virginia

Virginia Employment Commission
703 East Main Street
Richmond, VA 23219

Mr. Dan Jones
804-786-8308

Virgin Islands

University of the Virgin Islands
Caribbean Research Institute
Charlotte Amalie
St. Thomas, VI 00802

Dr. Frank Mills
809-776-9200

Washington

Office of Financial Management
Estimation and Forecasting Unit
450 Insurance Bldg., MS: AQ-44
Olympia, WA 98504-0202

Ms. Sharon Estee
206-586-2504

West Virginia

Community Development Division
Governor's Office of Community & Industrial
Development
Capitol Complex
Building 6, Room 553
Charleston, WV 25305

Ms. Mary C. Harless
304-348-4010

Wisconsin

Demographic Services Center
Department of Administration
101 S. Webster St., 6th Floor
P.O. Box 7868
Madison, WI 53707-7868

Mr. Robert Naylor
608-266-1927

Wyoming

Department of Administration and Fiscal
Control
Research and Statistics Division
Emerson Building 327E
Cheyenne, WY 82002-0060

Mr. Kreg McCollum
307-777-7504

Bibliography □□□

Adams, J. S. (1970) "The Residential Structure of Midwestern Cities," *Annals, Association of American Geographers* 60:37–62.

Adams, J. S. (1987) *Housing America in the 1980s* (Census Monograph Series). Russell Sage Foundation, New York.

Alonso, W., and P. Starr, eds. (1987) *The Politics of Numbers* (Census Monograph Series). Russell Sage Foundation, New York.

Anderson, M.J. (1988) *The American Census: A Social History.* Yale University Press, New Haven.

Apgar, W. C., G. S. Masnick, and N. McArdle (1991) *Housing in America: 1970–2000.* Joint Center for Housing Studies, Harvard University, Cambridge.

Baer, W. C. (1976) "The Evolution of Housing Indicators and Housing Standards," *Public Policy* 24:361–393.

Baer, W. C. (1990) "Aging of the Housing Stock and Components of Inventory Change," pp. 249–273 in Myers, D., ed., *Housing Demography: Linking Demographic Structure and Housing Markets.* University of Wisconsin Press, Madison, WI.

Batutis, M. J. (1988) "How to Understand Tables," pp. 31–34 in Wickham, P., ed., *The Insider's Guide to Demographic Know-How.* American Demographics Press, Ithaca, NY.

Bean, F. D., and M. Tienda (1987) *The Hispanic Population of the United States* (Census Monograph Series). Russell Sage Foundation, New York.

Burchell, R. W., and D. Listokin (1978) *The Fiscal Impact Handbook: Estimating Local Costs and Revenues of Land Development.* Center for Urban Policy Research, Rutgers University, New Brunswick, NJ.

Butz, W. P. (1987) "Discussion" [of Diemer and Tippett housing papers], pp. 477–479 in *Proceedings, Third Annual Research Conference.* U.S. Bureau of the Census, Washington, D.C.

Butz, W. P. (1991) "Evaluation of the 1990 Census." Paper presented at the annual meeting of the Population Association of America, Washington, D.C.

355

Carn, N., J. Rabianski, R. Racster, and M. Seldin, eds. (1988) *Real Estate Market Analysis: Techniques and Applications.* Prentice-Hall, Englewood Cliffs, NJ.

Cervero, R. (1989) "Jobs–Housing Balancing and Regional Mobility," *Journal of the American Planning Association* 55:136–150.

Choldin, H. M. (1991) Looking for the Last Percent: Science and Politics in the Census. Unpublished manuscript.

Clark, W. A. V. (1986) *Human Migration.* Sage Publications, Beverly Hills, CA.

Clemmer, R. B., and J. C. Simonson (1983) "Trends in Substandard Housing 1940–1980," *Journal of the American Real Estate and Urban Economics Association* 10:442–464.

COMSIS Corporation (1983) *Transportation Planners' Guide to Using the 1980 Census.* Federal Highway Administration, U.S. Department of Transportation, Washington, D.C.

Conk, M.A. (1987) "The 1980 Census in Historical Perspective," pp. 155–186 in Alonso, W., and P. Starr, eds., *The Politics of Numbers* (Census Monograph Series). Russell Sage Foundation, New York.

Coulson, M. R. C. (1970) "The Distribution of Population Age Structures in Kansas City," pp. 408–430 in Demko, G. J., H. M. Rose, and G. A. Schnell, eds., *Population Geography: A Reader.* McGraw-Hill, New York.

County of Los Angeles Department of Regional Planning (1990) *Bulletin.* Los Angeles.

Dahmann, D. C. (1983) "Racial Differences in Housing Consumption During the 1970s: Insights from a Components of Inventory Change Analysis," *Urban Geography* 4:203–222.

Davis, J. A. (1974) "Hierarchical Models for Significance Tests in Multivariate Contingency Tables: An Exegesis of Goodman's Recent Papers," pp. 189–230 in Costner, H., ed., *Sociological Methodology.* Jossey-Bass, San Francisco.

Davis, J. A., and A. M. Jacobs (1968) "Tabular Presentation," pp. 497–509 in Sills, D. L., ed., *International Encyclopedia of the Social Sciences, Vol. 15.* Crowell Collier and Macmillan, New York.

Deming, W. E. (1948) *Statistical Adjustment of Data.* John Wiley and Sons, New York.

Diemer, W. D. (1987) "Micro-Evaluation of the 1980 Census of Housing," pp. 437–476 in *Proceedings, Third Annual Research Conference.* U.S. Bureau of the Census, Washington, D.C.

Ehrenberg, A. S. C. (1977) "Rudiments of Numeracy," *Journal of the Royal Statistical Society* 140 (3):277–297.

Fienberg, S. E. (1977) *The Analysis of Cross-Classified Categorical Data.* MIT Press, Cambridge, MA.

Freedman, D. A. (1991) "Adjusting the 1990 Census," *Science* 252:1233–1236.

Gober, P. (1990) "The Urban Demographic Landscape: A Geographic Perspective," pp. 232–248 in Myers, D., ed., *Housing Demography: Linking Demographic Structure and Housing Markets.* University of Wisconsin Press, Madison, WI.

Guest, A. M. (1974) "Neighborhood Life Cycles and Social Status," *Economic Geography* 50:228–243.

Hamilton, C. H., and J. Perry (1962) "A Short Method for Projecting Population by Age from One Decennial Census to Another," *Social Forces* 41(2):160–170.

Hausrath, L. C. (1988) "Economic Basis for Linking Jobs and Housing in San Francisco," *Journal of the American Planning Association* 54(Spring):210–216.

Hedderson, J. (1987) *SPSSx Made Simple.* Wadsworth Publishing, Belmont, CA.

Hogan, H. (1990) "The 1990 Post-Enumeration Survey: An Overview," *Proceedings of the Section on Survey Research Methods of the American Statistical Association* (August): 518–523.

Innes, J. E. (1990) *Knowledge and Public Policy: The Search for Meaningful Indicators.* Expanded second edition. Transaction Publishers, New Brunswick, NJ.

Jencks, C. (1987) "The Politics of Income Measurement," pp. 83–131 in Alonso, W., and P. Starr, eds., *The Politics of Numbers* (Census Monograph Series). Russell Sage Foundation, New York.

Jenkins, R. (1987) *Procedural History of the 1940 Census of Population and Housing.* University of Wisconsin Press, Madison, WI.

Jones, C. D. (1991) "Taking the Census: Lessons from 1990." Paper presented at the annual meeting of the Population Association of America, Washington, D.C.

Kaplan, C. P., and T. L. Van Valey (1980) *Census '80: Continuing the Factfinder Tradition.* U.S. Government Printing Office, Washington, D.C.

Kobrin, F. E. (1976) "The Fall in Household Size and the Rise of the Primary Individual," *Demography* 13:127–138.

Lieberson, S., and M. C. Waters (1988) *From Many Strands: Ethnic and Racial Groups in Contemporary America* (Census Monograph Series). Russell Sage Foundation, New York.

Long, L. H. (1988) *Migration and Residential Mobility in the United States* (Census Monograph Series). Russell Sage Foundation, New York.

Long, L. H., and P. C. Glick (1976) "Family Patterns in Suburban Area: Recent Trends." pp. 39–67 in Schwartz, B., ed., *The Changing Face of the Suburbs.* University of Chicago Press, Chicago.

Lowry, I. S. (1964) *Migration and Metropolitan Growth: Two Analytical Models.* Chandler Publishing, San Francisco.

Lowry, I. S. (1983) "Designing Readable and Persuasive Tables." Series P-6945. Rand Corporation, Santa Monica, CA.

Lowry, I. S. (1991) Personal communication. May 15 memorandum.

Lucerna, R., and J. Landis (1989) *Desktop Mapping for Planning and Strategic Decision-Making.* Strategic Mapping, San Jose, CA.

Masnick, G. S., and M. J. Bane (1980) *The Nation's Families: 1960–1990.* Auburn House Publishing, Boston.

Masnick, G. S., J. R. Pitkin, and J. Brennan (1990) "Cohort Housing Trends in a Local Housing Market: the Case of Southern California," pp. 157–173 in Myers, D., ed., *Housing Demography: Linking Demographic Structure and Housing Markets.* University of Wisconsin Press, Madison, WI.

Mitroff, I. I., R. O. Mason, and V. P. Barabba (1983) *The 1980 Census: Policymaking and Turbulence.* Lexington Books, Lexington, Mass.

Mosbacher, R. A. (1991) Statement of Secretary Robert A. Mosbacher on Adjustment of the 1990 Census. July 15, 1991. U.S. Department of Commerce, Washington, D.C.

Mosteller, F. (1968) "Association and Estimation in Contingency Tables," *Journal of the American Statistical Association* 63(March):1–29.

Murray, M. P. (1991) "Census Adjustment and the Distribution of Federal Spending," Appendix 15 to *Decision on Whether or Not a Statistical Adjustment of the 1990 Decennial Census of Population Should Be Made for Coverage Deficiencies Resulting in an Overcount or Undercount of the Population; Explanation.* U.S. Department of Commerce, Washington, D.C.

Myers, D. (1978) "Aging of Population and Housing: A New Perspective on Planning for More Balanced Metropolitan Growth," *Growth and Change* 9:8–13.

Myers, D. (1985) "Wives' Earnings and Rising Costs of Homeownership," *Social Science Quarterly* 66:319–329.

Myers, D. (1990a) "Introduction: The Emerging Concept of Housing Demography," pp. 3–31 in Myers, D., ed., *Housing Demography: Linking Demographic Structure and Housing Markets.* University of Wisconsin Press, Madison, WI.

Myers, D. (1990b) "Filtering in Time: Rethinking the Longitudinal Behavior of Neighborhood Housing Markets," pp. 274–296 in Myers, D., ed., *Housing Demography: Linking Demographic Structure and Housing Markets.* University of Wisconsin Press, Madison, WI.

Myers, D. (1990c) "The Contribution of Demographic Methods to Real Estate Market Analysis," pp. 45–73 in Kapplin, S. D., and A. L. Schwartz, Jr., *Research in Real Estate. Vol. 3*, JAI Press, Greenwich, CT.

Myers, D., and A. Doyle (1990) "Age-Specific Population-per-Household Ratios: Linking Population Age Structure with Housing Characteristics," pp. 109–130 in Myers, D. ed.,

Housing Demography: Linking Demographic Structure and Housing Markets. University of Wisconsin Press, Madison, WI.

Nathan, R. P. (1987) "The Politics of Printouts: The Use of Official Numbers to Allocate Federal Grants-in-Aid," pp. 331–342 in Alonso, W., and P. Starr, eds., *The Politics of Numbers* (Census Monograph Series). Russell Sage Foundation, New York.

Oppenheimer, V. K. (1982) *Work and the Family: A Study in Social Demography.* Academic Press, New York.

Patton, C. V. (1988) "Jobs and Commercial Office Development: Do New Offices Generate Jobs?" *Economic Development Quarterly* 2(November):316–325.

Petersen, W. (1987) "Politics and the Measurement of Ethnicity," pp. 187–233 in Alonso, W., and P. Starr, eds., *The Politics of Numbers* (Census Monograph Series). Russell Sage Foundation, New York.

Pisarski, A. E. (1987) *Commuting in America.* Eno Foundation for Transportation, Westport, CT.

Pittenger, D. B. (1976) *Projecting State and Local Populations.* Ballinger Publishing, Cambridge, MA.

Putnam, S. H. (1983) *Integrated Urban Models: Policy Analysis of Transportation and Land Use.* Pion, London.

Rives, N. W., and W. J. Serow (1984) *Introduction to Applied Demography: Data Sources and Estimation Techniques.* Sage Publications, Beverly Hills, CA.

Robey, B. (1989) "Two Hundred Years and Counting: The 1990 Census," *Population Bulletin* 44(1).

Robinson, J. G. (1988) "Perspectives on the Completeness of Coverage of Population in the United State Decennial Censuses." Paper presented at the annual meeting of the Population Association of America.

Rogers, A. (1990) "Requiem for the Net Migrant," *Geographical Analysis* 22 (4): 283–300.

Roseman, C. C. (1971) "Migration as a Spatial and Temporal Process," *Annals of the American Association of Geographers* 61(September):589–598.

Roseman, C. C. (1991) "Cyclical and Polygonal Migration in a Western Context," forthcoming in Wardwell, J. M., P. C. Jobes, and W. F. Stinner, eds., *Community, Society, and Migration.* University Press of America.

Schmitt, R. C. (1954) "A Method of Projecting the Population of Census Tracts," *Journal of the American Institute of Planners* 20(2):102.

Shyrock, H. S., J. S. Siegel and Associates (1976) *The Methods and Materials of Demography* (Condensed Edition by E. G. Stockwell). Academic Press, New York.

Smith, S. K., and B. B. Lewis (1983) "Some New Techniques for Applying the Housing Unit Method of Local Population Estimation: Further Evidence," *Demography* 5:475–484.

Sweet, J. A., and Bumpass, L. L. (1987) *American Families and Households* (Census Monograph Series). Russell Sage Foundation, New York.

Thernstrom, A. (1987) "Statistics and the Politics of Minority Representation: The Evolution of the Voting Rights Act Since 1965," pp. 303–327 in Alonso, W., and P. Starr, eds., *The Politics of Numbers* (Census Monograph Series). Russell Sage Foundation, New York.

Tippett, J. A. (1987) "Housing Data: The Quality of Selected Items." *Proceedings, Third Annual Research Conference.* U.S. Bureau of the Census, Washington, D.C.

Tufte, E. R. (1983) *The Visual Display of Quantitative Information.* Graphics Press, Cheshire, CT.

Tufte, E. R. (1990) *Envisioning Information.* Graphics Press, Cheshire, CT.

U.S. Bureau of the Census (1982) *Users' Guide: Part B. Glossary* (PHC80-R1-B). U.S. Government Printing Office, Washington, D.C.

U.S. Bureau of the Census (1983a) *Users' Guide: Part C. Index to Summary Tape Files 1 to 4* (PHC80-R1-C). U.S. Government Printing Office, Washington, D.C.

U.S. Bureau of the Census (1983b) *Public Use Microdata Samples, Technical Documentation.* U.S. Government Printing Office, Washington, D.C.

U.S. Bureau of the Census (1983c) *Census Tracts* (PHC80-2). U.S. Government Printing Office, Washington, D.C.

U.S. Bureau of the Census (1983d) *Detailed Population Characteristics* (PC80-1-D). U.S. Government Printing Office, Washington, D.C.

U.S. Bureau of the Census (1983e) *Metropolitan Housing Characteristics* (HC80-2). U.S. Government Printing Office, Washington, D.C.

U.S. Bureau of the Census (1984) *Gross Migration for Counties: 1975 to 1980* (PC80-S1-17). U.S. Government Printing Office, Washington, D.C.

U.S. Bureau of the Census (1988) *The Coverage of Population in the 1980 Census*, Evaluation and Research Reports (PHC80-E4). U.S. Government Printing Office, Washington, D.C.

U.S. Bureau of the Census (1989a) *Current Population Reports* (Series P-23, No. 162) *Studies in Marriage and the Family*. U.S. Government Printing Office, Washington, D.C.

U.S. Bureau of the Census (1989b) Census of Population and Housing, 1990:*Tabulation and Publication Program*. U.S. Government Printing Office, Washington, D.C.

U.S. Bureau of the Census (1990a) *TIGER: The Coast-to-Coast Digital Map Data Base*. U.S. Government Printing Office, Washington, D.C.

U.S. Bureau of the Census (1990b) *Current Population Reports* (Series P-60, No. 166) *Money Income and Poverty Status in the United States: 1988*. U.S. Government Printing Office, Washington, D.C.

U.S. Bureau of the Census (1990c) Census of Population and Housing, 1990:*Content Determination Reports* (CDR-1 to 13). U.S. Government Printing Office, Washington, D.C.

U.S. Bureau of the Census (1991a) *Current Population Reports* (Series P-20, No. 447) *Household and Family Characteristics: March 1990 and 1989*. U.S. Government Printing Office, Washington, D.C.

U.S. Bureau of the Census (1991b) *Current Population Reports* (Series P-20, No. 450), *Marital Status and Living Arrangements: March 1990*. U.S. Government Printing Office, Washington, D.C.

U.S. Bureau of the Census (1991c) *Census Catalog and Guide*. U.S. Government Printing Office, Washington, D.C.

U.S. Bureau of the Census (1991d) Census of Population and Housing, 1990: *Summary Tape File 1 Technical Documentation*. U.S. Government Printing Office, Washington, D.C.

White, M. J. (1984) "On the Estimation of Specific Net Intercensal Migration for Small Areas," pp. 68–89 in Bogue, D. J., and M. J. White, eds., *Essays in Human Ecology*, The Community and Family Study Center, University of Chicago, IL.

White, M. J. (1987) *American Neighborhoods and Residential Differentiation* (Census Monograph Series). Russell Sage Foundation, New York.

Wolter, K. M. (1991) "Accounting for America's Uncounted and Miscounted," *Science* 253: 12–15.

Index □□□